Josef Rom

High Angle of Attack Aerodynamics

Subsonic, Transonic, and Supersonic Flows

With 289 Illustrations

Springer-Verlag
New York Berlin Heidelberg London Paris
Tokyo Hong Kong Barcelona Budapest

Josef Rom
Department of Aerospace Engineering
Technion-Israel Institute of Technology
Haifa 32000
Israel

Library of Congress Cataloging-in-Publication Data
Rom, Josef, 1932–
High angle of attack aerodynamics : subsonic, transonic, and supersonic flows / Josef Rom.
p. cm.
Includes bibliographical references and index.
ISBN-13:978-1-4612-7686-9
1. Vortex-motion. 2. Aerodynamics. I. Title.
TL574.V6R66 1991
629.132'3—dc20 91-36124

Printed on acid-free paper.

Softcover reprint of the hardcover 1st edition 1992

Production managed by Henry Krell; manufacturing supervised by Robert Paella.
Camera ready copy supplied by the author.

9 8 7 6 5 4 3 2 1

ISBN-13:978-1-4612-7686-9 e-ISBN-13:978-1-4612-2824-0
DOI: 10.1007/978-1-4612-2824-0

Foreword

The aerodynamics of aircraft at high angles of attack is a subject which is being pursued diligently, because the modern agile fighter aircraft and many of the current generation of missiles must perform well at very high incidence, near and beyond stall. Much of the past and present research and development efforts in aerodynamics are devoted to investigations of the numerous flow phenomena which are accentuated as the angles of attack of complex configurations increase. The extensive research on the aerodynamics phenomena at high angles of attack, experimental, analytical and numerical, are reported in numerous reports and papers from the 1930's to the present date. Certain aspects of the aerodynamics at high angles of attack were included in various books and in the proceedings of conferences devoted to this subject. However, a comprehensive presentation of the methods and results applicable to the studies of the complex aerodynamics at high angles of attack has not been covered in monographs or textbooks nor in any specific course. I first attempted teaching a graduate course on High Angle of Attack Aerodynamics in the winter of 1985, while on sabbatical at the Department of Aeronautics and Astronautics of Stanford University. Having prepared notes and reference material, it was Prof. Holt Ashely who suggested that I should integrate all of this material into a book. Since then, I have taught such a course in the Department of Aerospace Engineering at the Technion and found that the material is interesting, important and specialized, warranting its publication. The work of preparing the book for publication became possible by the generous support of the Bloomfield Family, through a grant by the El-Dee Foundation. The support of the El-Dee Foundation, the Bloomfield Family and, particularly, Mrs. N. Bloomfield is gratefully acknowledged.

The book is not the usual textbook, in that it goes beyond just presenting the basic theoretical and experimental knowhow, and contains reference material to practical calculation methods and technical and experimental results which can be useful to the practicing aerospace engineer and scientist. It can be certainly used as a text-reference book for graduate courses on subjects related to high angles of attack aerodynamics and for topics related to three-dimensional separation in viscous flow courses. Each instructor can select the appropriate topics, chapters and sections, which fit his particular syllabus. In addition, the book is addressed to the aerodynamicist interested in a comprehensive reference to methods of analysis and computations of high angle of attack flow phenomena. I hope that it will also find its place in the working place of aerospace scientists and engineers, to be used as a reference book for understanding the flow phenomena and the methods for analysis and calculations of flows at high angles of attack on various configurations.

The book is written for the aerospace scientist and engineer who is familiar with the basic concepts of viscous and inviscid flows and with computational methods used in fluid dynamics. It may be helpful to various readers to refer back to some of the "classical" fluid mechanics textbooks, including those referenced in the book, to refresh their knowledge of some of the basic subjects required for the understanding of various methods for analysis or calculations. I have tried to present the material in an orderly and consistent way, using basic concepts, so that it can be followed also by a non-specialist, but reference to the basic fluid mechanics textbooks is recommended. An extensive list of references is presented following each chapter in order to enable interested readers to further study in details each subject.

The book presents the many aspects of the flow phenomena associated with the aerodynamics of high angles of attack, with particular emphasis on the methods for analysis and numerical calculations. The complexity of the flow requires reliance on experimental measurements for useful data which can also be used for verification of the calculation methods. The reader will find references to experimental results in the present book used mainly for the verification of the numerical methods. However, it should be clear that, in many cases, we must rely on experimental data not only for the verification of numerical calculations but also as the principal means of evaluation of the aerodynamic characteristics. Certain complex phenomena, for which our theoretical and numerical knowhow is limited, can be studied mainly on the basis of systematic experimental programs. In spite of the large advancements in the capabilities of the numerical methods to obtain reasonable values for the aerodynamic characteristics, most of the "fine tuning" of the aerodynamic design of configurations for operations at high angles of attack must rely on experimental measurements in wind tunnels and in flight tests. Still, the advantages of the simple panel methods, such as the NLVLM, for the fast evaluation of the aerodynamic coefficients in the preliminary design stage is most striking.

Therefore, fast and simple calculation methods using the various panel methods are presented as means for reasonable evaluation in the preliminary design stage. Then, more complex methods based on the solutions of the Euler equations can be used, both in the preliminary design and in full design calculations, as a reasonable guide for selecting configurations for detailed wind tunnel testing. The progress in the solutions of the Navier-Stokes equations promises to enable such solutions to be used in aerodynamic designs. The length of time needed for the computations and the large supercomputer resources required for the solution of the three-dimensional Navier-Stokes equations makes the numerical solution of the Navier-Stokes equations very long and costly. It seems that, for the foreseeable future, aerodynamic design will continue to rely on simple, fast and less accurate calculation methods for preliminary design and then on longer and more complex calculations solving the Euler equations. In some cases, the thin-layer Navier-Stokes equations can be solved in a long but still reasonable length of time, and these results can be used in the full design case. The full design requires a fairly extensive calculation effort using the Euler or thin-layer Navier-Stokes equations, eventually followed by the solution of the full Navier-Stokes equations. All of these calculations do not eliminate the need for the testing of the configurations, first in wind tunnels and then in the final stages of development in flight tests.

A book of this kind is a culmination of knowledge gathered over many years, and many of my students, friends and colleagues share my gratitude for their discussions and comments which directly or indirectly contributed to the formation of this book. In particular I am most grateful to my graduate students who participated, contributed and initiated many of the developments of the Nonlinear Vortex Lattice Method: Dr. Carlos Zorea, Dr. Rachel Gordon, Dr. Dan Almosnino, Mr. Mr. Beny Melamed and Mr. Gabriel Tennenbaum, the latter under the direction of Dr. Asher Sigal. Particular appreciation and thanks goes to Prof. Henry Portnoy who contributed to the initial formulation of the NLVLM and to the studies of the roll-up of the vortex wake. It is my pleasure to acknowledge the contributions of Prof. A. Seginer, Prof. E. Wasserstrom, Dr. Z. Rusak and Prof. J. Katz and Dr. D. Levin who used our basic NLVLM program for further developments of the steady and nonsteady linear and nonlinear Vortex Lattice applications. The development of the NLVLM program was supported, in part, by grants from the US AFOSR and the US Army. I wish to express my appreciation for this support and particularly to the directors and the officers of these organizations for their continued encouragement. In particular, it is my pleasure and duty to acknowledge the friendship and advice of Mr. Milton Rogers of the AFOSR and Dr. Max Scherberg of the ARL at Wright Patterson AFB.

Writing the manuscript is an individual effort. However producing a book is a team effort. It is therefore, my pleasure to acknowledge with many thanks the contributions of Judy Golub for editing the manuscript, Dr. E. Rakotch and I. Abramovich for formatting the text into TEX, Irit Nitzan for preparing the line drawings, Yair Naor for preparing the photographic copies and Alice Goodman for typesetting. It is also my pleasure to acknowledge many excellent suggestions by the Publisher's reviewers which certainly improved the final manuscript.

The support of all I love, my family, my children - Dalit, Rami-Yair and Vered - made this writing a reality.

Contents

Chapter 1

Introduction

Slender elongated configurations are being used in many flight applications, such as the Shuttle Spacecraft and the National Aerospace Plane (NASP), the highly agile supersonic – transonic military fighter aircraft, as well as the various types of controlled, guided and maneuverable missiles. Such configurations are able to generate considerable aerodynamic forces and moments even at relatively high angles of attack. Large part of the aerodynamic forces at these high angles of attack are due to the induced flow generated by the vortices which are shed from the lifting surfaces and the body of the configuration. The strength and position of these vortices depend on the lifting surfaces, planform and leading-edges' shapes and on the body geometrical shape and their local attitude in relation to the direction of the velocity vector (i.e. local angle of attack). Practical experience teaches us that the generation of these vortices can be controlled by variations in the planform shapes, such as double delta, canard-wing combinations, leading-edge extensions (LEX), sawtooth leading-edges, and others. These configurations, which utilize the strong vortex interactions, give the aerodynamic designer new tools for achieving controlled flight at very high angles of attack required for sharp maneuvers. Furthermore, introduction of small fins, acting as vortex generators at "strategic" positions, can affect the complete flow field because of the interaction of the generated vortex filament from the fin with the vortex structure established on the configuration before the addition of the fin. Therefore, it may be said that properly controlled vortex interactions may enable significant variation of the overall lift on the aircraft with or without change of its attitude. Thus, controlled vortex interactions can be used as a form of direct lift control.

It is, therefore, expected that the calculations of the aerodynamic characteristics at high angles of attack must involve evaluation of the effects of free vortices which are shed from the wing planforms and body planforms. For these vortical flow structures, it is expected that the generation of separated free shear layers is caused by the viscous shear and viscous-inviscid interactions. However, once these free shear layers roll-up and form the concentrated vortices, it may be assumed that effects of viscosity in the free flow field, outside the boundary layers on the surfaces, are confined to the limited zones in the cores of the rolled up vortices. Outside of these zones, the flow can be treated as irrotational nondisipative flow that is governed by the potential flow equation s and can be solved using the mathematical tools applicable to the solution of the inviscid flow equations. The equations governing these flows must account for the fact that these vortices are present in the flow field. Disregarding the fact that these vortices are generated by the separating viscous layers from the planform edges and also from its surfaces, and that the separation process is certainly dominated by viscous-inviscid flow interactions, it is expected that most of the flow field is dominated by the inviscid flow characteristics. Therefore, such flows can be calculated by various methods using approximate models of inviscid flow. One approach is to consider potential flow into which the vortices are introduced arbitrarily, such as the potential equation solutions andd the panel methods solutions discussed in Chapters 4 to 7. Another approach is based on the direct solutions of the inviscid conservation equations, the Euler equations, where the shock waves and, similarly also, the vortices are "captured", as presented in Chapter 8.

These introductory remarks indicate that the concentrated vortices, generated at the sharp leading-edge and/or on the separation line on the surface, due to the roll-up of the free shear layer or separating boundary layer, have a dominating effect on the flow field at high angles of attack. It is of interest to note the structure of such a line vortex. In the case of potential flow, that is inviscid, incompressible and thus, irrotational flow, assuming an infinitely long line vortex of uniform strength Γ, the potential for the vortex flow is $\phi(r,\theta) = (\Gamma/2\pi)\theta$. The streamlines are circles around the vortex line and the tangential velocity $u_\theta(r) = (\Gamma/2\pi)\cdot(1/r)$. Therefore, there is a singularity as $r \Rightarrow 0$, and the vorticity is concentrated in the vortex line which is positioned at $r = 0$. This singularity can be eliminated by assuming a vortex "core" of a finite radius a around $r = 0$. The flow inside the core can be assumed to be a solid body rotation, where $u_\theta(r) \propto r$ for $0 < r < a$, and outside the core, for $r > a$, the tangential velocity is given by the potential vortex solution $u_\theta \propto 1/r$.

The introduction of the free vortex with an inner core, which has a solid body rotation inside and potential vortex outside the inner core – such a concentrated vortex structure can be justified on the basis of experimental observations and by the theory of viscous flow. Most calculation methods based on potential flow models use either the potential vortex or the free vortex with an inner core, as will be discussed in Chapters 4 to 7. This vortex structure is an over-simplification, since in a three-dimensional flow reality there is an axial velocity to the flow inside the vortex inner core in addition to the solid body rotation. The concentrated vortex flow is then very sensitive to axial pressure gradients and the vortex inner core tends to burst under certain flow conditions. The bursting of the inner core leads to the vortex breakdown phenomenon. Furthermore, there are very intricate effects of compressibility at increased Mach numbers, where the flow velocity inside the core may become supersonic. The methods of calculation based on potential flow approximations are not capable to include these effects, and it remains to the solutions of the full Navier-Stokes equations to be able, in principle, to simulate these flow phenomena.

The initiation and the position of the separation of the free shear layers from the surfaces of the configuration, which then roll-up into concentrated vortices, depend on the state of the boundary layer whether laminar, transitional or turbulent. In these cases, the position of separation on a given configuration is generally determined by semi-empirical correlation relating the position of separation to the Reynolds and Mach numbers of the flow. An example of such semi-empirical correlations of the separation positions on elongated axisymmetrical bodies at various Reynolds and Mach numbers is presented in Section 7.1.3.3. The interactions between the flow characteristics, represented by the Reynolds and Mach numbers, with the configuration geometry and the state of the boundary layer, and their effects on the position of separation and the state of the separating shear layer, is most complex. However, once the position of separation is established (by experimental data, for example), the external flow with its concentrated vortices can be assumed to be governed by the inviscid flow equations, and the appropriate methods of solution, such as panel methods, can be used. In the case of configurations where the separation positions are fixed by the geometry, with no dependence on the state of boundary layer or flow Reynolds number, such as in the cases of sharp leading-edges, the inviscid calculation methods are expected to give reasonable results. For this reason, the reader will note that, in most of the chapters dealing with the inviscid potential solutions, there will be no reference to the flow Reynolds number or to the boundary layer on the surface. This can be justified by reference to results of experimental investigations that indicate that in many cases of such flows (once the separation position is fixed), the Reynolds number and the state of the boundary layer on the configuration have a very small effect on their aerodynamic characteristics. This consideration also justifies the comparison of the results of the calculations using the various inviscid methods with experimental results. In all such cases the effects of Reynolds numbers of the experimental data, are ignored. Although, in many cases, this approximation of using the inviscid calculation methods and ignoring the Reynolds number effect gives reasonable results which can be compared with experimental data, in general, it is known that three-dimensional separations on complex configurations depend both on the flow Reynolds number, the state of the boundary layer and the scale effects which may be characterized by requiring also simulation of a unit Reynolds number. Ignoring all these effects by using any of the inviscid methods is a very significant simplifying approximation which must be remembered whenever these inviscid methods are used.

The correct viscous effects can be, in principle, accounted for by solving the full Navier-Stokes equations. It is expected that solving the viscous conservation equations, the full Navier-Stokes equations, will enable full simulation of the complex viscous-inviscid interactions in the three-dimensional separated flows at high angles of attack, even on complex configurations. This is provided, of course, that we have the correct algorithm and an appropriate grid generation scheme which can be solved on a large and fast supercomputer. However, there are still fundamental difficulties in modeling the turbulent shear and of predicting the transition from laminar to turbulent flows. Therefore, even the present "exact numerical" solutions of the full Navier-Stokes equations for transitional and turbulent flows are limited by the approximations for the turbulent model and transition. The progress that has been achieved in the solutions of the Navier-Stokes equations for laminar and turbulent flows over complex configurations at high angles of attack are discussed in Chapter 9.

In order to familiarize ourselves with the type of flows that may occur due to the development of the three-dimensional separated flow over a slender configuration as the angle of attack increases, let us observe the evolution of such flows which can be described as comprising the following steps. At low angles of attack, the viscous shear layers are confined to separation bubbles attached to the surfaces with only small effects on the aerodynamic characteristics. As the angle of attack is increased to moderate and high angles, the separated layers detach from the surface into the flow as vortex sheets which then roll up into concentrated vortices. For symmetric configuration at zero sideslip, the generated vortex structure is also symmetric up to a certain angle of attack. At higher angles of attack, this symmetric vortex structure may be transformed into an asymmetric structure on the symmetric configuration with zero sideslip. At still higher angles of attack, the cores of the concentrated vortices may "burst" at a certain point on the lifting surface. The position of the vortex "burst" is known to move upstream as the angle of attack is further increased. This is the well known phenomenon of "vortex breakdown". At even higher angles of attack the vortex breakdown occurs nearer to the wing apex, and then the flow may break up into nonsteady turbulent wake.

Thus, for an incoming steady flow of known Mach and Reynolds numbers, we can distinguish the following flow regimes on slender configurations as a function of the increasing angle of attack:

I. Very low angles of attack – attached, symmetric, steady flow. Linear lift variation with angle of attack.

II. Low angles of attack – attached, symmetric, steady flow with closed separation bubbles. Nearly linear lift variation with angle of attack.

III. Moderate and high angles of attack – separated, symmetric, rolled-up vortices in steady flow. Nonlinear lift variation with angle of attack.

IV. High angles of attack – separated, asymmetric, rolled-up vortices in steady flow. Nonlinear lift variation with angle of attack.

V. Higher angles of attack – vortex breakdown causing loss of lift and nonsteady flow conditions.

VI. Very high angles of attack – separated, nonsteady turbulent wake. Post stall aerodynamic characteristics.

Linear aerodynamic theories and numerical methods (discussed in Chapter 4) can be used to evaluate the aerodynamic characteristics of configurations in categories I and II at low angle of attack. The rolling-up of the vortex sheet behind the wing at low angles of attack is presented in Chapter 5. For the calculations of the nonlinear aerodynamic characteristics in categories III and IV, we must use the nonlinear methods, as discussed in Chapters 6 to 9. It should be noted that on various configurations at certain flow Mach numbers, vortex breakdown (category IV) may begin on the lifting surface at moderate angles of attack, thus eliminating for that flow case the asymmetric vortex flow postulated in category III. There are some indications that the occurrence of vortex breakdown and that the flow characteristics in category V can be studied using the Euler equations solution, discusse d in Chapter 8. We may hope that the solutions to the Navier-Stokes equations, as discussed in Chapter 9, may be useful for studies of vortex breakdown and flows in the stall and post stall categories.

A good illustration of the type of flows that can be expected on aircraft and missile configurations at moderate and high angles of attack can be obtained from the observation of flows over slender sharp edge delta wings. The flow configurations over a delta wing with leading-edge sweep angle Λ, which is positioned at an angle of attack α at Mach number M_∞, can be classified in accordance to the angle of attack measured normal to the leading-edge and the normal component of the Mach number $\alpha_N = \tan^{-1}(\tan\alpha/\tan\Lambda)$ and $M_N = [1 - \sin^2\Lambda \cdot \cos^2\alpha]^{1/2}$, presented in Fig. 1.1.

It is seen that when the Mach number normal to the leading-edge M_N is less than 1, the subsonic rolled-up vortex structure is obtained even when the external flow is supersonic. Thus, vortical flow structures which are identical to those obtained at subsonic speeds are also observed over a very slender delta wing in supersonic flow, including the transition to asymmetric vortex structure. However, for the supersonic Mach number normal to the leading-edge, $M_N > 1$, there is a Prandtl-Meyer expansion at the leading-edge on the leeward side of the wing, causing an over-expansion which requires a terminating shock wave. As the angle of attack increases, the terminating shock wave strength increases, causing separation of the boundary layer on the wing surface. At moderate angles of attack, the separated boundary layer reattaches to form a separation bubble. At higher angles of attack, the separating boundary layer detaches from the surface and rolls up to concentrated vortices with the terminating shock waves.

This brief introduction should convince the reader that we are dealing with very complex flow phenomena which are associated with high angles of attack aerodynamics. Therefore, the present book is arranged in accordance with the following structure. In order to initiate some appreciation, and provide a basis, for analytical and numerical approximate methods for

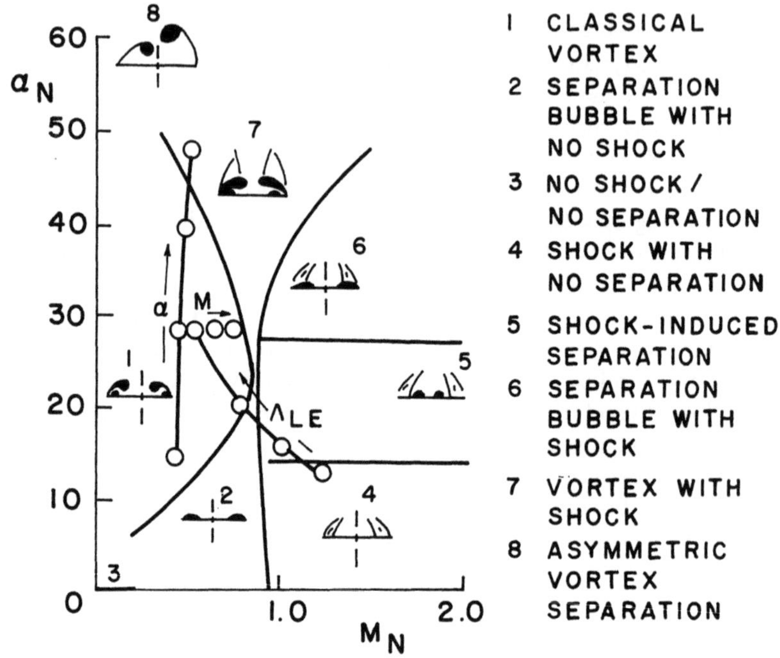

Fig. 1.1. Classification of flow structures over slender sharp-edge delta wings

these complicated flow structures that do develop on simple and, more so, on complex configurations at increasing angles of attack, Chapter 2 is devoted to descriptive introduction of these complicated three-dimensional separated flows un many cases by the use of visualization photographs. The selected results of the visualization experiments serve to illustrate the various important flow phenomena which are characteristic of the viscous vortical flow field over various configurations. The film technique has been used to visualize the surface skin-friction line patterns generated by the three-dimensional separated flow over the configuration. Interpreting these patterns and relating them to the overall vortical flow structure can be done by the utilization of topological concepts and rules. An introduction to the application of the topological concepts and rules to flow structures at high angles of attack is presented in Chapter 3. Thus, the phenomenological descriptions of the high angles of attack flows in Chapter 2 and the topological concepts and rules in Chapter 3 will be useful for the understanding and interpretation of the complex flow structures at high angles of attack. This will give a good fundamental understanding of the physical phenomena which will be very useful when applying the various analytical and numerical methods for the calculations of the aerodynamic characteristics.

Since the calculations of the nonlinear aerodynamic characteristics of configurations at high angles of attack are based on the methods developed for the calculations of the linear characteristics, it is required and recommended to begin by presenting some of the linear aerodynamic methods, which is the subject matter of Chapter 4. The characteristics of the rolling-up of a vortex sheet in the wake of a lifting surface is discussed in Chapter 5. In this presentation, the structure of the rolled-up concentrated vortex, including its core, is detailed. Some analytical and semi-empirical methods for calculating the nonlinear aerodynamic characteristics of wings and bodies at high angles of attack are described in Chapter 6. The panel methods for the calculations of the nonlinear aerodynamic characteristics of aircraft and missile configurations are presented in Chapter 7. The capability of the solution of the inviscid Euler equations to capture the three-dimensional vortical flow structure and to compute the nonlinear aerodynamic characteristics is discussed in Chapter 8. The recent progress in the developments of methods of solution for the Navier-Stokes equations is presented in Chapter 9.

Chapter 2

Description of Flows at High Angles of Attack

The distinguishing features of flows at high angles of attack are caused by the generation of free shear layers at sharp leading edges, by separation of the viscous layers from the surfaces of wings and bodies and by the flow in the wakes of the wings and bodies. These types of flow structures, which are in general the result of three-dimensional separation, induce velocity fields which result from the strong interactions between the generated vortical flows, usually including concentrated vortex cores, and the outer flow governed by the configuration's geometry. This complicated flow field is inherently three-dimensional in its structure. The vortical flow is generated either by the separated viscous shear layers at the sharp leading and side edges (in addition to the "classical" separation at the trailing edge as expressed by the Kutta condition), or by the three-dimensional boundary layer separation from the surfaces of the wings and bodies. As can be anticipated, many unexpected flow structures may occur in these complicated flow fields. Therefore, it is necessary to study and classify the various elements of flow structures which can occur over the various aerodynamic configurations at increasing angles of attack.

Theoretical analyses of such complicated three-dimensional flows with strong viscous-inviscid interactions are unattainable, certainly, at present and, probably, also in the foreseeable future. Even numerical solutions for the governing three-dimensional equations, the Navier-Stokes equations with three-dimensional boundary conditions, are prohibitive, at the present and, probably, will remain so in the near future and even beyond the next (one or two) generations of supercomputers. Consequently, at present and in the near future, we still must rely on approximate theories and simplified models to calculate the variations of the aerodynamic characteristics with a reasonable degree of approximation, knowing that these must be substantiated in experiments both in wind tunnels and in flight tests, hoping to approach the true values to within "engineering accuracy". It is well known that practical design of aircraft and missile configuration must rely very heavily on experimental results from both wind tunnel and flight tests. This is particularly so in the evaluation of the aerodynamic characteristics at high angles of attack.

In order to study, understand and catalogue the three-dimensional flow structures at increasing angles of attack, it is best to begin by the study of flow visualization photographs obtained in experiments and to analyze the corresponding aerodynamic effects. In the following presentations, we chose those photographs of visualization experiments that illustrate, as clearly as possible, the various important flow phenomena which are characteristic of the viscous-vortical flow fields on the different configurations; we indicate the corresponding aerodynamic characteristics as the angles of attack increase.

The concentrated vortex lines-tubes, rotating around their viscous cores, are known to "breakdown" under certain flow conditions. Flow visualization can be used to observe the phenomena associated with this vortex breakdown and to study the conditions under which breakdown occurs. Vortex breakdown affects the induced flow fields and causes loss of aerodynamic lift forces, leading to the stall of the lifting surfaces. The breakdown may not be symmetrical and/or steady, causing the appearance of asymmetric loading, lateral forces and moments, and nonsteady phenomena such as wing rock. The subject of vortex breakdown is not discussed in details in the present book, however, a brief description of some of the flow elements which are observed when "breakdown" occurs is included in this phenomenological chapter for completeness.

These visualization experiments can also be used to define the flow regimes in terms of the steadiness of the flow, i.e. the appearance of periodic fluctuations or the breakdown of the "organized" structures. It is observed that in many practical cases, the process of the flow development, even up to reasonably high angles of attack, is steady, with no time-dependent variations. Therefore, these flows can be analyzed using the time-independent steady flow equations. However, there are certain values for angles of attack under which the separated vortex flow becomes nonsteady, either with periodic time variation or with chaotic flow structure. Nonsteady flow conditions are also obtained at varying attitudes of the configurations when high turning rates maneuvers or oscillatory motions of the complete configuration, or of certain parts thereof (such as oscillations of control surfaces), are performed.

In order to present a complete overview of the phenomena observed in flows at high angles of attack, we include a brief description of the results of visualization experiments focusing on vortex breakdown; some effects of the oscillatory variation of angle of attack on aerodynamic characteristics will also be described. However, the emphasis of this book, both in the descriptive presentation in this chapter and, later, in the presentation of the various analysis and calculation methods will be on the study of the aerodynamics characteristics at steady flow conditions.

The description of the various flow structures will be best accomplished by studying the photographs of visualization experiments on various configurations at varying attitudes, beginning with simple lifting surface planforms, then moving to more involved geometries and, finally, to full aircraft and missile configurations.

2.1 The Medium and High AR Wings at Low Subsonic Speeds

A clear example of a flow field which is dominated by the induced vortical flow is seen in the flow field established by the vortices shed in the wake of the finite lifting wing of medium and high aspect ratios at small and up to moderate angles of attack. These vortices, which are shed from the wing side edges and trailing edges, roll up downstream of the trailing edge into the well-defined tip vortices. The generation of rolled-up "tip vortices" starts with the lift generation already at relatively small angles of attack and, as the angle of attack is increased, the strength and size of these tip vortices also increase. An excellent illustration of the vortical wake and the rolling up of the vortices shed from the side edges of the rectangular wing is presented in a top view in Fig. 2.1 and in a rear view at various cross-sections behind the trailing edge in Fig. 2.2. Both the top view (Fig. 2.1) and particularly the rear view of the cross-sections (Fig. 2.2) clearly show the essentially inviscid rolled-up vortex wrapped around a well defined viscous vortex core.

It is also clear from these photographs that the diameter of the core of these vortices increases only very gradually, reaching an asymptotic core diameter as the downstream distance is increased. It is well known that these tip vortices persist in the wing's wake for very long distances (thousands of span lengths in an undisturbed atmosphere) and that the actual deterioration of the pair of the tip vortices is due to the onset of an instability mechanism (Crow 1970), destroying the linear structure of the pair, and *not* by viscous dissipation.

The aerodynamic characteristics of the wing and its near-vortex wake can be calculated by assuming that these tip vortices have a viscous core (considered as solid body rotation) which reaches an asymptotic size within a few span lengths behind the trailing edges and from there on, the tip vortices extend far downstream keeping the core diameter constant. Therefore, outside of the viscous vortex cores, the flow is dominated by the rotational vortex flow which may be considered mainly inviscid. At small and moderate angles of attack, the effect of the rolled-up tip vortices on the aerodynamic characteristics of the high AR wing is small, so that at low subsonic speeds, the variation of these coefficients with angle of attack remains linear. As the angle of attack is increased, first, a closed separation bubble is developed over certain parts of the wing surface which will result in aerodynamic deformation of the wing profile shape and then, further increase of the angle of attack will cause the separation bubble to open up resulting in

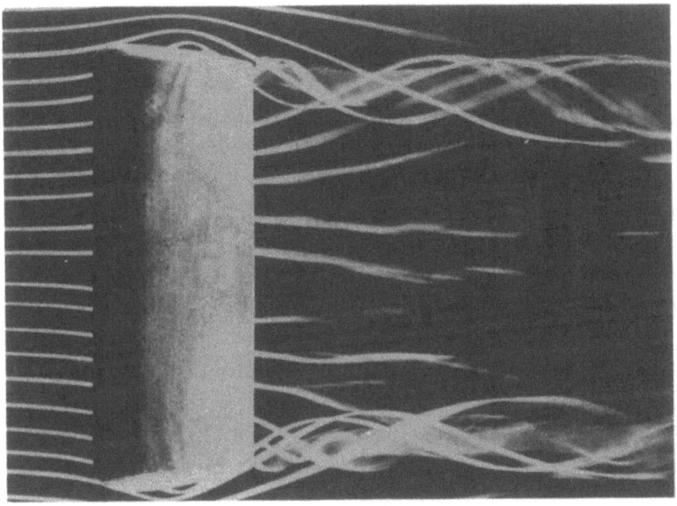

Fig. 2.1. The rolled-up tip vortices behind a rectangular wing-top view (Head 1982)

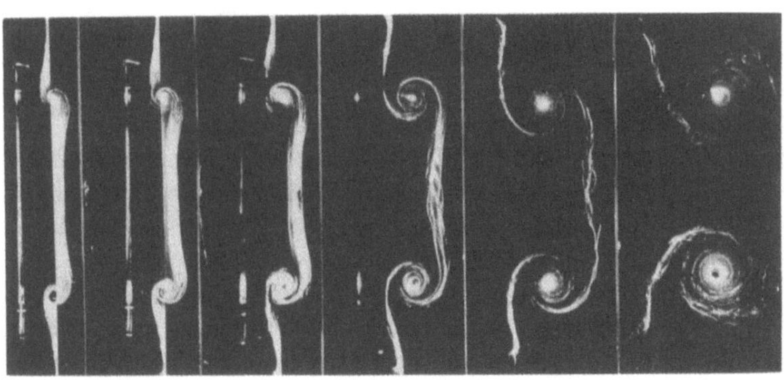

Fig. 2.2. The rolled-up tip vortices behind a rectangular wing – cross-section rear view (Bippes 1972)

a large separated region over the wing involving a large loss of the lift force. Further increase in the angle of attack will increase the separated region. As a result of this increase in the loss of lift force – the wing is "stalled". At stall and at angles of attack beyond stall, for the post stall attitudes, three-dimensional separations and various types of rolled-up vortices may be generated. The local flow structure of the vortices separated near the tips of the high aspect ratio wing is similar to the vortex flow over low aspect ratio wing and since configurations for high angles of attack flight are in general slender with low aspect ratio wings, we will not dwell on the post-stall flow over high aspect ratio wings. It is important to remind the reader that the stall can be delayed by various high lift devices. High lift devices, such as the various types of flaps, jet flaps, etc., are well known aerodynamic tools described in many textbooks and will not be described in the present book. However, the separated vortical flows over low aspect ratio wings, which will be discussed in this book, can be used as a good qualitative guide to the types of flows that can be seen also on high aspect ratio wings for post-stall conditions.

2.2 The Low AR Rectangular Wing at Low Subsonic Speeds

As the span of the rectangular wing is decreased, resulting in a low AR case, the vortices, which are shed from the side edges, roll up over the wing surface. For the low AR wing, these rolled-up vortices affect a large portion of the wing. The induced flow due to these rolled-up vortices causes the generation of additional aerodynamic forces, resulting in the nonlinear variation of the aerodynamic coefficients with increasing angles of attack. The rolling up of the vortices shed from the side edge of a rectangular wing is shown in the visualization photograph in Fig. 2.3, which also includes an illustration of the vortex structure over the low AR wing.

At higher angles of attack, another type of flow is observed: the viscous boundary layer flow, attached to the aerodynamic surfaces, tending to separate from these surfaces at increasing angles of attack in regions of adverse pressure gradients. These separations are inherently *three-dimensional* phenomena, even when the attached flow over the aerodynamic configuration indicated essentially *two-dimensional* characteristics. Initially, at small angles of attack, the separated flow from a flat surface reattaches further downstream so as to form a "separation bubble" near the leading edges of the flat configuration. As the angle of attack is further increased, the separated shear layer does not reattach and flows into the free stream. These free shear layers tend to roll up into vortices and form various *three-dimensional* vortical flows. These may take the form of the rolled up leading edge vortices or other vortex types, depending on the specific characteristics of the corresponding flow fields. As the angle of attack is further increased, secondary vortices, as well as more complicated separated flows, are observed.

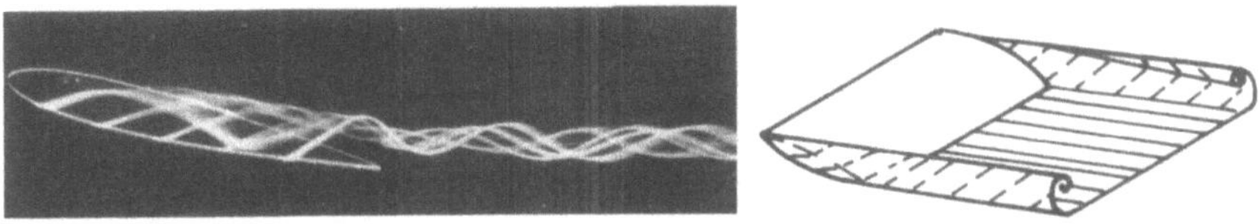

Experiment (Werle 1974) Illustration

Fig. 2.3. The rolling up of the vortices from the side edge of a rectangular wing

An example of such an intriguing three-dimensional separated flow structure is observed at low speeds on the low AR rectangular wing at high angles of attack, as illustrated in Fig. 2.4. At the high angles of attack, the flow becomes more "complicated", as local three-dimensional separations cause the separated shear layers to roll up into vortex "spirals" that detach from the planar surface of the wing. The visualization technique utilizes an oil film which is forced to follow the skin friction lines on the surface by the flow friction forces. Regions of attached flow and various types of separation pattern are observed on the wing surface, as shown in Fig. 2.4.

The skin friction line pattern, which includes several singular points, is the basis for a topological interpretation of such flows. In this case, the particular interpretation for this flow was proposed by Peake and Tobak (1980) in the perspective picture in Fig. 2.5, which is reconstructed from the oil flow pattern on the surface. The complicated structure of the primary and secondary vortices, which separate and reattach at various positions on the slender rectangular wing at the high angles of attack, are shown both in the top view of the lee-side of the wing and in the perspective in Fig. 2.5.

2.3 The Slender Delta Type Wings

One way to visualize the flow over a slender delta wing, which was suggested by Rehbach (1973), is to consider the flow established on a deformed slender rectangular wing as its leading edge span is deformed towards zero span, i.e. a pointed nose. The final stage of this deformation is a delta wing where the two side edges of the slender rectangular wing become the leading edges of the triangular wing. The vortices, shed from these leading edges, roll up over the wing surface and cause the nonlinear variation of the lift as the angle of attack is increased.

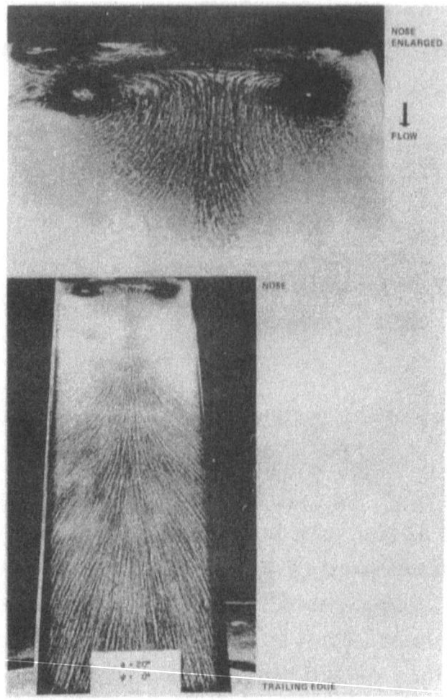

Fig. 2.4. Oil flow visualization of the skin friction lines on the top surface of a rectangular wing at high angles of attack (Wickens 1966)

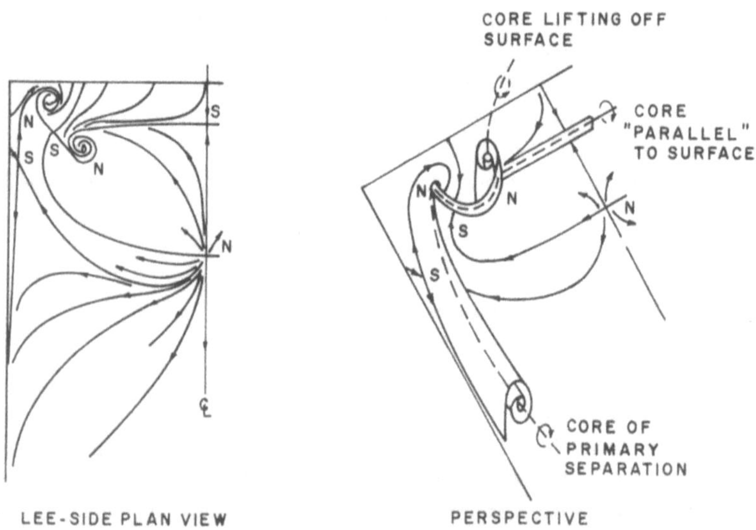

Fig. 2.5. Topological singularities in the skin friction lines observed on the surface of the low AR rectangular wing at high angles of attack, according to Peake and Tobak (1980)

2.3.1 Sharp Edges Delta Wings at Low Subsonic Speeds

In some respects, this rolled-up flow is a much simpler flow structure than that observed on the low AR rectangular wing at reasonably high angles of attack. Some early flow visualization photographs, illustrating the rolling-up of the leading edge vortices over delta wings with sharp edges starting at low angle of attack, are shown in Fig. 2.6. The separation of a vortex sheet from the sharp leading edges, already at small angles of attack, and the rolling-up of the vortex sheet at moderate angles of attack, are very well illustrated in this figure. The process of the generation of the leading edges rolled-up vortices and the associated flow field, including the appearance of the secondary vortices as the angle of attack increases, is described graphically in Fig. 2.7.

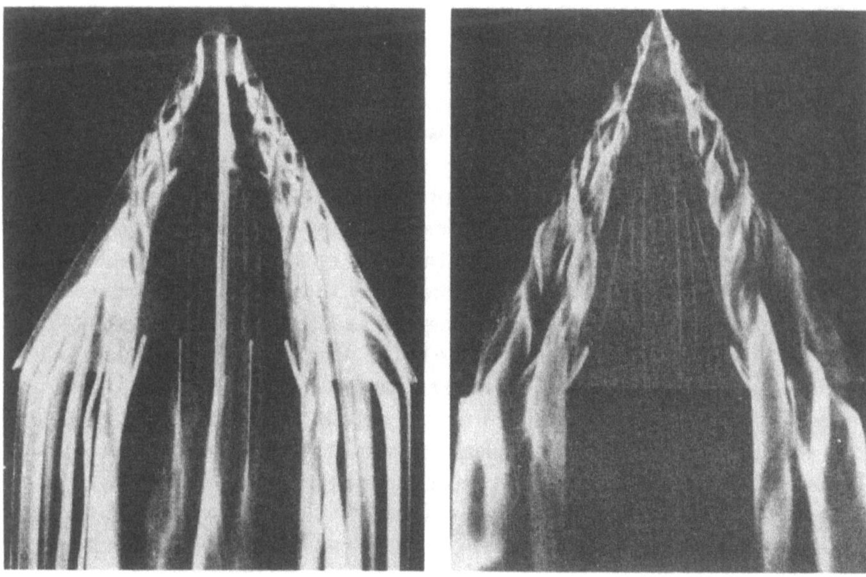

a. Small angle of attack – bubble separation case, $\alpha = 3^o$

b. Moderate angle of attack – rolled-up vortices case, $\alpha = 12^o$

Fig. 2.6. Flow over delta wings with sharp leading edges (Werle 1958)

Illustration of the fully developed flow over the low AR delta wing at reasonably high angles of attack, $\alpha \cong 20^o - 25^o$, is presented in Figs. 2.8 and 2.9. The streamline pattern above the delta wing is illustrated in Fig. 2.8. The turning of the streamlines into the wing surface, due to the induced velocities of the rolled-up leading edge vortices, is clearly seen in the side elevation and the separation and attachment lines of the primary and secondary vortices are seen in the plan view from the top. These streamline patterns are in accordance with the illustrated flow field presented in Fig.

2.7c. It is important to note that, although the initial separation of the vortex sheets at the sharp leading edges is due to viscous separation as the angle of attack increases, these vortex sheets roll up into fully developed leading edges vortices. These rolled-up vortices, which are clearly seen in the cross section view in Fig. 2.9, feature a viscous core, with the vortex flow around it being essentially inviscid. The experimental evidence that the viscous effects are not important and that the flow over the wing is dominated by the inviscid vortex structure is illustrated in Fig. 2.10, where the experimental data on the vortex flow structure over slender delta wings at various Reynolds numbers are shown. It is indicated there that the main features of the vortical flow vary only slightly, while the flow over the wing moves from being initially a laminar viscous flow ($Re_c = 0.8 \cdot 10^6$) through being transitional ($Re_c = 2.5 \cdot 10^6$) and finally becoming fully turbulent ($Re_c = 4.1 \cdot 10^6$). These photographs show that, even though the viscous flow in these experiments spans the range of laminar, transitional and turbulent flows, it is evident that the leading edge vortices in all three cases are very similar in their shape and position over the wing. The variation of the Reynolds number in these tests is seen to have only a small effect on the shape and location of these vortices. Thus, for these three Reynolds numbers, spanning flows which are laminar, transitional and turbulent, the basic flow structure, featuring the leading edge vortices, is very similar. This experimental evidence can be taken as one of the justifications for considering such flows as being dominated by the inviscid vortex flows.

The vortex flow over a delta wing is also affected by the introduction of a vertical fin on the leeward surface of the wing. The vertical fin actually defines a corner-type flow. The interaction between the rolled-up leading edge vortices and the corner flow near the normal vertical fin, positioned on the rear part of a delta wing, causes the appearance of secondary vortices. This typical corner flow with its secondary vortices is shown in Fig. 2.11. The effect of the vertical fin and the corner flow is made clear by comparing this flow (Fig. 2.11) with the flow over a "clean" delta wing presented in Fig. 2.9.

2.3.2 Spiral Type Separations on Delta Wings at Very High Angles of Attack at Low Speeds

As the angle of attack is inreased, new separation phenomena are observed. In addition to the separation from the leading edge and/or side edges generating the free vortex sheets which curl up into the concentrated rolled-up vortices (already discussed in the previous section), as the angle of attack is further increased, the boundary layer on the leeward side of the lifting surface may separate at certain points. This separation has generally the nature of a focus-type separation, in which the separated shear layer rolls up in a spiral motion into a "tornado" type vortex stem, as illustrated in Fig. 2.12. The spiral flow separation is observed in this case in a water

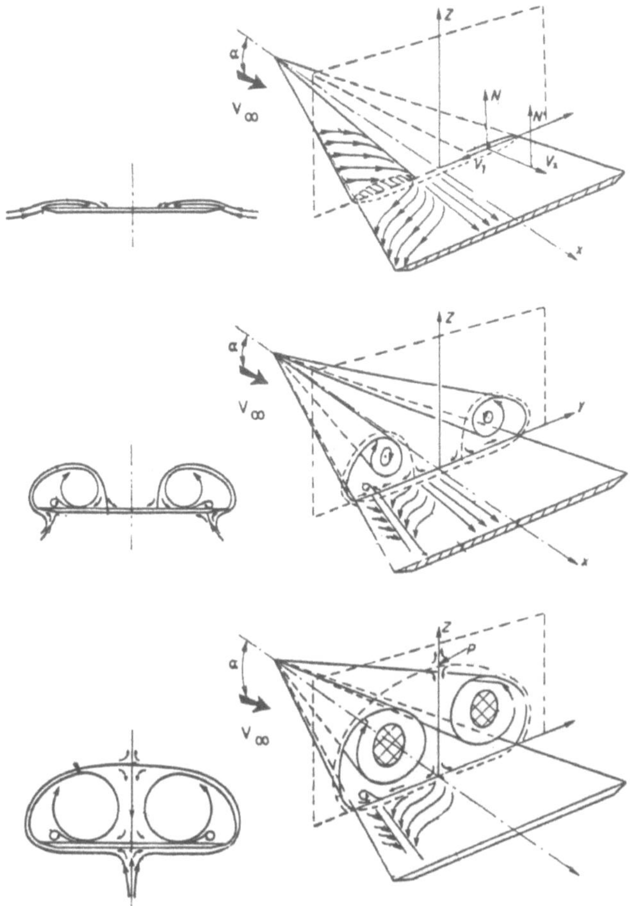

Top – Small α – separation bubble at leading edges

Medium – Medium α – two rolled-up vortices reattaching on surface

Bottom – High α – large rolled-up vortices with one reattachment line on surface and one singularity in free stream

Fig. 2.7. Structure of flows over delta wings with sharp leading edges (Werle 1958)

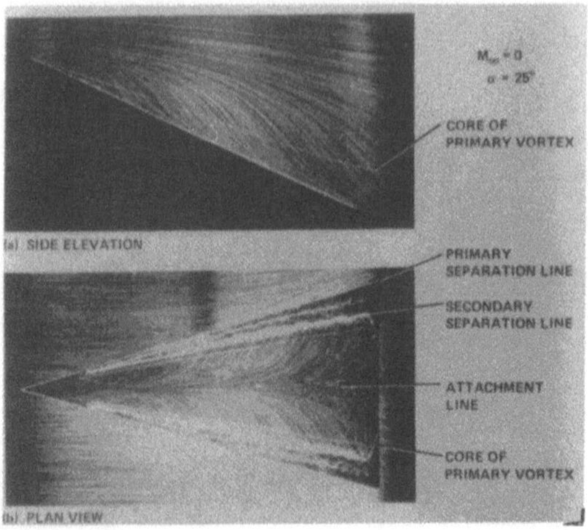

Fig. 2.8. Visualization of streamlines and skin friction lines on a 75° delta wing (Monnerie and Werle 1968)

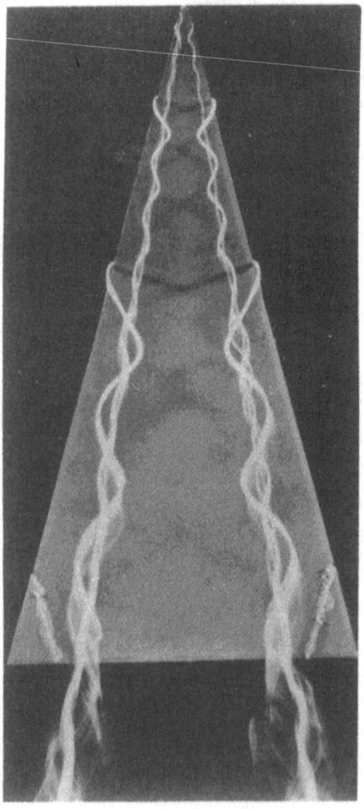

Fig. 2.9. Visualization of the rolled-up vortices over a delta wing – top view and cross-section view (Werle 1958)

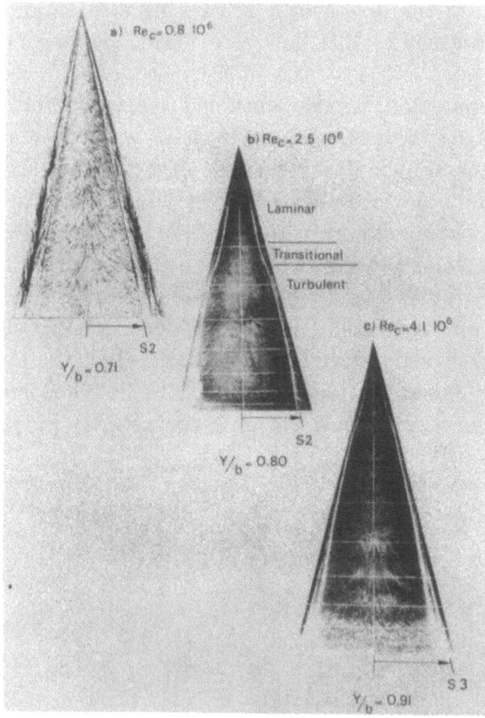

Fig. 2.10. Oil flow visualization on delta wings at various Reynold numbers – effects of laminar, transitional and turbulent flows (Carcaillet et al. 1986)

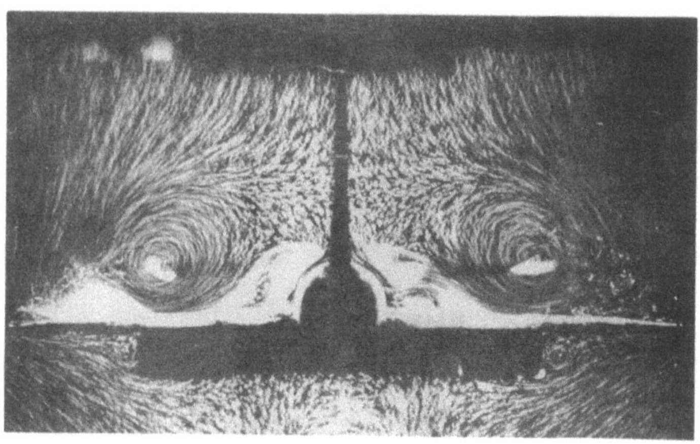

Fig. 2.11. The effect of a vertical fin on the flow over a delta wing (courtesy of Aeronautical Journal)

tunnel visualization of the flow over a delta wing at high angles of attack. The topological structure of this flow, postulated by Legendre (1965), is also shown.

Spiral type separation is also obtained on thick delta wings with rounded leading edges. The flow over the delta wing with rounded leading edges and thickness, at increasing angles of attack, is illustrated in Fig. 2.13. In this case, the more complicated flow is affected by the strong interaction between the viscous layers (which include local separated zones) and the outer flow. The vortex layers, which are separated from the surface of the delta wing with rounded leading edges, roll up into a three-dimensional spiral-type separation structure. The rounded leading edge also causes the formation of secondary separation layers, which include "bubble" type separation, and, as the angle of attack increases, the flow becomes more complicated by the generation of new vortex structures, as illustrated in Fig. 2.13.

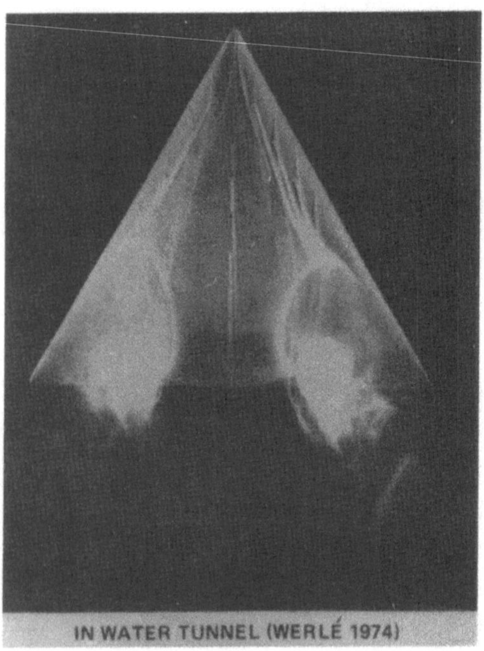

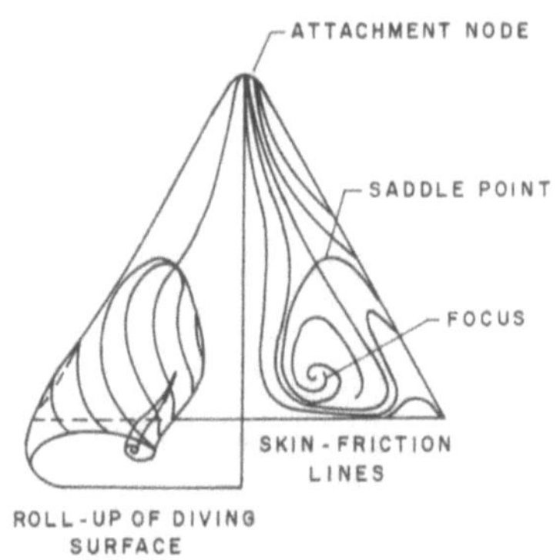

POSTULATED PATTERNS OF LEGENDRE 1965

Fig. 2.12. Three-dimensional flow separation over a delta wing at high angles of attack (Werle 1974, Legendre 1965)

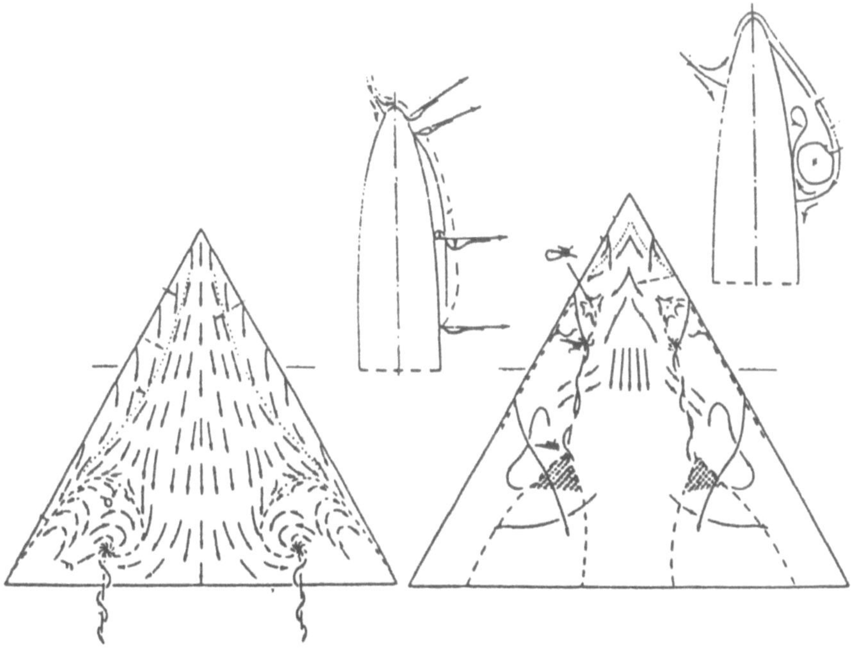

Fig. 2.13. Flow structure on a thick delta wing with rounded leading edges at moderate and high angles of attack (Werle 1958)

2.3.3 Slender Delta Type Wings in Supersonic Flows

It is interesting to note that vortex structures, similar to those observed in subsonic flow, are also observed at supersonic speeds in the case of flows over slender configurations at high angles of attack. This is due to the fact that for very slender configurations, the cross flow component of the velocity in the direction normal to the lifting surface is much smaller than the undisturbed supersonic free stream velocity. Thus, vortices can occur in the cross flow similar to those in the subsonic flow case.

A new phenomenon is observed in the vortex structure over very slender configurations at increasing angles of attack – the onset of asymmetric vortex structures. In supersonic flow, the onset of asymmetric flow is observed already at relatively low angles of attack, at about 15° to 17°, as seen in Fig. 2.14. It should be noted that flow asymmetry also develops over slender configurations in subsonic flows. However, in the subsonic flow case, the flow asymmetry appears only at relatively large angles of attack, over 35°, and will be discussed in a later section. The asymmetric flow structure over a slender delta wing in supersonic flow is demonstrated in Fig. 2.14. It is seen that up to moderate angles of attack (up to $\alpha = 15^o$), the flow field,

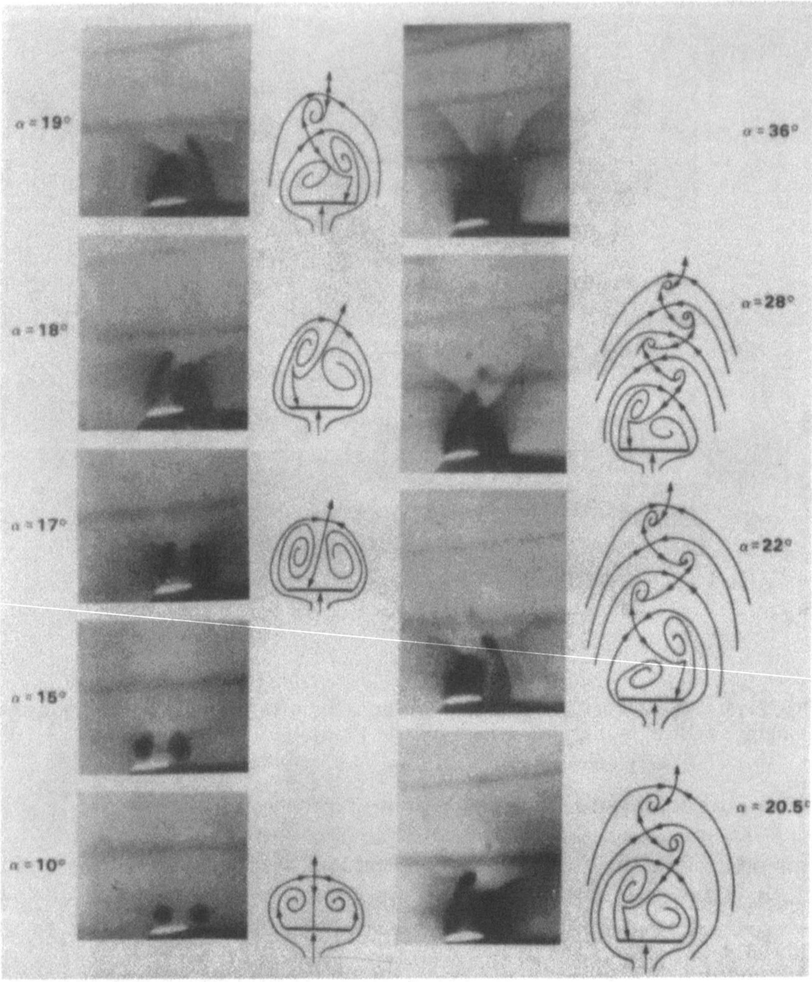

Fig. 2.14. The development of vortices over a slender delta wing at supersonic speeds – the symmetric and asymmetric separation structure variations at increasing angles of attack at $M_\infty = 2.8$ (Fellows and Carter 1969)

which includes the separation zones and the rolled-up vortices, is symmetric and very similar to the symmetric flow over the delta wing in subsonic flow. As the angle of attack is increased to $\alpha = 17^o$, the vortex flow is "suddenly" deformed into an asymmetric flow structure. This asymmetric flow structure is stable; i.e. at the specific angle of attack, the flow remains steady at the specific asymmetrical positions of the rolled up vortices. As the angle of attack is further increased, the flow asymmetry may increase or, at certain angles of attack, "suddenly" the flow structure may "jump" into an opposite asymmetry (in this case, the first jump from left to right occurs between $\alpha = 18^o$ and $\alpha = 19^o$) as is well demonstrated in Fig. 2.14. The most intriguing aspect of this phenomenon is the fact that initial direction of the flow asymmetry is unpredictable. Similar models under similar flow conditions may start with either left or right asymmetry. However, once this asymmetry is established (to the right or to the left) at the specific angle of attack, a further increase of the angle of attack will result in the "jump" to the opposite asymmetry at the same angles of attack, in a predictable and repeatable order. The topological structure of the flow at increased angle of attack, proposed by Peake and Tobak (1980), are also illustrated in this figure. At still higher angles of attack, multiple vortex pairs are observed, "jumping" from the left to the right sides, alternatively, as the angle of attack increases. Eventually, at a certain high angle this vortex structure will break down and become a turbulent wake region.

2.4 The Flow Over Elongated Slender Bodies

2.4.1 The Symmetric Vortex Flows over Slender Bodies in Subsonic Flows

The aerodynamic characteristics of bodies at high angles of attack are dependent on the shape and position of the separation line. For rounded bodies, slender or not, the position of separation is a result of the viscous-inviscid interaction of the flow over the body. Therefore, the line of separation of the vortex sheet will depend on the body's geometry and the specific flow conditions. But once the separation position is established, the vortex sheet, which separates from the body surface, curls up into the well familiar rolled-up concentrated vortex structure. This is well illustrated for the low speed flow over an Ogive-Cylinder model positioned at $\alpha = 30^o$ in the excellent visualization photograph presented in Fig. 2.15. The details of the rolled-up symmetric vortices and particularly, the well-defined vortex cores in the flow over Ogive-Cylinders in the cross flow plan are seen in Fig. 2.16.

The effect of the nose geometry (sharp elongated or blunt rounded) on the type of separated flows which can be observed on such bodies is illustrated in Figs. 2.17a, 2.17b and 2.17c, which also include the topological structure of these flows. The separation lines on an elongated sharp-nosed body are drawn out in Fig. 2.17a, showing the "regular" separation with the symmetric rolled-up vortices. The flow structure for a blunt spherical-nosed body at the higher angle of attack, is shown in Fig. 2.17b, where, in addition to the primary rolled-up symmetric vortices, secondary vortices are also generated with the corresponding reattachment lines. In the case of higher angles of attack on a blunt nosed body, a local separation zone funneling the shear layer into a spiral type vortex is obtained, as shown in Fig. 2.17c.

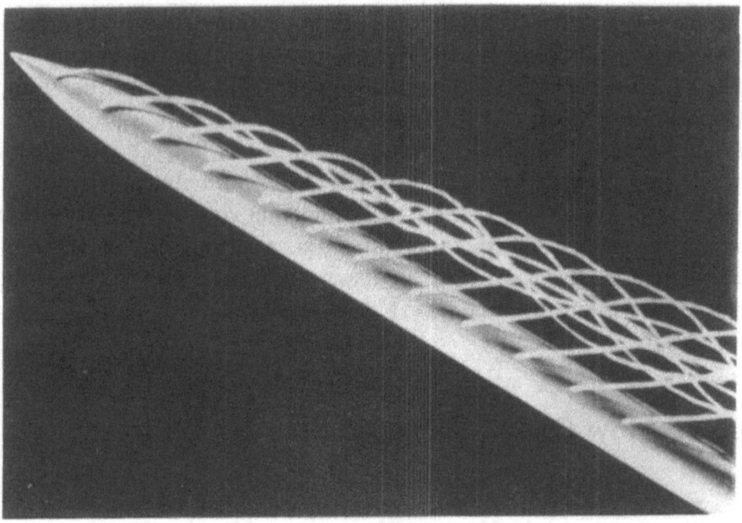

Fig. 2.15. The vortex structure on a slender Ogive-Cylinder body at high angles of attack (Fiechter 1969)

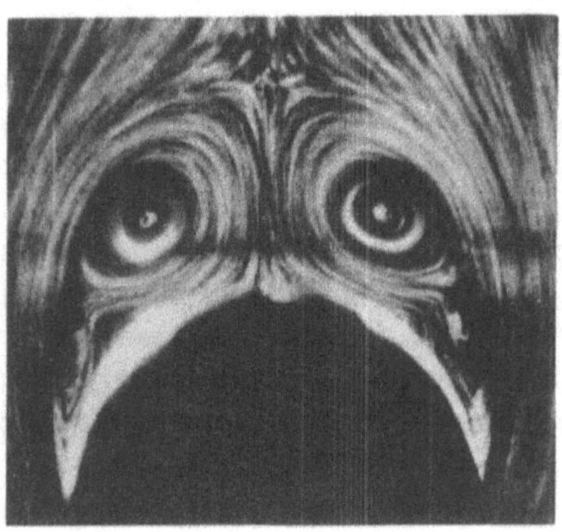

Fig. 2.16. A cross-section rear view of the rolled-up vortices over the slender Ogive-Cylinder at high angles of attack (Werle 1974)

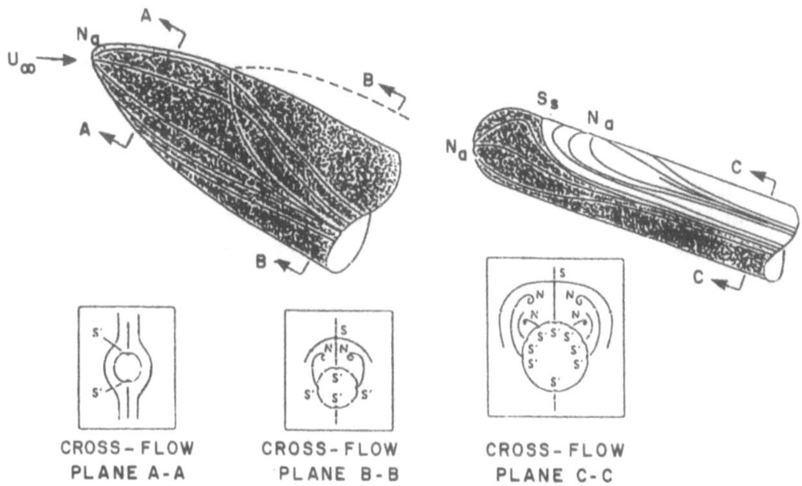

a. Regular primary separation structure at low α

b. Primary and secondary vortex separation at moderate α

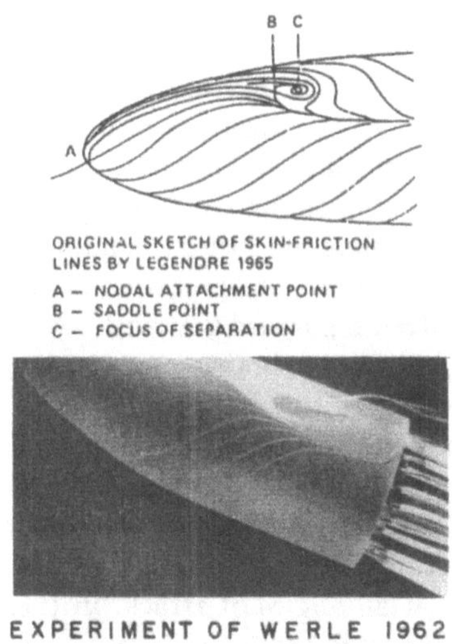

c. Foci type separation at high α

Fig. 2.17. The separated vortical flows over blunt elongated bodies

2.4.2 The Asymmetric Vortex Flow over Slender Bodies in Subsonic Flow

The symmetric vortex structure established on the elongated wing or body can be transformed into an asymmetric structure as the angle of attack is increased also at subsonic speeds. At subsonic speeds, the angle of attack for which the flow becomes asymmetric is relatively high in comparison to that observed in the case of supersonic flow. The visualization of the vortex flow in the wake over a slender Ogive-Cylinder body is presented in Fig. 2.18. The rolled-up vortices, seen in a side view in this figure, show a steady symmetrical vortex structure at $\alpha = 25^o$. The flow becomes asymmetric but still remains a steady vortex wake flow at $\alpha = 48^o$. At a higher angle of attack, at $\alpha = 60^o$, this vortex wake breaks up into a nonsteady chaotic turbulent structure. The corresponding variation of the aerodynamic forces and moments with angle of attack is of interest. For the symmetric flow range, when the vortex wake is composed of the steady rolled-up vortices, the lift force increases with increasing angles of attack while the lateral forces and moments are negligible. At increasing angles of attack, as the flow becomes asymmetric but steady, the lift is still increasing at a lower rate, but a new and very important effect is observed, the appearance of a significant side force and moment. The direction of the side force and the yawing moment may be reversed at various angles of attack. The relatively large size and the direction reversal of these lateral aerodynamic coefficients is a phenomenon that will be of importance for the performance of slender aircraft with long nose sections.

The asymmetric vortices at various stations on an Ogive-Cylinder model at an angle of attack of 40° are shown in Fig. 2.19. The visualization photographs present the positions of the cores of the vortices above the cylindrical body. Included in the figure is a topological structure of the flow at various cross-sections along the body axis. These structures, illustrating the limiting streamlines and the streamlines about the various types of singular point, enable better interpretation of these flows.

A much more involved separated flow with asymmetric vortex structure can be obtained for noncircular bodies, even at small or moderate angles of attack. The vortex shapes that can be seen with a square cross-section elongated body at moderate angle of attack and at various roll attitudes are shown schematically in Fig. 2.20. The topological structure of these flows is indicated for each roll attitude of the square body. It can be seen that asymmetric structure can be obtained for roll angles which result in an asymmetric geometry at all angles of attack. In these cases of asymmetric geometries, it can be expected that significant side forces and moments will be generated even at small angles of attack. These effects are of importance for configurations with noncircular bodies, used for some missile bodies or for some aircraft fuselages. Some available results for noncircular bodies can be found in the publications of Polhamus (1959), Lijewski et al. (1982) and Clarkson et al. (1982) and in a summary paper of Sigal (1990).

a. Symmetric and steady vortices at $\alpha = 25^o$

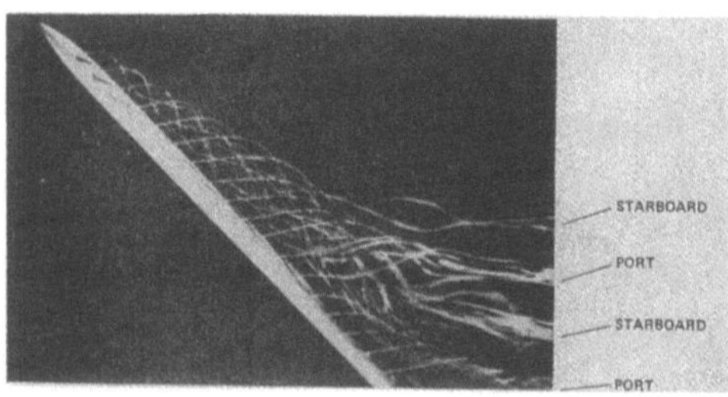

b. Asymmetric and steady vortices at $\alpha = 48^o$

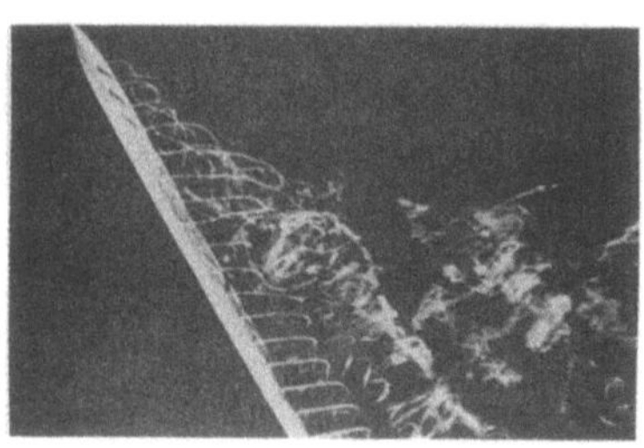

c. Asymmetric and nonsteady vortex flow at $\alpha = 60^o$

Fig. 2.18. Development of vortex flows over elongated bodies at increasing angles of attack (Fiechter 1966)

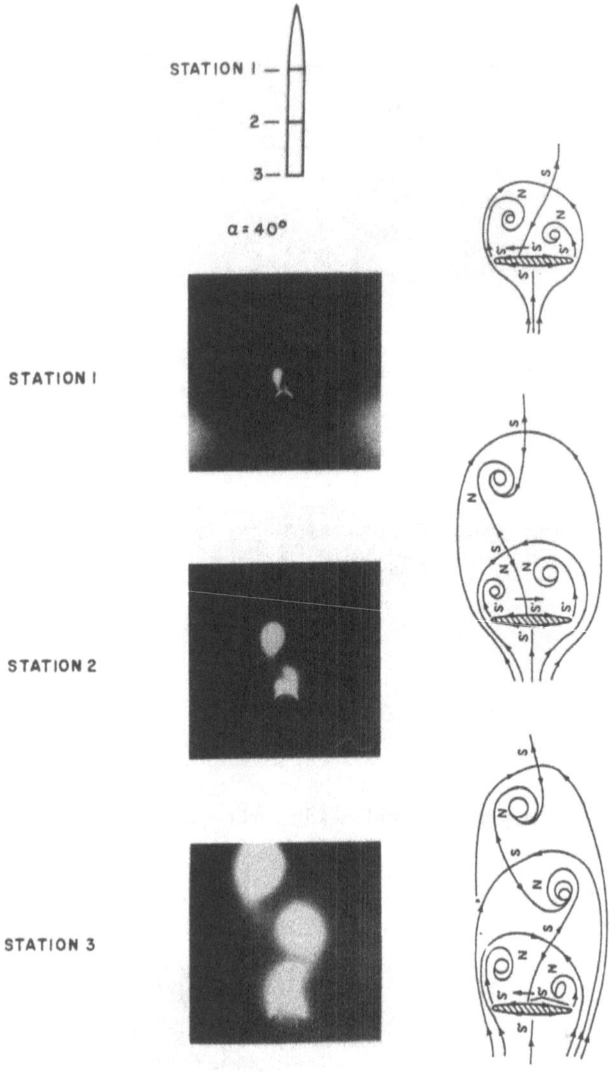

Fig. 2.19. Views at various cross-sections along the elongated body of the asymmetric vortices and the corresponding topological interpretations (Jorgensen 1977, Peake and Tobak 1980)

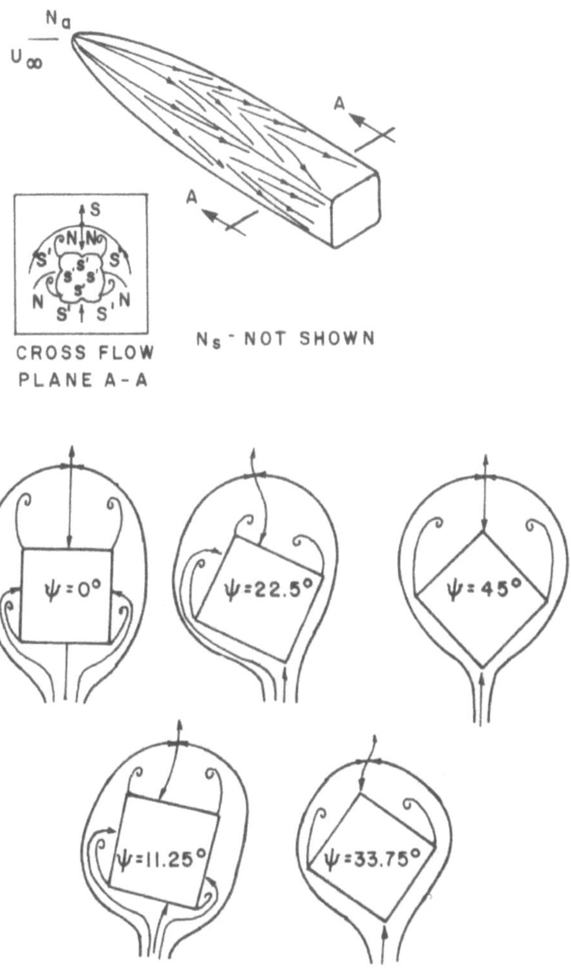

Fig. 2.20. Flow structure over a square cross-section body at various roll attitudes and at moderate angles of attack

2.4.3 Symmetric and Asymmetric Vortex Flows over Elongated Bodies in Supersonic Flows

Supersonic flows over elongated bodies at angles of attack also (like in the case of slender wings) generate rolled-up vortices in the flow behind the body, as seen in the schlieren photograph presented in Fig. 2.21. It is noted that at supersonic speed of $M = 4$ and at a 5° angle of attack, the boundary layer is already separated and rolled-up vortices are already observed.

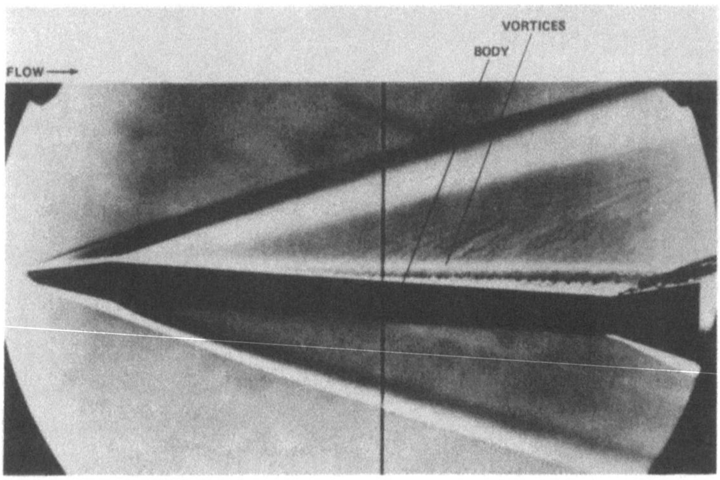

Fig. 2.21. Schlieren photo of the flow over a 10° blunted cone-cylinder with a flare at $\alpha = 5^o$ and $M = 4$ (Peake and Tobak 1980)

The detailed structure of the separated supersonic flow over the inclined cone is described in Fig. 2.22. This flow, which represents the flow at high angles of attack, includes the conical shock wave which envelopes the cone, the internal shock waves generated at the separation lines from which the primary vortices are established, and the secondary vortices, all shown schematically in Fig. 2.22. At lower angles of attack, the flow structure is somewhat simpler, since only the primary vortices are seen. An illustration of such flows by oil flow visualization is shown in Fig. 2.23 for $M = 2.94$ and for angles of attack of $\alpha = 12^o$ and $\alpha = 18^o$. In these cases, the separated vortex structure is symmetric and at $\alpha = 12^o$, there is only a primary vortex separation, while at $\alpha = 18^o$, the secondary vortices are also observed. At the higher angle of attack of $\alpha = 34^o$, the observed vortex structure is asymmetric, but still steady, as seen in Fig. 2.24. This figure indicates that both the primary and secondary vortices are asymmetrically positioned in respect to the cone axis.

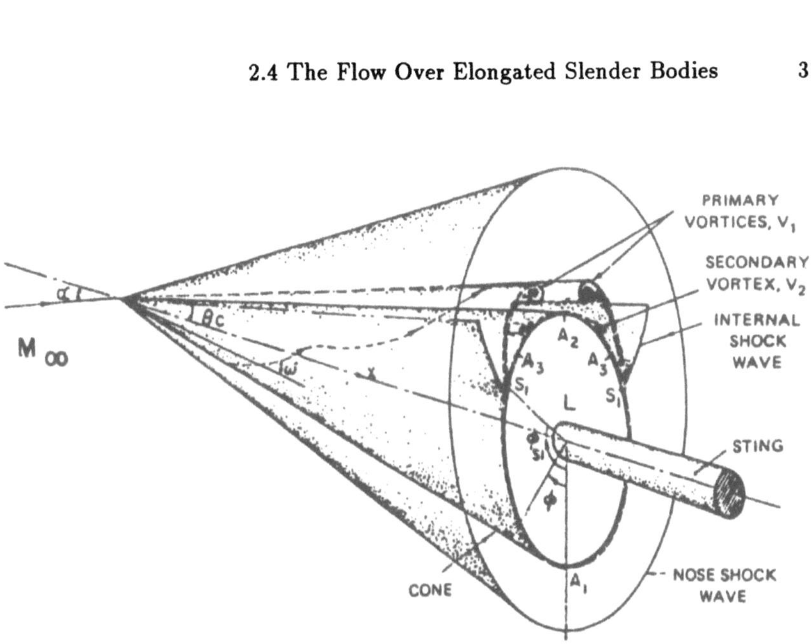

Fig. 2.22. Schematic of the shock wave and symmetric vortex structure on a cone at high angles of attack (Rainbird 1968a,b)

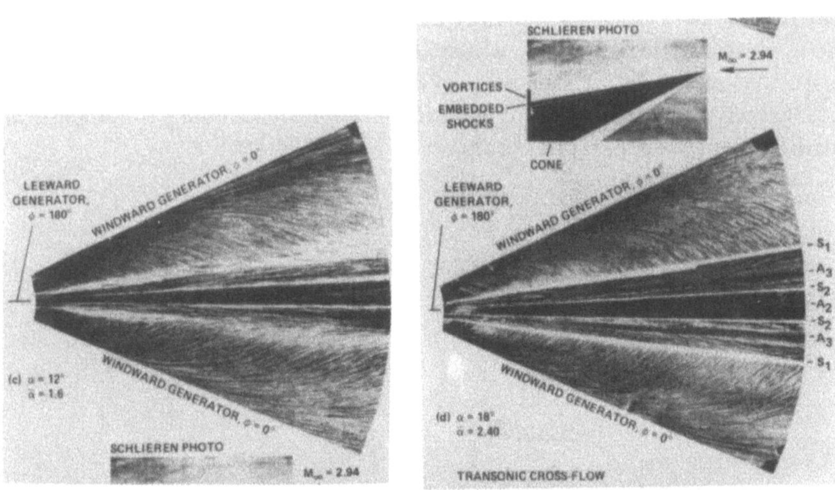

a. Primary vortex separation and reattachment at $\alpha = 12^o$

b. Primary and secondary vortices separation and reattachment at $\alpha = 18^o$ illustrated in Fig. 2.22

Fig. 2.23. Oil flow visualization of the skin friction lines over a cone (Bannink and Nebbeling 1978)

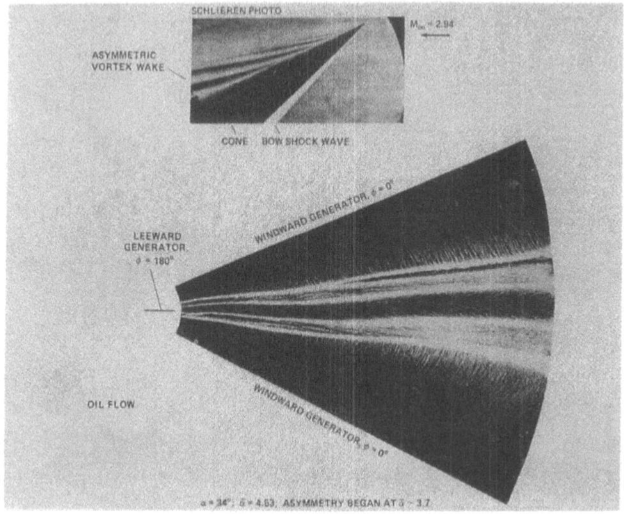

Fig. 2.24. Oil flow visualization of the skin friction lines over a cone with asymmetric separation and reattachment (Bannink and Nebbeling 1978)

An interesting structure is obtained for the flow over a cone-cylinder-flare model, a configuration which has a conical nose, a cylindrical body and a flared conical aft-body. Here, a very intricate "foot print" for the separated flow is obtained, already at small angles of attack in a supersonic flow, as shown in Fig. 2.25, where the flow at $\alpha = 4^o$ can be compared with that at $\alpha = 0^o$. A schematic of the flow structure is presented in Fig. 2.26a, and the pattern of the skin-friction lines (including a comparison with computational results) is shown in Figs. 2.26b and 2.26c.

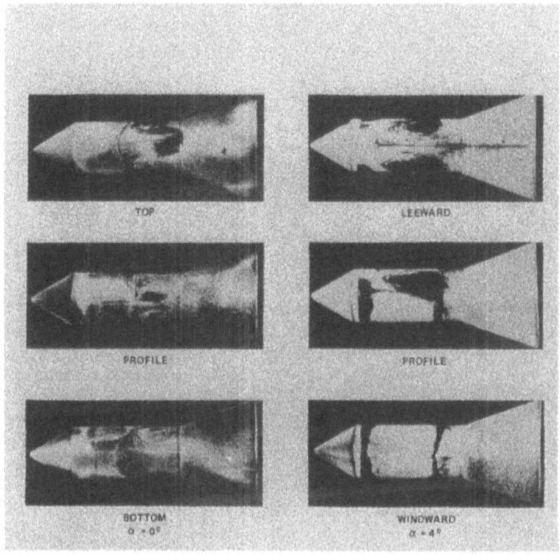

Fig. 2.25. Oil flow pattern over a cone-cylinder-flare body at small α (Ericsson, Reding and Guenther 1969)

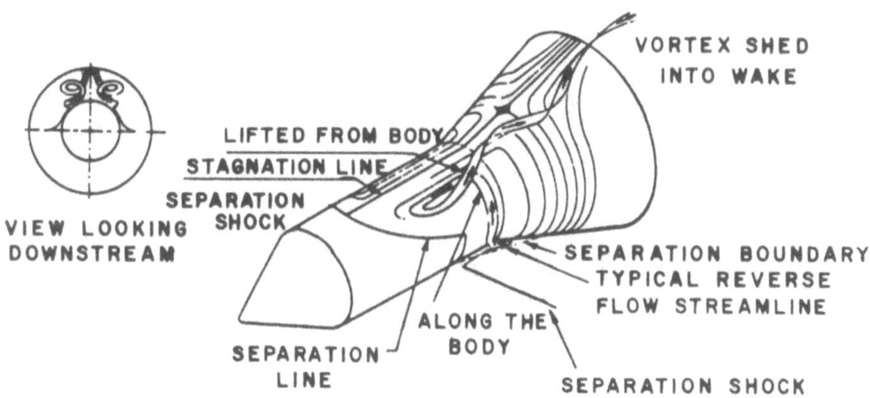

Fig. 2.26a. A three-dimensional schematic of flow structure over a cone-cylinder-flare body with foci type separation (Peaké and Tobak 1980)

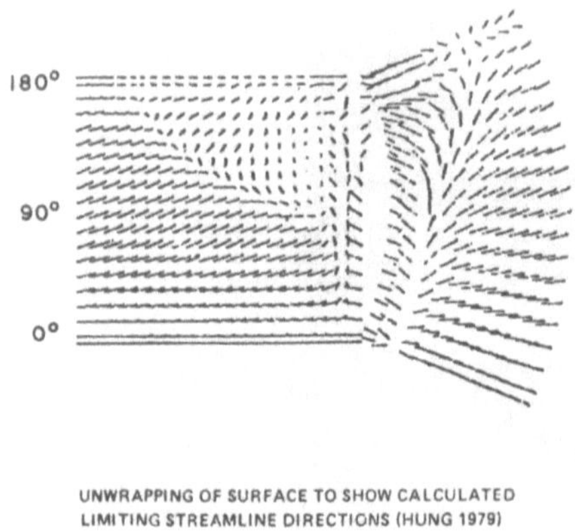

Fig. 2.26b. Calculated streamline directions on the surface of a cone-cylinder-flare configuration (Hung 1979)

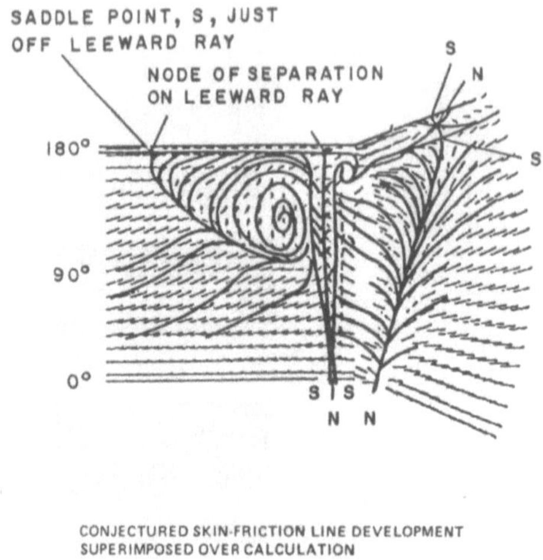

Fig. 2.26c. Conjectured skin friction lines superimposed over calculations (Peake and Tobak 1980)

2.5 Aircraft Type Configurations

The characteristics of the flow over aircraft configurations, which include combinations of bodies of various shapes and lifting surfaces, (wing, canard, tail etc.) with various planforms and profile shapes, result in very involved and interesting flow fields. This flow is strongly affected by the free rolled-up vortices, which are present over the lifting surfaces, and by additional vortices, generated at different regions by various devices such as LEX. Some visualization photographs of representative aircraft configurations will be presented in order to illustrate various features of such flows.

An illustration of the flow with separated regions, already obtained at relatively small angles of attack, is seen in the oil-flow pattern on the Shuttle Orbiter at the transonic speed of $M = 1.2$ and at $\alpha = 0^o$ in Fig. 2.27. At this low angle of attack, there are no shed vortices and only the local separation zones at corners are seen, as well as the "foot print" of the transonic shock waves on the mid chord of the wing.

Another flow pattern is illustrated by the flow structure on a fighter aircraft with relatively high AR swept-back wings, as seen in Fig. 2.28. As this aircraft is placed at high angle of attack, a large swirling vortex flow is generated near the root of the wings. This vortex is similar to the foci type separation, observed on the delta wing at high angles of attack, illustrated in Fig. 2.12. A topological interpretation of the visualization photograph, illustrating the vortex structure on the top surface of the wing, is also shown.

The oil flow pattern for fighter aircraft with a long slender nose, and long narrow strakes and twin fins is shown in Fig. 2.29. The "foot prints" of vortices generated by the strakes, as well as the separation and reattachment lines from the vortices generated at the leading edges and at the intersection points between the strakes and the wings' leading edges, are also clearly seen.

An oil flow pattern on another fighter aircraft with short strakes and a single fin, and with both leading edge and trailing edge flaps deflected, is presented in Fig. 2.30. The skin-friction lines are visible on the wing surfaces and on the flaps, indicating the flow directions on these surfaces.

2.5.1 Aircraft Aerodynamic Characteristics

The appearance of separated regions and the shedding of the free rolled-up vortices from the various parts of the aircraft affect the flow field around these configurations and, therefore, also the aerodynamic forces and moments. At low angles of attack, the aerodynamic forces and moments vary in accordance with well-known linear aerodynamic behavior. However, as the angles of attack increase, drastic changes in the aerodynamic characteristics are observed.

2.5.1.1 Effects of Symmetrical Vortex Flows at Subsonic Speeds. An illustration of the significant changes, due to increasing the angles of attack on the pitching moment of an aircraft configuration at various elevator deflection angles, is presented in Fig. 2.31. The pitching moment variation is reasonably well behaved up to $\alpha = 20^o$. However, as the angle of attack is further increased, there is a clear loss of elevator effectiveness up to the complete loss of elevator control above $\alpha = 45^o$. Another result of the separated flow at high angles of attack is the loss of roll control by ailerons on a delta type configuration at increasing angles of attack, as shown in Fig. 2.32. Here the aileron effectiveness is reasonably constant for angles of attack up to 20°–25°, as seen in Fig. 2.32, beyond which the rolling moment due to aileron deflection falls off drastically.

2.5.1.2 Effects of Asymmetric Vortex Flows at Zero Sideslip in Subsonic Flow. It was shown that slender configurations, such as low AR delta wings and slender elongated bodies, develop asymmetric vortex flow, as illustrated in Figs. 2.14 and 2.19, respectively. Similar asymmetric vortex flows are also observed on slender aircraft configurations at high angles of attack, even without any sideslip. The appearance of asymmetric vortex pattern on the aircraft forebody is illustrated schematically in Fig. 2.33. The asymmetric vortex flow which is generated on the nose induces lateral forces and moments on the configuration. The effect of the various forebody shapes on

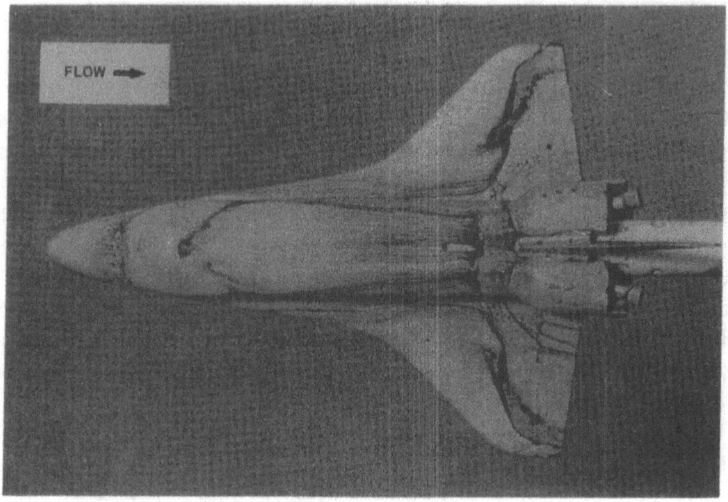

Fig. 2.27. Oil flow pattern on the shuttle-orbiter at $\alpha = 0$ (Ericsson and Reding 1976)

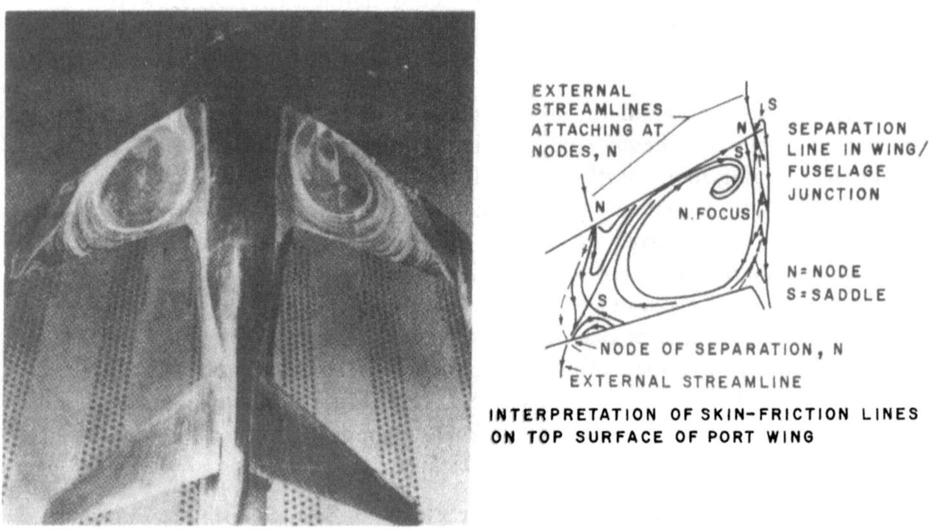

Fig. 2.28. The flow pattern over a swept back wing of a fighter aircraft with its topological interpretation (Peake and Tobak 1980)

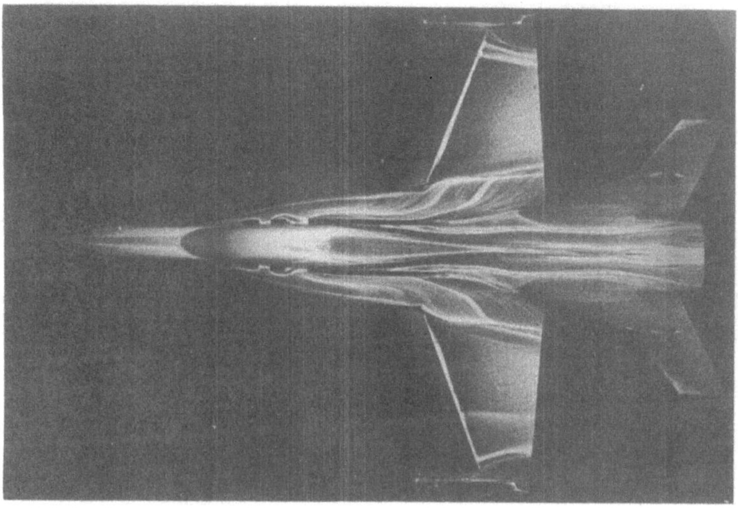

Fig. 2.29. Oil flow pattern on a fighter aircraft showing the effects of long leading edge strakes and two vertical fins (Beeman 1977)

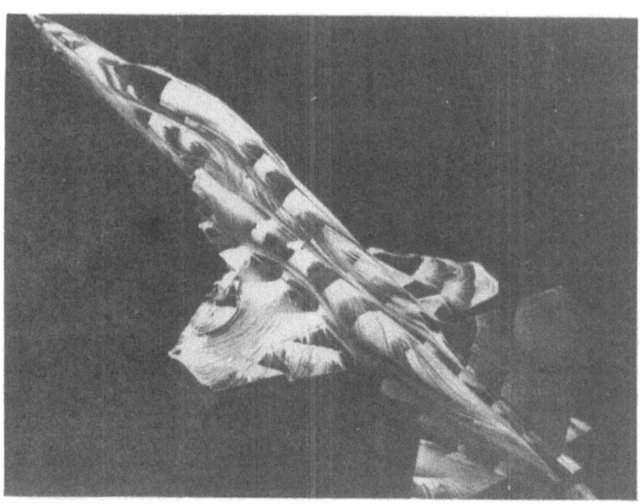

Fig. 2.30. Oil flow pattern on a fighter aircraft showing the effects of short leading edge strakes and a single fin (Beeman 1977)

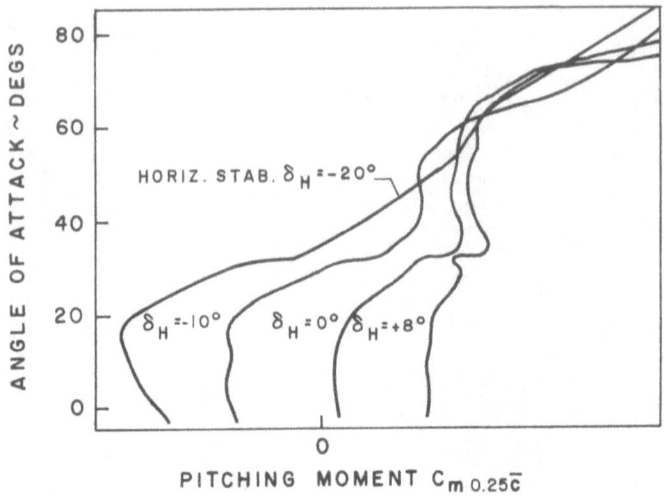

Fig. 2.31. The loss of pitch control as the angle of attack is increased for a low mounted aft tail fighter aircraft configuration (Skow 1983)

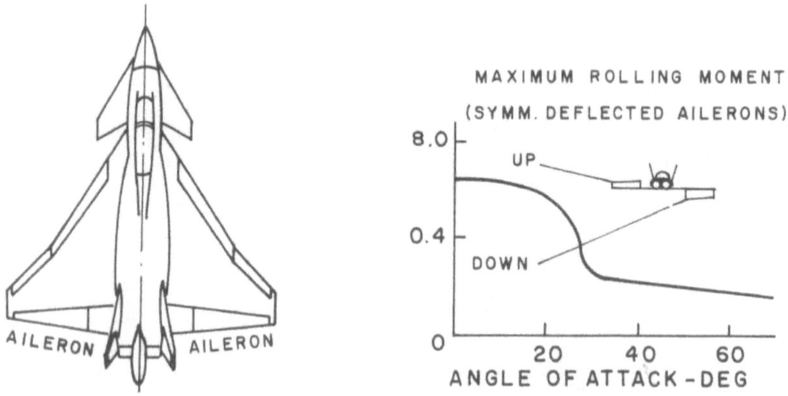

Fig. 2.32. Loss of ailerons effectiveness at increasing angles of attack for a delta-canard fighter configuration (Skow 1983)

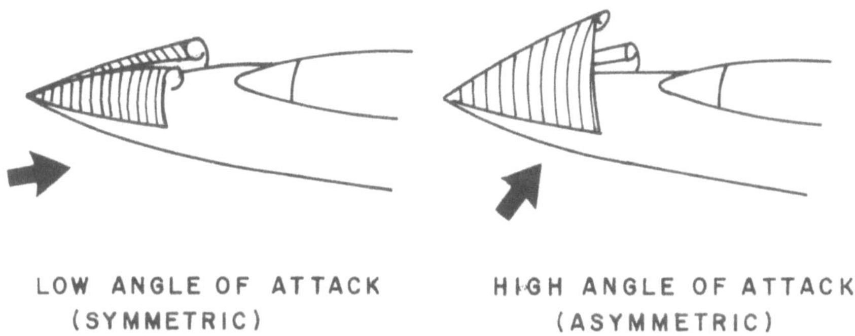

Fig. 2.33. Schematic of the symmetric and asymmetric separation of vortices on the slender nose of fighter aircraft

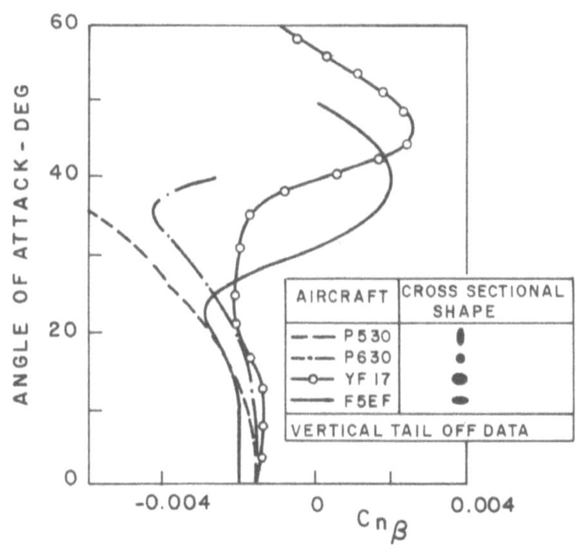

Fig. 2.34. The reversal of the yawing moment coefficient at increasing angles of attack due to variations of the elongated nose shapes of various fighters (Skow 1983)

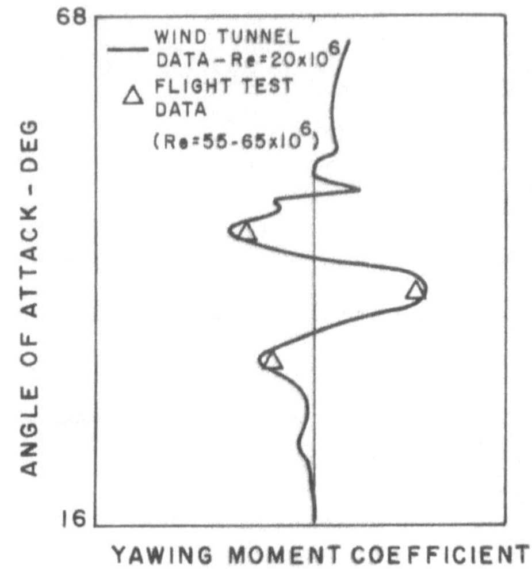

Fig. 2.35. Wind tunnel and flight test data illustrating the yawing moment reversal at increasing angles of attack

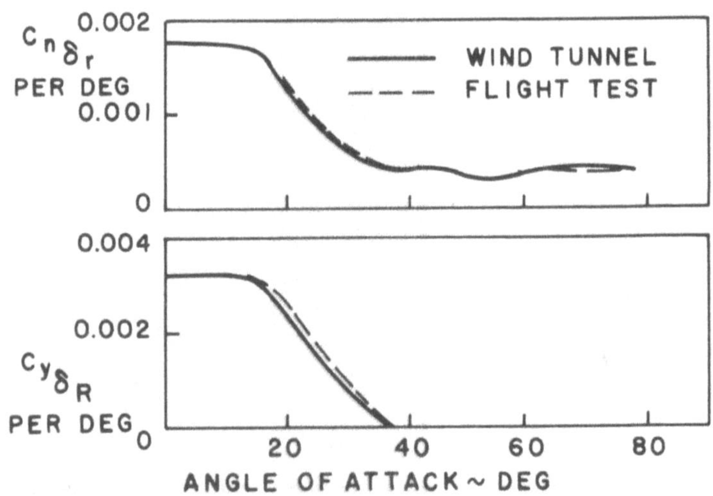

Fig. 2.36. The loss of lateral control effectiveness due to rudder deflections at increasing angles of attack (Skow 1983)

the lateral stability moment coefficient is illustrated in Fig. 2.34. It is seen that for angles of attack above 20°–25°, there is a large change of the lateral moment coefficient, whereas at higher angles of attack, the variation of the pitching moment reverses to the other direction. The lateral moment coefficient reversals resulting from the effects of these aerodynamic asymmetries, as measured in both wind tunnels and flight tests, is shown in Fig. 2.35. The extensive region of separated flows at high angles of attack results in the loss of lateral control effectiveness. The loss of effectiveness of the lateral directional control forces and moments due to these vortex asymmetries is shown in Fig. 2.36.

2.5.1.3 Effect of LEX at High Angles of Attack at Subsonic Flows. The dominant effects of the interaction between vortices, generated by the addition of the Leading Edge Extension in front of the wing, is very well illustrated in a series of visualization photographs, obtained in a water tunnel on a 3% model of the F-18 aircraft. The water tunnel data correlated very well with flight test data on the NASA F-18 high alpha research vehicle as reported by Del Frate and Zuniga (1990).

We will begin by observing the flow structure at $\alpha = 30^o$ (at zero sideslip) shown in Fig. 2.37a, where the forebody vortices and the LEX vortices and their interactions over the LEX and the wing-root surfaces are clearly indicated. The LEX vortex core breakdown occurs in this case at the point of intersection of the wing and the LEX. At $\alpha = 35^o$, as shown in Fig. 2.37b, the LEX vortex core breakdown occurs at about 2/3 of the length of the LEX chord. This breakdown must affect the forebody vortices upstream of the wing. At increasing the angles of attack to 40° and 45°, as presented in Figs. 2.37c and 2.37d respectively, the LEX vortex core breakdown moves very close to the LEX leading edge, as most of the LEX and the wing are affected by the turbulent mixed flow caused by the break-up of the vortices.

2.5.1.4 Closed Coupled Canard-Wing Configuration at High Angles of Attack at Subsonic Flows. Enhancement of the leading edge vortices on a slender delta wing can also be achieved by generating strong vortices on a slender canard, as shown in Fig. 2.38. It can be seen that the induced flow field pulls the canard vortices inwards and below the wing leading edge vortices. The interaction of these vortices, due to their approach towards one another over the rear part of the wing, strengthens the wing vortices and delays their breakdown. The side view of the vortices obtained by a schlieren photograph (Er-El and Seginer 1987) shows the pulling down of the canard vortices into the wing vortices, while the top view, which is a water tunnel visualization (Hummel 1988), shows that the canard vortices are drawn inside the cores of the wing leading edge vortices.

2.5.1.5 Effect of Sideslip Angle at Subsonic Speeds. The introduction of a sideslip angle, while the configuration is at moderate or high angles of attack, results in an asymmetric vortex flow structure. The asymmetric vortex flow on a Concorde model at $\alpha = 12^o$ and at $\beta = 10^o$ is seen in Fig.

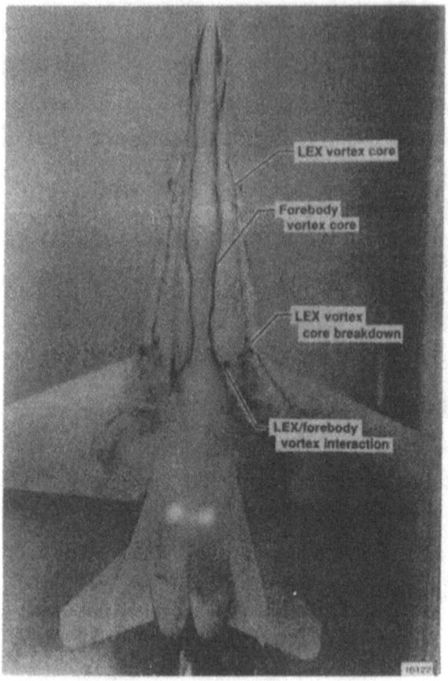

a. $\alpha = 30^o$

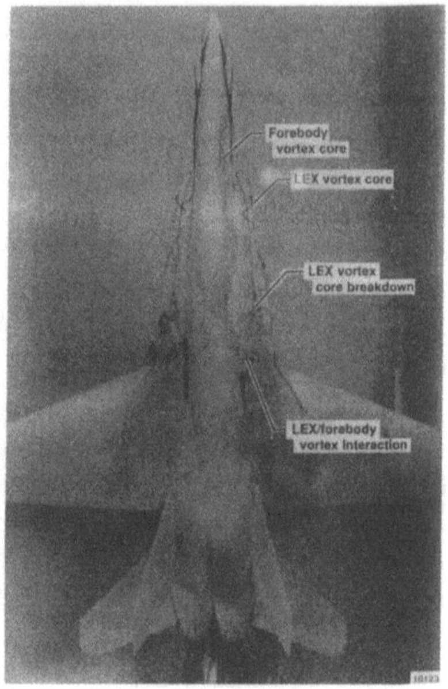

b. $\alpha = 35^o$

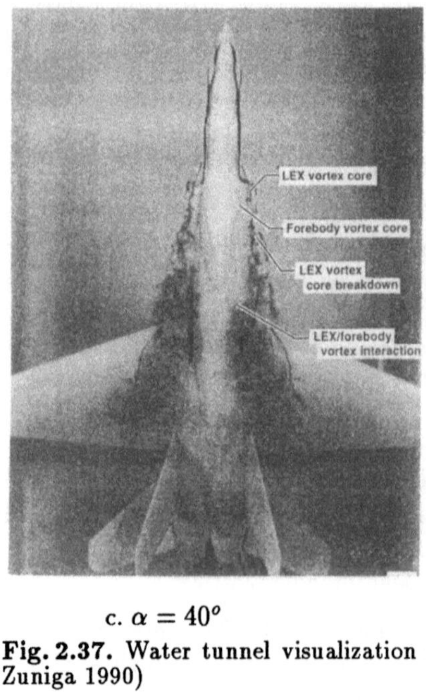

c. $\alpha = 40^o$

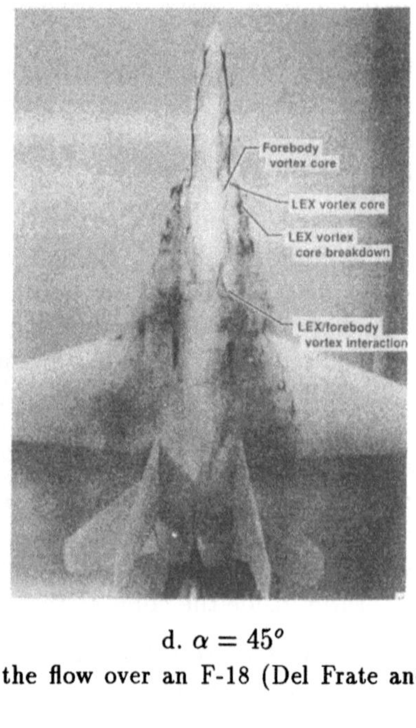

d. $\alpha = 45^o$

Fig. 2.37. Water tunnel visualization of the flow over an F-18 (Del Frate and Zuniga 1990)

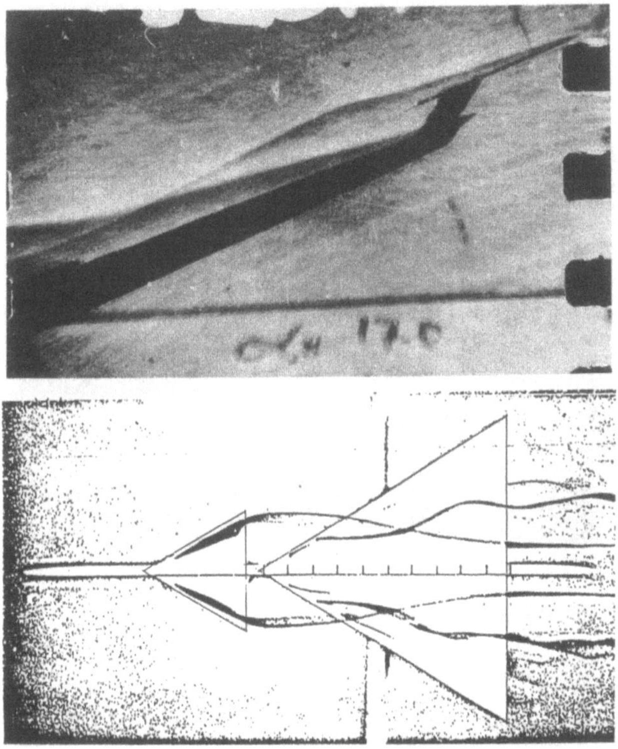

Top – Schlieren, side view (Er-El and Seginer 1987)

Bottom – oil flow, top view (Hummel 1988)

Fig. 2.38. The flow over a wing-canard configuration

2.39. Two counter rotating vortices are seen (in the top view and, more clearly, in the rear view) on the left wing; while only one is present on the other wing. The interactions of the vortices, generated on a slender aircraft configuration with LEX, is shown in a side view in Fig. 2.40. A top view of the vortex trajectories, obtained on a elongated nose aircraft with sideslip, is shown in Fig. 2.41. The effect of this asymmetry on the loss of rudder effectiveness is shown in Fig. 2.42 and on the loss of the aileron effectiveness on Fig. 2.43.

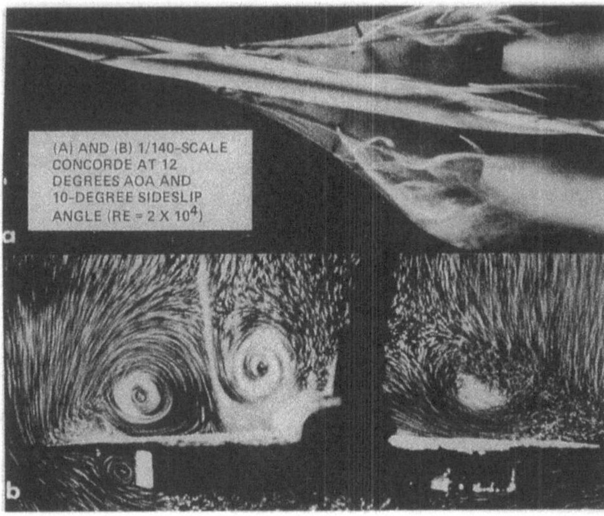

Fig. 2.39. The asymmetric vortex structure on the concorde due to combined angles of attack and sideslip (Skow 1983)

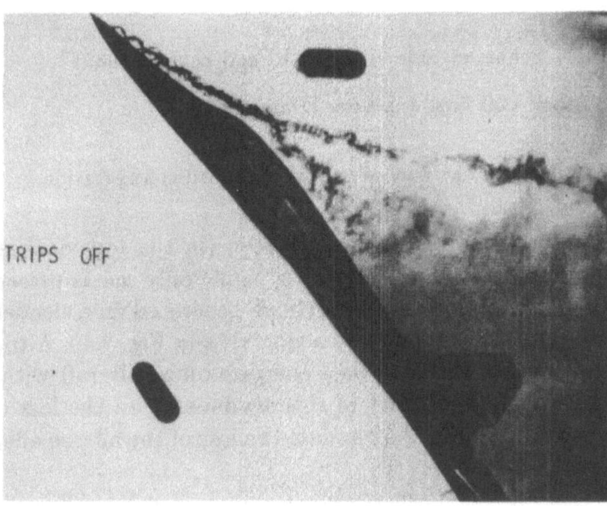

Fig. 2.40. Side view of the asymmetric vortex trajectories due to sideslip on a fighter aircraft showing the trajectories of the forebody and LEX vortices (Skow 1983)

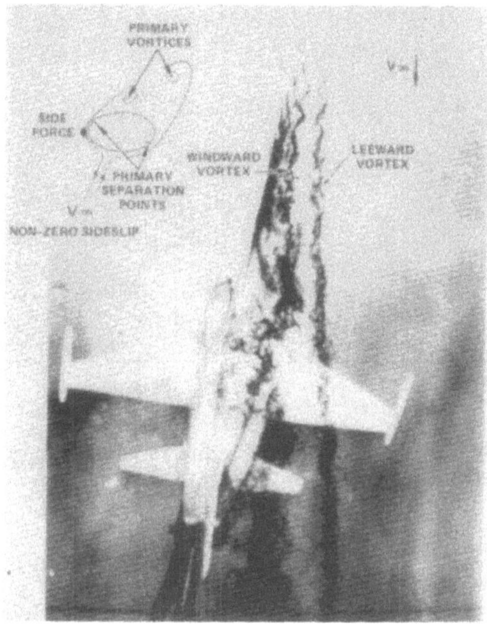

Fig. 2.41. Top view of the trajectories of the asymmetric vortices due to sideslip of a fighter aircraft (Skow 1983)

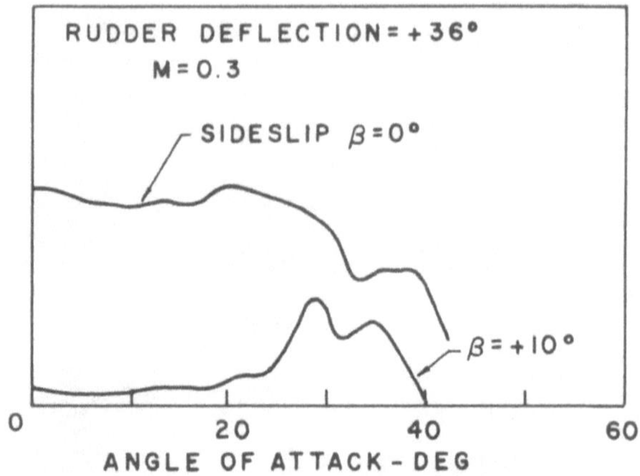

Fig. 2.42. Loss of rudder effectiveness due to combined variations of angle of attack and sideslip (Skow 1983)

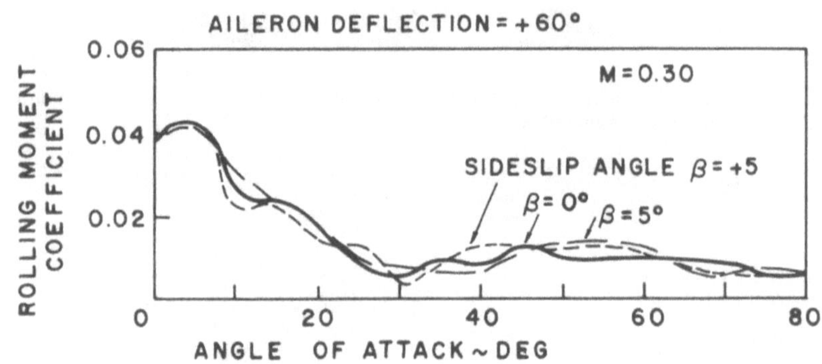

Fig. 2.43. Loss of aileron effectiveness at increasing angles of attack in the presence of sideslip for a fighter aircraft (Skow 1983)

2.6 Vortex Breakdown

It is well established that concentrated vortex filaments suddenly break down at certain flow conditions. A view of breakdown of vortices on the delta wing at increasing angles of attack is shown in Fig. 2.44. The breakdown can occur symmetrically, as in Fig. 2.45a or asymmetrically, as a result of sideslip, as in Fig. 2.45b. A side view of the asymmetric vortex breakdown on an aircraft configuration is shown in Fig. 2.46.

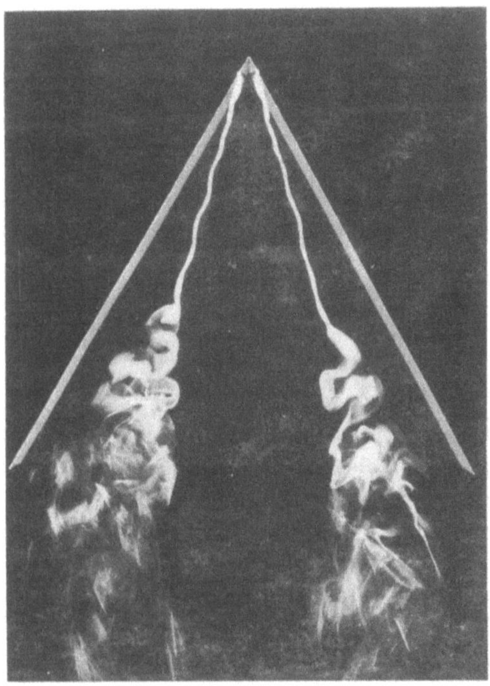

Fig. 2.44. Top view of the breakdown of the leading edges vortices on a delta wing – symmetric case, no sideslip (Werle 1960)

The effect of the asymmetric vortex breakdown on the roll stability of a delta wing with sideslip is illustrated in Fig. 2.47. The vortex breakdown results in the loss of the effectiveness of the vortex to generate lift, first, on one side of the wing. This results in a rolling moment caused by increase of angle of attack. As the breakdown position moves up the wings' planform, the lift on the wing is destroyed and the wing stalls. The irregular unsteady mixing, which is generated in the breakdown zone of the concentrated vortex, affects the post-stall flow field. The unsteady mixing in the wake is dissipated and, finally, the turbulent wake is established.

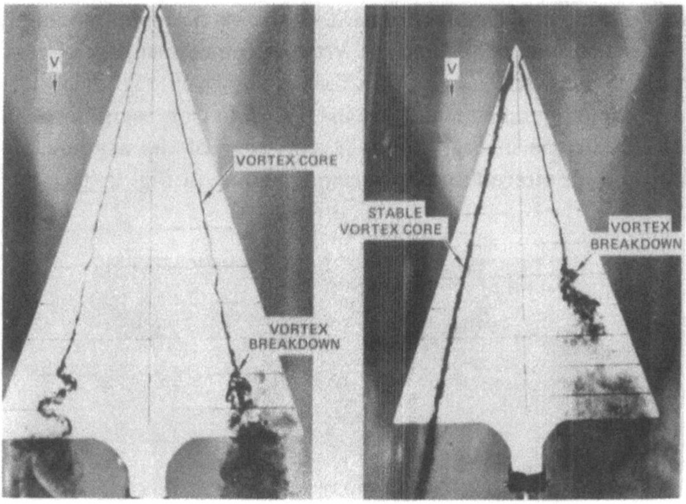

a. Symmetric, $\beta = 0^o$ b. Asymmetric, $\beta \neq 0^o$

Fig. 2.45. Effect of sideslip on vortex breakdown on a 70° delta wing at 25° angle of attack (Skow 1983)

The vortex breakdown mechanism is rather involved and is one of the unresolved problems in fluid mechanics. Therefore, we will include here only some descriptive experimental results which will demonstrate some of the physical characteristics observed when the vortex breakdown occurs. Two types of breakdown are observed; the "bubble" type, shown in Fig. 2.48 and the "spiral" type, shown in Fig. 2.49. The flow parameters that affect these two types of vortex breakdown, and the physical process of the breakdown mechanism, are still among of the most challenging fundamental research problems in fluid dynamics. Modern high performance aircraft with high thrust-to-weight ratio are capable of partially controlled performance in the post-stall flight regime; therefore, the aerodynamics of the vortex breakdown and the stall and the post-stall characteristics are of increasing interest.

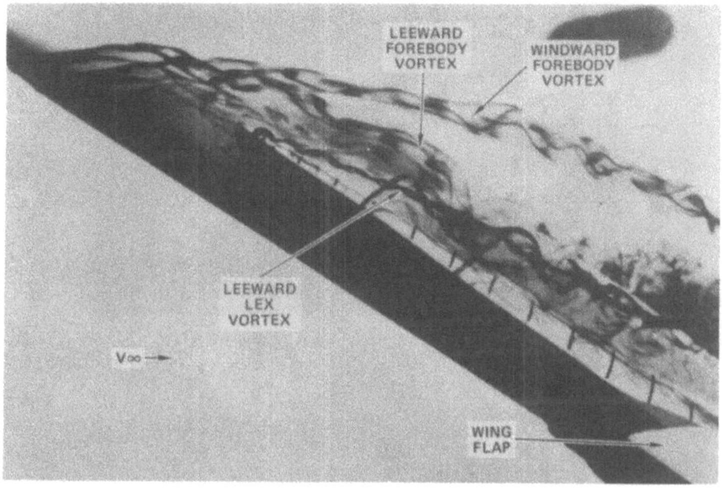

Fig. 2.46. Side view of the trajectories of the asymmetric vortex breakdown on a fighter configuration (Erickson 1979)

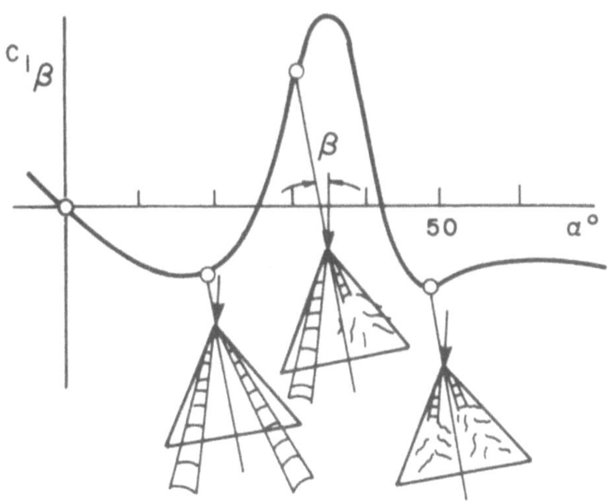

Fig. 2.47. The rolling moment reversal due to asymmetric vortex breakdown at increasing angle of attack (Skow 1983)

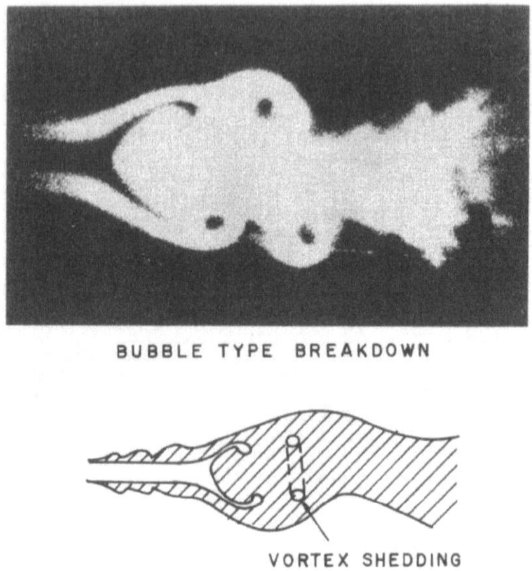

a. Side view visualization and illustration drawing

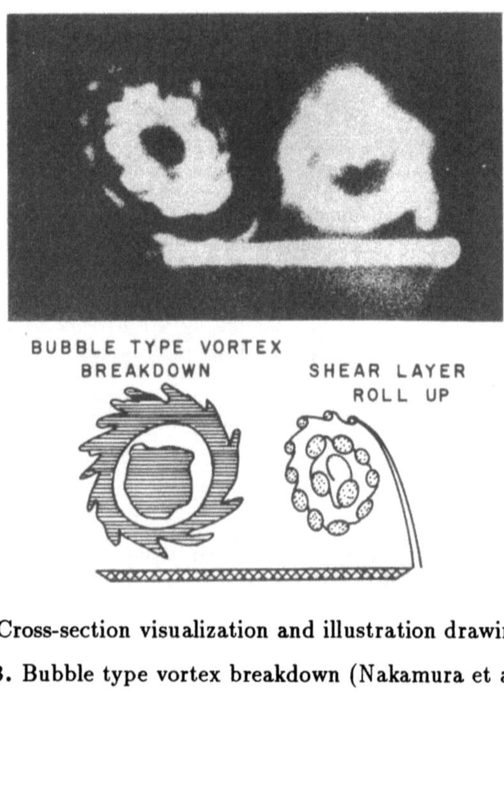

b. Cross-section visualization and illustration drawing

Fig. 2.48. Bubble type vortex breakdown (Nakamura et al. 1986)

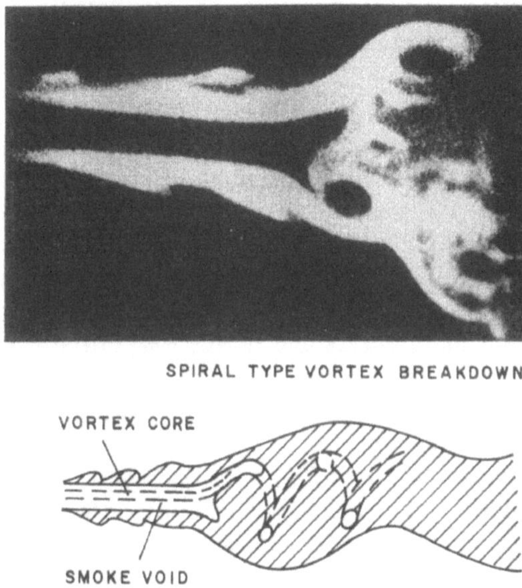

a. Side view visualization and illustration drawing

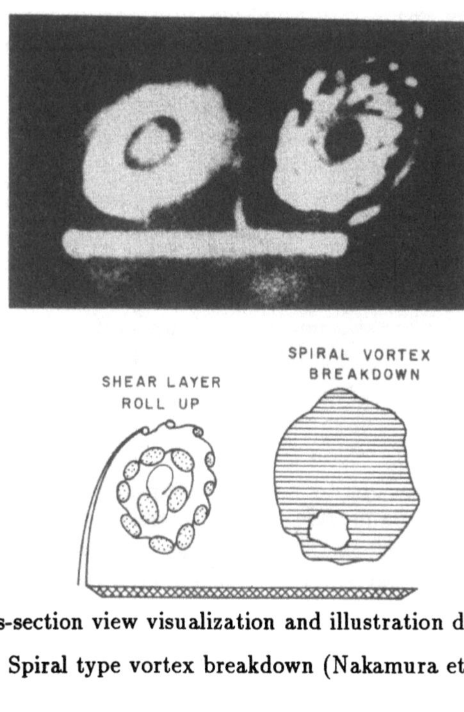

b. Cross-section view visualization and illustration drawing

Fig. 2.49. Spiral type vortex breakdown (Nakamura et al. 1986)

2.7 Nonsteady Aerodynamics at High Angles of Attack on Slender Configurations

Modern fighter-type aircraft and missiles are designed for very high rate maneuvers. In such maneuvers, the attitude of the configuration, the angles of attack, sideslip and roll may change at such a fast rate as to affect the aerodynamic characteristics of the vehicle. The fast rate of change of a slender configuration's angle of attack from a moderate or a high angle of attack causes time variations of both the position of vortex separation and of the vortex trajectories. In general, there is a difference in the vortex movements during the upstroke and downstroke of the angle of attack sweep. This difference causes a type of hysteresis loop for both vortex positions and, therefore, also a hysteresis loop for the aerodynamic forces and moments. The visualization experiments on an oscillating delta wing, presented in Fig. 2.50, show the vortex formation on a delta wing during a full upstroke and downstroke cycle, which illustrates both the movement of the separation pattern as the angle of attack is changing and also the differences in the positions of the vortices during the upstroke compared to their position during the downstroke. The hysteresis loop for the height of the vortex above the oscillating delta wing during a full upstroke-downstroke cycle is presented in Fig. 2.51. Obviously, these variations in the positions of the vortices will cause corresponding time-dependent variations in the aerodynamic forces and moments on the delta wing. The hysteresis loop for the position of vortex separation on the body of an ogive-cylinder during an upstroke-downstroke cycle is presented in Fig. 2.52. The vortex structure over an elongated body during an upstroke-downstroke cycle of oscillation is shown in Fig. 2.53.

The generation of nonsteady forces during high rate maneuvers which are demonstrated in the experimental results of Figs. 2.50 to 2.53 represent non-steady aerodynamic forces obtained by controlled rate and stroke of the movement of the configuration. However, nonsteady aerodynamic forces may be generated by nonstationary vortex separation or nonsteady vortex breakdown which may occur in the vortex wake behind a body when the orderly rolled-up vortex wake is transformed into a turbulently mixed wake. The case of the vortex breakdown leading to turbulent mixing may occur at post-stall flow conditions. In this case, the turbulent mixing may result in nonsteady time variation of the aerodynamic forces on the configuration.

Nonsteady aerodynamic effects due to unsteady vortex separation and breakdown may cause effects such as wing rock and buffet and may also couple to the structural dynamics in various aeroelastic effects such as flutter. These are extremely interesting and complex phenomena which, as discussed before, are only reviewed here for the completeness of the presentation.

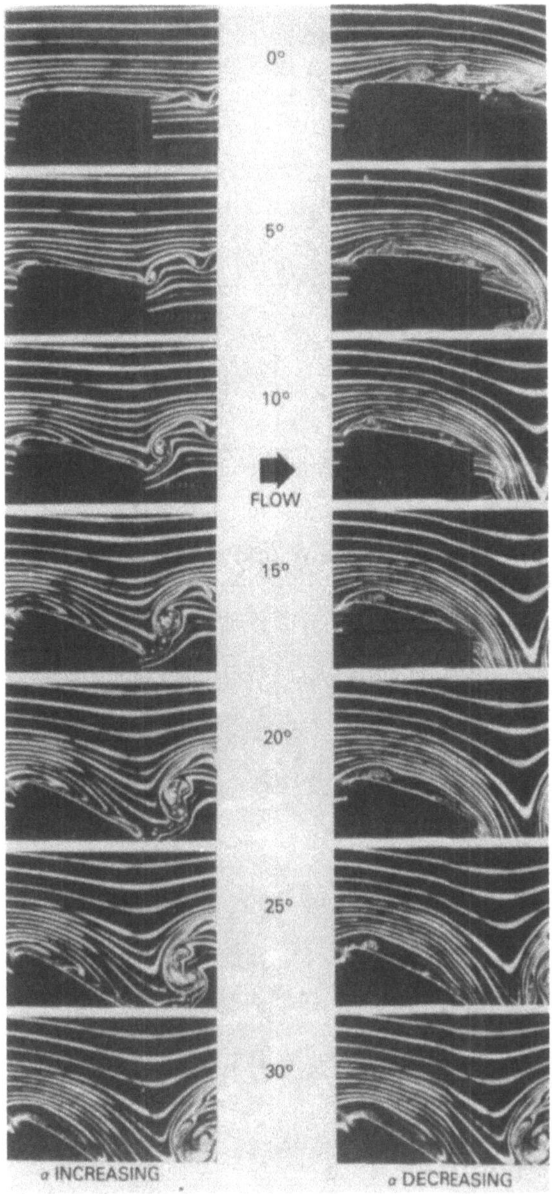

Fig. 2.50. Vortex structure on an oscillating delta wing during the upstroke and downstroke pitching motions (Gad-El-Hak and Ho, C.M. 1985)

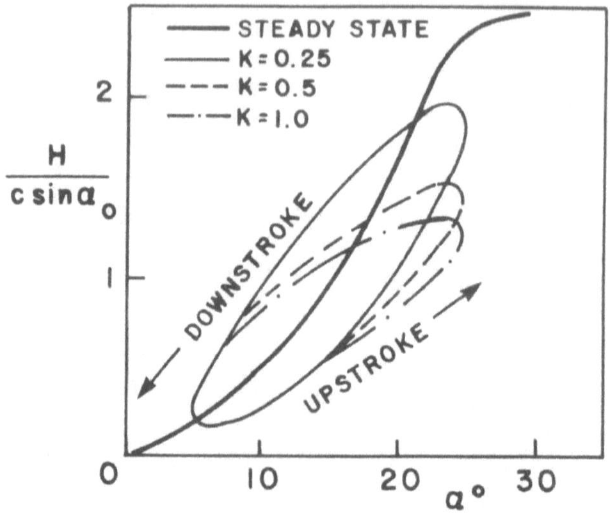

Fig. 2.51. The hysteresis loop of the vertical position of the vortex over an oscillating delta wing (Gad-El-Hak and Ho, C.M. 1985)

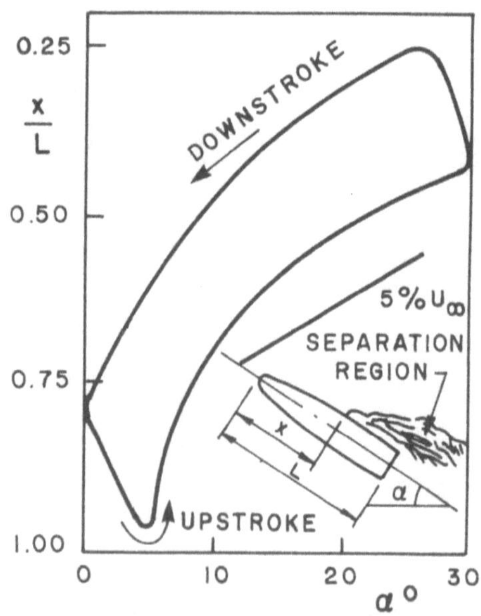

Fig. 2.52. The hysteresis loop of the position of the separation on an oscillating elongated body (Gad-El-Hak and Ho, C.M. 1986)

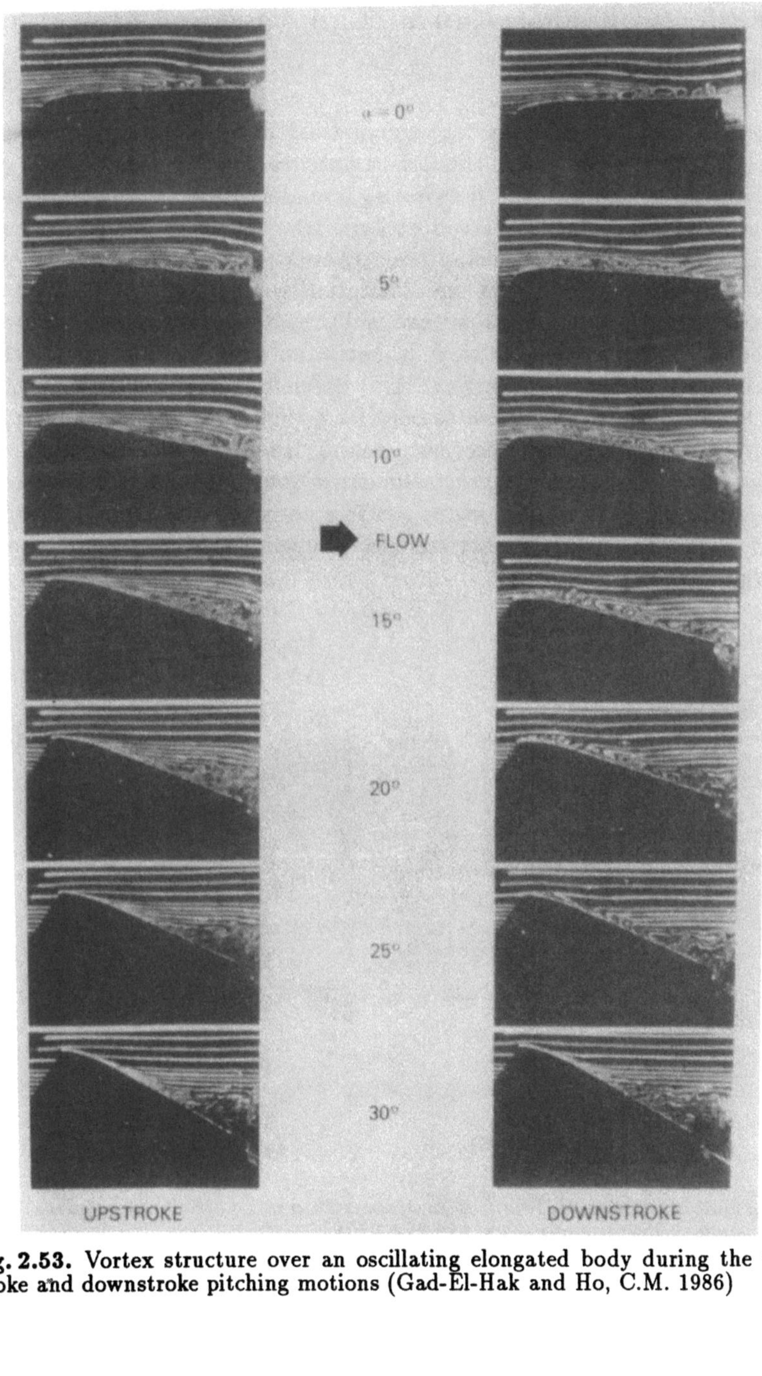

Fig. 2.53. Vortex structure over an oscillating elongated body during the upstroke and downstroke pitching motions (Gad-El-Hak and Ho, C.M. 1986)

2.8 Effects of Separation at High Angles of Attack in Hypersonic Flows

Viscous-inviscid interactions have dominating effects in hypersonic flows. It is very interesting to relate the flow structures which are associated generally with separations of the viscous boundary layer at high angles of attack with the effects observed at hypersonic speeds on the surfaces – the high local shear stresses and heat transfer rates. The main features of three-dimensional separation are illustrated by the flow generated by the interaction of the oblique shock wave and the viscous layers for the flow obtained at the corner by a vertical fin, set at an angle to the flow direction. The interaction of the shock wave and the viscous layer causes separation of the boundary layer on the plate surface, well ahead of the wedge root, and the separated shear layer that, leaving the surface, reattaches to the wedge face. The flow at the reattachment region causes high shear stresses and also high heating rates on the surfaces ahead of the wedge and on the wedge face. The flow on the surface and the shear stress lines are illustrated in Fig. 2.54.

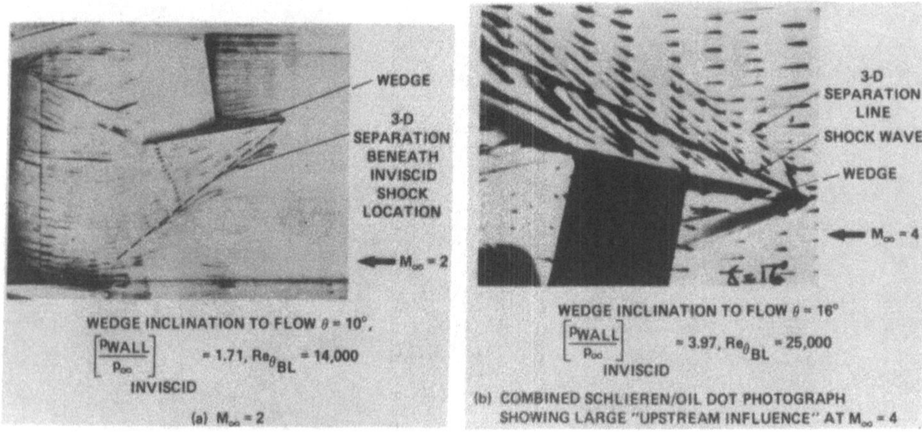

Fig. 2.54. Experimental skin friction line pattern in a turbulent boundary layer oblique shock interactions (Peake et al. 1970)

The pressure distribution and the heat transfer rate distribution are shown in Fig. 2.55. The sharp rise in the heat transfer rate at the reattachment zone on the face of the fin is observed.

Particularly high heat transfer rates are obtained, as the shock wave impinges on the surface, and the interaction there may result in severe damage to the surface structure. One of the most striking examples of the peak heating rates is seen in Fig. 2.56 which, with the burned surface of the ventral fin of the X-15A-2 aircraft, shows the damages associated with shock wave impingement at hypersonic speeds.

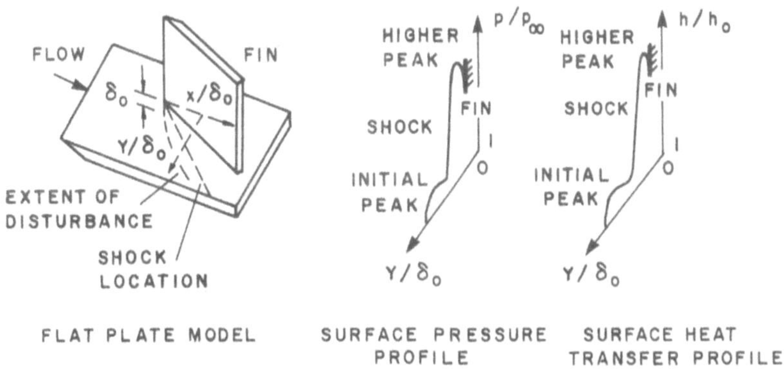

Fig. 2.55. Surface heat transfer rate and pressure profiles on an inclined fin (Scuderi 1978)

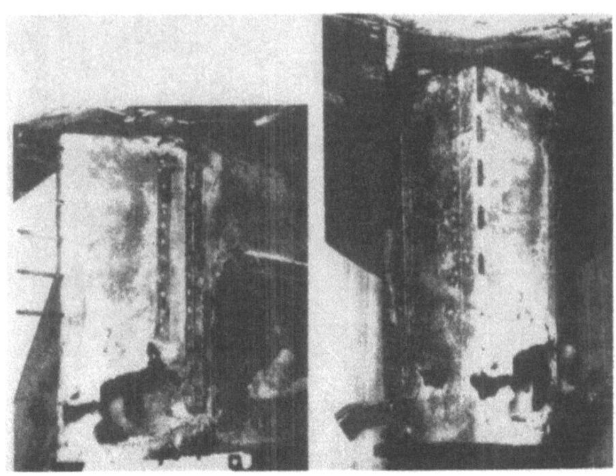

Fig. 2.56. Side and front views of the damage to the ventral fin of the X-15A-2 aircraft due to shock impingement (courtesy of NASA)

Another example of the high heating rates in regions of separation and reattachment is the case of the blunted cone at increasing angles of attack, as shown in Fig. 2.57. The flow structure and the "hot spot" at the local reattachment point are clearly seen in the oil flow visualization presented in this figure. An interpretation of the separation flow structure, including the assumed forms of the singularities, is proposed by Peake and Tobak (1980) and shown in Fig. 2.58. This high heat rate peak is similar to the peak in heat transfer observed at the reattachment point in the center of the flat base obtained in the measurement of the heat transfer distribution on the blunt base in the shock tube (Rabinowicz-Rom 1958).

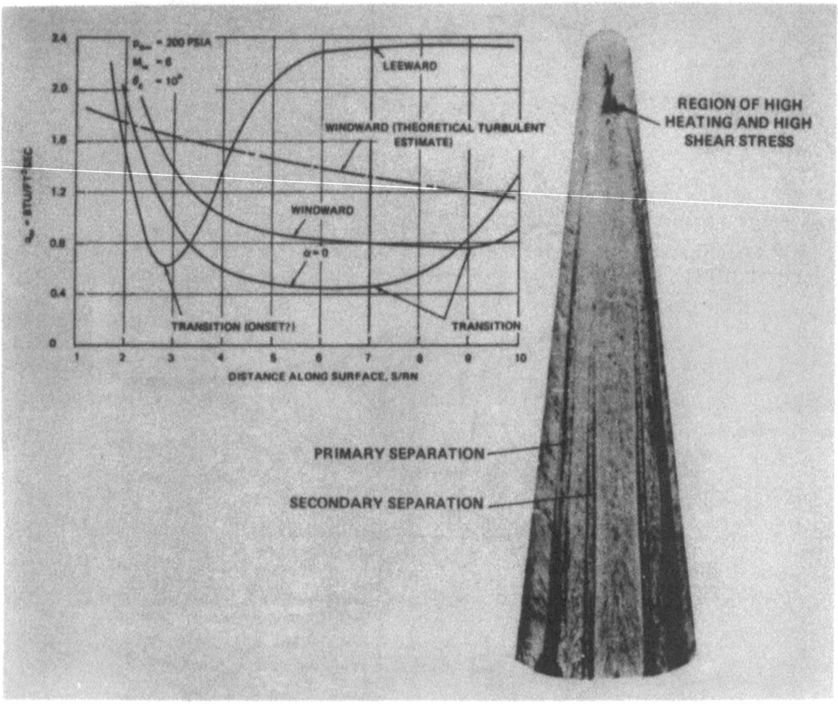

Fig. 2.57. Heat transfer rates and oil flow on the surface of a 10° blunt cone at M=6 and $\alpha = 20°$ (Zakkay et al. 1972)

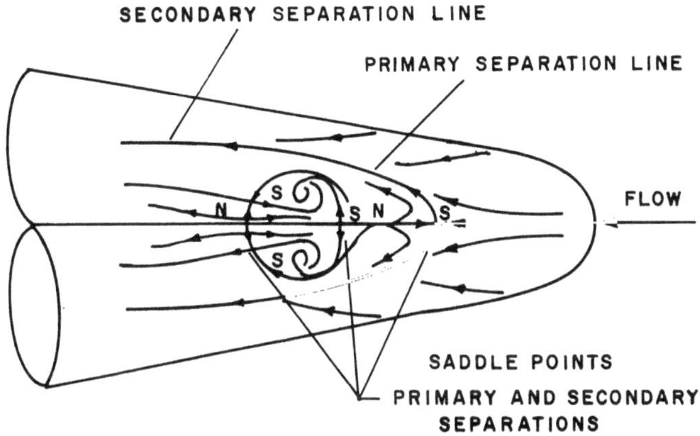

Fig. 2.58. Interpretation of the singularities in the skin friction line pattern near the separation point of the blunt cone (Peake and Tobak 1980)

References

1. Bannink, W.J. and Nebbeling, C. (1978), "Measurements of the Supersonic Flow Field Past a Slender Cone at High Angles of Attack", AGARD CP-247.

2. Bippes, H. (1978), NASA TM 75243.

3. Boersen, S. (1975), "Reynolds Number Effects on Pressure Normal Force and Distributions along Conically Pointed Circular Cylinder at Mach Number 2.3", NLR TN 7512U.

4. Carcaillet, R., Manie, F., Pagan, D. and Solignac, J. (1986), "Leading Edge Vortex Flow over a 75 Degree Swept Delta Wing – Experimental and Computational Results", Proceedings of the 15th ICAS Congress, Paper No. 86-1.5.1, pp. 589–603.

5. Clarkson, H.M., Malcolm, G.N. and Britam, V.A. (1982), "Aerodynamic Characteristics of Bodies with Noncircular Cross-Sections at High Angles of Attack", AIAA Paper 82-0056.

6. Crow, S.C. (1970), "Stability Theory for a Pair of Trailing Vortices", AIAA Journal, Vol.8, No. 12, pp. 2172–2179.

7. Er-El, J. (1985), Private Communication.

8. Ericsson, L.E., Reding, J.P. and Guenter, R.A. (1969), "Effects of Shock-Induced Separation", Lockheed Missiles and Space Co., Sunnyvale, California, L-87-69-1.

9. Ericsson, L.E. and Reding, J.P. (1976), "Unsteady Aerodynamic Flow Field Analysis of the Space Shuttle Configuration, Part I, Orbiter Aerodynamics", Lockheed Missiles and Space Co., Sunnyvale, California, LMSC-DO57194.

10. Erickson, G.E. (1979), "Water Tunnel Flow Visualization: Insight into Complex Three-Dimensional Flow Fields", AIAA Paper 79-1530.

11. Fellows, K.A. and Carter, E.C. (1969), "Results and Analysis of Pressure Measurements on Two Isolated Slender Wings and Slender Wing-Body Combinations at Supersonic Speeds, Vol. 1 Analysis", ARA Report 12.

12. Fiechter, M. (1966), Über Wirbelsysteme an schlanken Rotationskörpern und ihren Einfluss auf die aerodynamischen Beiwerte", Deutsch-Französisches Forschungs-Institut Saint-Louis, Report 10/66.

13. Fiechter, M. (1969), Jahrb. 1969 DGLR, pp. 77–85.

14. Gad-el-Hak, M. and Ho, Ching-Ming (1985), "The Pitching Delta Wing", AIAA Journal, Vol. 23, No. 11, pp. 1660–1665.

15. Gad-el-Hak, M. and Ho, Chin-Ming (1986), "Aerodynamics of a Missile in Unsteady Flight", AIAA Paper 86-0572.

16. Hummel, D. and Oelker, H.C. (1986), "Vortex Interference Effects on Close-Coupled Canard Configurations in Incompressible Flow", in the Proceedings of "International Vortex Flow Experiment and Euler Code Validation", Stockholm, pp. 47–61.

17. Hung, C.M. (1979), "Numerical Solution of Supersonic Laminar Flow over an Inclined Body of Revolution", AIAA Paper 79-1547.

18. Jorgensen, L.H. (1977), "Prediction of Static Aerodynamic Characteristics for Slender Bodies Alone and With Lifting Surfaces to Very High Angles of Attack", NASA TR R-474.

19. Legendre, R. (1965), "Lignes de Courant d'un Ecoulement Continu", ONERA Rech. Aerosp. No. 105, pp. 3–9.

20. Lighthill, M.J. (1963), "Attachment and Separation in Three-Dimensional Flow", in *Laminar Boundary Layers,* Section II 2.6, Rosenhead, L., ed. Oxford Univ. Press, pp. 72–82.

21. Lijewski, L.E., Zollars, G.J., Yechout, T.R. and Haupt, B.F. (1982), "Experimental Flow Field Measurements of Missiles with Square Cross-Sections", AIAA Paper 82-0055.

22. McDevitt, J.B. and Mellenthin, J.A. (1969), "Upwash Patterns on Ablating and Non-Ablating Cones at Hypersonic Speeds", NASA TN D-5346.

23. Mendenhall, M.R., Perkins, S.C. and Lesieutre, D.J. (1986), "Prediction of the Aerodynamic Characteristics of Flight Vehicles in Large Unsteady Maneuvers", Proceedings of the 15th Congress of ICAS, Paper ICAS-86-2.5.3, pp. 662–675.

24. Monnerie, B. and Werle, H. (1968), "Etude de l'Ecoulement Supersonique et Hypersonique autour d'une Aile Elancée en Incidence", AGARD CP-30.

25. Murman, E.M., Powell, K.G., Miller, D.S. and Wood, R.M. (1986), "Comparison of Computations and Experimental Data for Leading Edge Vortices – Effects of Yaw and Vortex Flaps", AIAA Paper 86-0439.

26. Payne, F.M., Ng, T.T., Nelson, R.C. and Schiff, L.B. (1986), "Visualization and Flow Survey of the Leading Edge Vortex Structure on Delta Wing Planforms", AIAA Paper 86-0330.

27. Peake, D.J., Jones, D.J. and Rainbird, W.J. (1970), "The Half-Cone Flow and its Significance to Side-Mounted Intakes", AGARD CP-71.

28. Peake, D.J. (1978), "Phenomenological Aspects of Quasi-Stationary Controlled and Uncontrolled Three-Dimensional Flow Separations", AGARD LS 94.

29. Peake, D.J., Owen, F.K. and Johnson, D.A. (1980), "Control of Forebody Vortex Orientation to Alleviate Side Force", AIAA Paper 80-0183.

30. Peake, D.J. and Tobak, M. (1980), "Three-Dimensional Interactions and Vortical Flows With Emphasis on High Speeds", NASA TM 81169.

31. Polhamus, E.C. (1959), "Effect of Flow Incidence and Reynolds Number on Low Speed Aerodynamic Characteristics of Several Noncircular Cylinders with Application to Directional Stability and Spinning", NASA TR R-29.

32. Rabinowicz (Rom), J. (1958), "Measurement of turbulent heat transfer rates on the off portion and and blunt base of a hemisphere cylinder in the shock tube", Jet Propulsion, Vol. 28, No. 9, pp. 615–620.

33. Rainbird, W.J. (1968a), "The External Flow Field About Yawed Circular Cones", AGARD CP-30.

34. Rainbird, W.J. (1968b), "Turbulent Boundary-Layer Growth and Separation on a Yawed 12.5 Cone at Mach Numbers 1.8 and 4.25", AIAA J. Vol. 6, No. 12, pp. 2410–2416.

35. Scuderi, L.F. (1978), "Expressions for Predicting 3D Shock Wave Turbulent Boundary Layer Interaction Pressures and Heating Rates", AIAA Paper 78-0162.

36. Sigal, A. (1990), "Analysis Methods and Experiments for Missiles with Noncircular Fuselages", in *Tactical Missile Aerodynamics*, ed. Nielsen, J.N., Mendenhall, M.R. and Hemesch, J.

37. Skow, A.M. (1983), "Control of Advanced Fighter Aircraft", AGARD FDP-VKI Lecture Series.

38. Van Dyke, M. (1982), "An Album of Fluid Motion", The Parabolic Press, Stanford, California.

39. Werle, H. (1958), "Aperçu sur les Possibilités Expérimentales du Tunnel Hydrodynamique a Visualization de l'O.N.E.R.A.", ONERA Tech. Note 48.

40. Werle, H. (1960), ONERA Rech. Aeron. No. 74, pp. 23–30.

41. Werle, H. (1962), "Separation on Axisymmetrical Bodies at Low Speed", ONERA Rech. Aeron. No. 90, pp. 3–14.

42. Werle, H. (1973), "Hydrodynamic Flow Visualization", *Annual Review of Fluid Mechanics*, Vol. 5, pp. 361–382.

43. Werle, H. (1974), "Le Tunnel Hydrodynamique au Service de la Recherche Aérospatiale", ONERA Publication No. 156.

44. Werle, H. (1980), "Structures Décollements Ails en Fleche", ONERA Rech. Aérosp. 2.

45. Wickens, R.H. (1966), "The Vortex Wake and Aerodynamic Load Distribution of Slender Rectangular Plates (the Effects of a 20 Degrees Bend at Mid-Chord)", Nat. Res. Council of Canada Aero. Report LR-458.

46. Zakkay, V., Calarese, W. and Wang, C.R. (1972), "A Theoretical and Experimental Investigation of the Hypersonic Turbulent Boundary Layer Subject to Normal and Longitudinal Pressure Gradients and Cross Flow Along a Windward Plane of Symmetry", AIAA Paper 72-0187.

Chapter 3

The Topology of Separating and Reattaching Vortical Flows

3.1 Equations for Vortical flows

The Euler and the Navier-Stokes equations for inviscid and viscous flows, respectively, are the governing equations representing the conservation laws of mass, momentum and energy. Direct solutions of these equations are presently being developed, using realistic boundary conditions to represent aerodynamic configurations in flows with varying Mach numbers and angles of attack; with viscous flows, varying Reynolds numbers are also used. The considerable progress that has been achieved in this active area of research of numerical solutions of the Euler and Navier-Stokes equations will be discussed in Chapters 8 and 9. However, the capacity and speed of the present generation of supercomputers, and probably even those of the next one or two generations of future supercomputers, are below the capacity and speed which are required for accurate computations on complex three-dimensional configurations.

So, for many practical aerodynamic design applications, the method of calculations based on simplified models must be used. It is my opinion that even when future generations of supercomputers will make direct solution of the Euler and Navier-Stokes equations practical, there will remain a need for relatively fast calculations utilizing some of the presently developed simplified calculation methods, at least for the preliminary design stages of aircraft and missile development.

Some mathematical formulations which are useful for the understanding and can be useful also in the analysis and calculations of vortical flows about aerodynamic configurations at medium and high angles of attack are presented in this chapter. The first part of this chapter includes the presentation of the equations for vorticity transport and the Biot-Savart law for induced velocities caused by vortex flow. The second part of this chapter includes an introduction to the application of rules of topology towards the interpretation and the understanding of three-dimensional separated vortical flows. Since the application of topological discipline can become a useful tool in reconstructing the characteristics of separated three-dimensional flow fields which are encountered at high angles of attack.

3.1.1 The Vorticity Transport Equation

In a flow field where the velocity vector is $\mathbf{V} = \mathbf{V}(\mathbf{r}, t)$; the angular velocity $\omega = \omega(\mathbf{r}, t)$ is defined as $\omega = 1/2$ curl $\mathbf{V}$ and the vorticity $\Omega = \Omega(\mathbf{r}, t)$ is defined as Ω=curl $\mathbf{V}$. The vorticity transport equation for incompressible flow (with constant kinematic viscosity ν) is derived from the momentum equation to be

$$\frac{d\omega}{dt} = \frac{\partial \omega}{\partial t} + u\nabla\omega = \omega\nabla u + \nu\nabla^2\omega \tag{3.1}$$

The mathematical formulations for the vorticity field, based on the basic theorems of Helmholtz and Kelvin and their applications to fluid flows, have been presented in many textbooks. The book by Karamcheti (1960) is an excellent reference.

3.1.2 The Biot-Savart Law

The velocity induced by a vortex filament on the surface $\mathbf{s}$ at any point $\mathbf{r}(x, y, z)$ in the flow field, can be calculated by the *Biot-Savart law*, given by

$$\mathbf{V}(\mathbf{r}) = \frac{\gamma}{4\pi} \int \frac{\mathbf{dl} \times (\mathbf{r} - \mathbf{s})}{|\,\mathbf{r} - \mathbf{s}\,|^3} \tag{3.2}$$

As an example, the induced velocity due to an infinite line vortex of strength Γ is calculated to be $\mathbf{V}(\mathbf{r}) = (\Gamma/2)\pi h$, where h is the normal distance of the point $P(x, y, z)$ to the vortex line. The Biot-Savart law is used in most methods developed for the analysis and computations of the induced flow velocities due to vortices in the high angles of attack flow fields. An excellent review of the state of the art on the various methods of analysis and computations for vortex flows is presented by Sarpkaya (1989).

3.2 Topological Concepts for the Analysis of Vortical Flows

The application of concepts of topology provides a rational means for understanding the intricate three-dimensional separated flow fields which occur around aerodynamic configurations at high angles of attack. Techniques of flow visualization, particularly the oil streak technique, make it possible to obtain skin friction line patterns on the surfaces of models in wind tunnels as well as on aircraft in flight tests. The topological rules can be useful tools in reconstructing the characteristics of external three-dimensional flow fields using the patterns of skin friction lines on the surfaces of the particular configuration.

The "signature" of streamlines of flow on body surfaces was first discussed in terms of "limiting streamlines" by Sears (1948). It was Legendre (1956) who first viewed these "wall streamlines" trajectories as a continuous vector field which may include certain singular points. As the "limiting streamlines" remain parallel to the surfaces, the pattern of these streamlines may be viewed as a sheath surrounding the body, whose projection on the surfaces are the skin friction lines. The mathematical significance of the surface skin friction lines and, more precisely, the mathematical definition of the various types of singular points as well as their use in the analysis of separated flows, was clearly shown by Lighthill (1963). In this formulation, the skin friction lines and their associated surface vortex lines cover the surface completely, crossing each other everywhere at right angles. Lighthill (1963) also postulated that flow separation can be defined by the convergence of the wall streamlines into a specific wall streamline that originates from a singular point – a saddle point. It is shown that a necessary condition for flow separation is that the surface streamlines converge into a single line originating from a saddle point. Lighthill (1963) also showed that the number of the various types of singular points on the surface must obey topological rules.

Experimental methods for surface flow visualization utilizing oil streaks or various thin layers of paints, produce surface patterns which are accepted as the pattern of skin friction lines on that surface. Legendre (1977) observed that these patterns can be interpreted by using the streamline structure of the specified number of elementary singular points. In an extensive definitive paper on three-dimensional vortical flows, Peake and Tobak (1980) use these surface visualization patterns and topological concepts to decipher the intricate flow fields. They note that singular points in the patterns are consistent with the simplest possible hypotheses with regard to the characteristics of continuous vector fields. Into this hypothesis, Peake and Tobak (1980) integrate the results of Hunt et al. (1978) showing that the notions of elementary singular points and topological rules can be extended to apply as well to flows above the surfaces. They conclude that it is useful to consider the limited number of singular points and topological rules as components of an organizing principle – a flow grammar, whose finite number of elements can be combined in myriad ways to describe, understand, and connect together the properties common to all three-dimensional viscous flows. These studies are also summarized in a latter overview paper by Tobak and Peake (1982).

3.2.1 Attached and Separated Flows

It is useful to introduce the types of flows which constitute the basic elements for the construction of the intricate three-dimensional separated flow fields.

Consider the flow over a smooth slender body of revolution inclined at

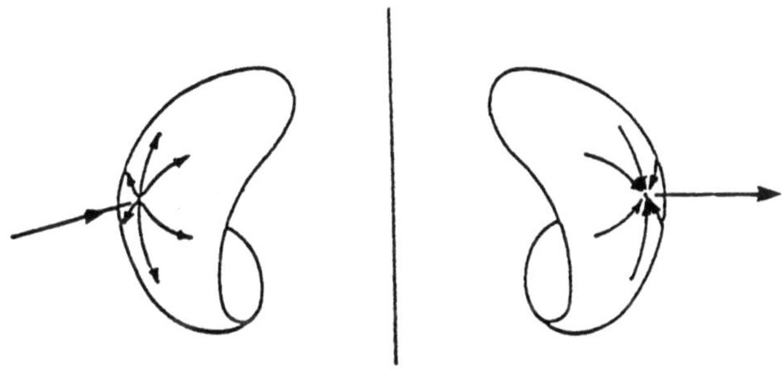

Fig. 3.1a. Skin friction line pattern at nodal points of attachment and separation

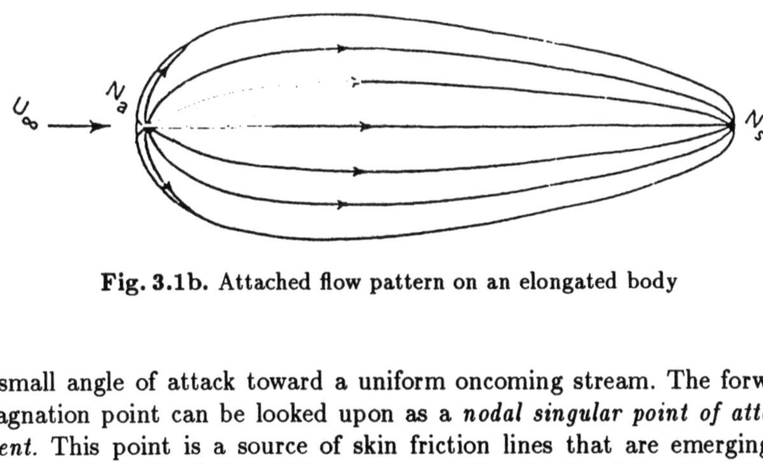

Fig. 3.1b. Attached flow pattern on an elongated body

a small angle of attack toward a uniform oncoming stream. The forward stagnation point can be looked upon as a *nodal singular point of attachment.* This point is a source of skin friction lines that are emerging to envelope the body. A favorable pressure gradient in the circumferential direction produces the skin friction lines which emanate from the nodal point of attachment (the forward stagnation point), sweep around the sides of the body, and converge on either side of the skin friction line running along the leeway ray. All the skin friction lines join at a single point (the rear stagnation point), near the aft end of the body, from which a single streamline departs into the external flow. This point, which can be considered a sink for the collection of all skin friction lines, is called a *nodal point of separation.* Thus, Peake and Tobak (1980) define the attached flow as a flow containing *two*, and *only two singular points*, in the pattern of skin friction lines; these are a *nodal point of attachment* and a *nodal point of separation.*

The surface line patterns at the nodal points of attachment and separation are illustrated in Fig. 3.1a and the attached flow pattern on an elongated body is shown in Fig. 3.1b.

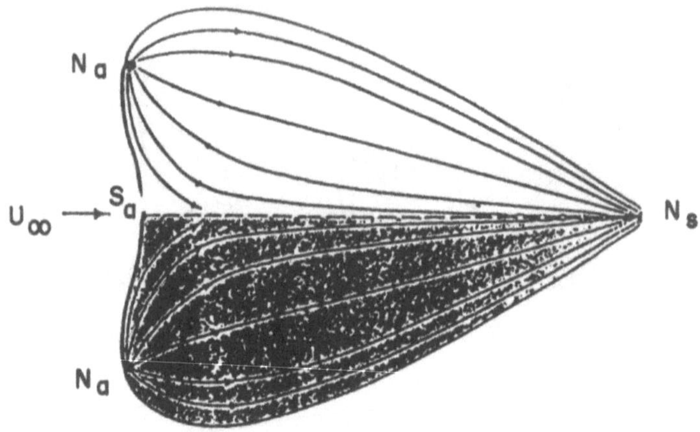

Fig. 3.2a. Illustration of skin friction pattern on bodies with two nodes of attachment, one node of separation and one saddle point of attachment

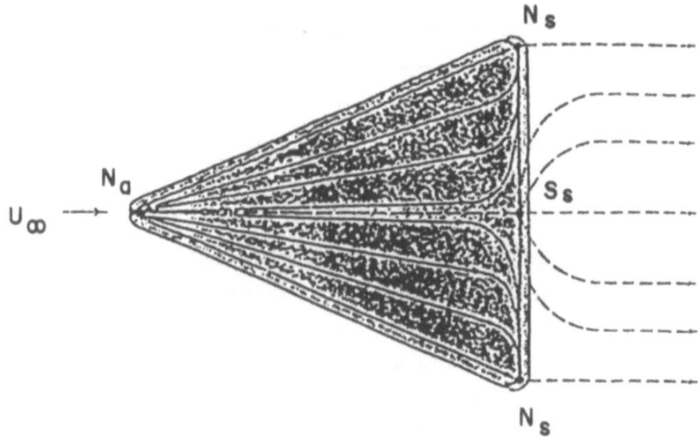

Fig. 3.2b. Illustration of skin friction pattern on bodies with two nodal points of separation, one nodal point of attachment and one saddle point of separation

This definition of attached flow leads to a definition of separated flow as any three-dimensional flow with skin friction patterns containing *more than two nodal points*. Since two of the nodal points must be of the same type, the skin friction lines emanating from each of these nodal points must be prevented from crossing; this is achieved by inserting a *saddle point* between them. Therefore, the simplest separated flow must contain three nodal points and one saddle point. Illustrations of such patterns are presented in Figs. 3.2a and 3.2b. Surface patterns for a flow with two nodal points of attachment, one node of separation and one saddle point of attachment is shown in Fig. 3.2a. Additionally, the surface line patterns for a flow with one node of attachment, two nodal points of separation and one saddle point of separation is shown in Fig. 3.2b. The skin friction line patterns at the saddle points of attachment and separation are illustrated in Fig. 3.3. The fact that the number of nodal points exceeds the number of saddle points by two, conforms to one of the rules of topology.

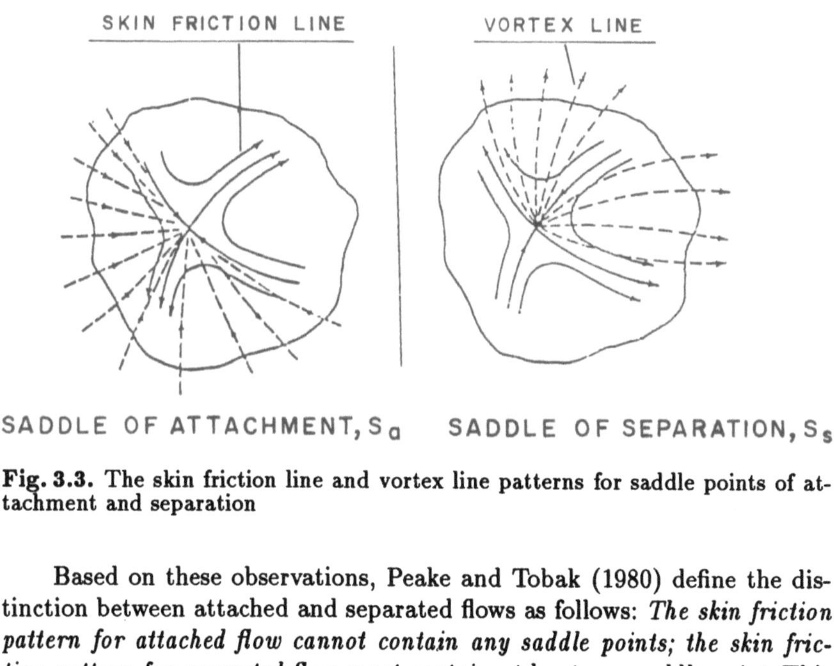

Fig. 3.3. The skin friction line and vortex line patterns for saddle points of attachment and separation

Based on these observations, Peake and Tobak (1980) define the distinction between attached and separated flows as follows: *The skin friction pattern for attached flow cannot contain any saddle points; the skin friction pattern for separated flow must contain at least one saddle point.* This is a necessary condition for defining separation, however, the full definition of the separation phenomena will be included in the discussion on the characteristics of the singular points in Section 3.2.4.

3.2.2 The Equations for Surface Stream Lines and Surface Vortex Lines of Three-Dimensional Flows

The flow near the surface can be defined by the mean velocity vector $\mathbf{v}(x, y)$ which defines the streamlines of the flow at a small distance z above the surface, the mean surface shear stress vector ϵ, and the mean surface vorticity vector ω. It can be shown that near the surface according to Lighthill (1963), (neglecting terms of z^2)

$$\mathbf{v} = \epsilon_w z; \quad \mu\epsilon_w = \tau_w \tag{3.3}$$

where

$$\mathbf{v} = (u, v); \quad \epsilon_w = (\epsilon_u, \epsilon_v); \quad \epsilon_w = \omega_w \times \mathbf{n} \tag{3.4}$$

The vertical velocity w is then

$$w = -\frac{1}{2}(\nabla \cdot \epsilon)z^2 \text{ and } \nabla \cdot \epsilon_w = \frac{\partial \epsilon_u}{\partial x} + \frac{\partial \epsilon_v}{\partial y} \tag{3.5}$$

Since at the singular point, by definition: $\epsilon_u = \epsilon_v = 0$, then the singular points are classified by the values of the $\nabla \cdot \epsilon_w$ and the value of the Jacobian

$$J = \frac{\partial \omega_u}{\partial x}\frac{\partial \omega_v}{\partial y} - \frac{\partial \omega_u}{\partial y}\frac{\partial \omega_v}{\partial x} = \frac{\partial \epsilon_u}{\partial x}\frac{\partial \epsilon_v}{\partial y} + \frac{\partial \epsilon_u}{\partial y}\frac{\partial \epsilon_v}{\partial x} \tag{3.6}$$

as follows:

1. For $J < 0$ the singularity is a ***saddle point.***
   It is a ***saddle point of attachment*** if div $\epsilon_w > 0$
   It is a ***saddle point of separation*** if div $\epsilon_w < 0$
2. For $J > 0$ the singularity is a ***nodal point.***

The structures of skin friction lines and vortex lines at the various types of the ***nodal points*** depend on the values

$$\Delta = \text{ div } \epsilon_w = \frac{\partial \epsilon_u}{\partial x} + \frac{\partial \epsilon_v}{\partial y} = \frac{\partial \omega_u}{\partial x} + \frac{\partial \omega_v}{\partial y} \tag{3.7}$$

and

$$\Omega = \text{ div } \omega_w = \frac{\partial \omega_u}{\partial x} + \frac{\partial \omega_v}{\partial y} \tag{3.8}$$

The corresponding skin friction line patterns and the vortex lines which are perpendicular to these skin friction lines are included in the discussions of the characteristics of the singular points and are shown in Fig. 3.4.

3.2.3 The Characteristics of the Singular Points

In a regular flow, only one skin friction line and a single vorticity line, pass through a given point on the surface. Singular points in this pattern may occur at isolated positions where the surface vorticity and skin friction (which are orthogonal vector quantities tangential to the surface) become simultaneously zero. These singular points can be classified into two main types: *nodes* and *saddle points*. The nodes can be further divided into two sub-classes: *nodal points* and *foci*, both for either attachment or separation. The patterns of the skin friction lines and the vorticity lines at these various singular points are shown in Fig. 3.4.

The various types of nodal points (Fig. 3.4a) are distinguished by the fact that the nodal singular point is common to an infinite number of skin friction lines. At the node, all the skin friction lines except one, labeled AA in this figure, are tangential to the line BB while the angle between these lines can be varied, as indicated in Fig. 3.4a. At a nodal point of attachment, all the skin friction lines are directed outwards away from the singular point, while at the nodal point of separation the lines are aimed in the opposite direction.

The *focus* is distinguished from the "regular" nodal point by lack of a common tangent line, as indicated in Fig. 3.4b. An infinite number of skin friction lines spiral around the singular point, either leading away from it (a *focus of attachment*) or leading into it (a *focus of separation*). Foci of attachment generally occur in the presence of rotation, either in the flow or rotation of the surface.

The *saddle point* has only two skin friction lines that pass through the singular point, lines CC and DD in Fig. 3.4c. The directions on either side of the singular point are inward on one particular line and outward on the other particular line. All of the other skin friction lines do not pass through the singular point and follow directions relating to those of the adjacent lines. These particular lines act as field barriers for the skin friction lines.

The classification of these singularities according to the values of the Jacobian J and the values of Ω and Δ (defined in Equations 3.6, 3.7 and 3.8) according to nodes of various types, foci and saddle points are illustrated in Figs. 3.4a, 3.4b and 3.4c, respectively.

3.2.4 Skin Friction Lines Near Separation in Three-Dimensional Flows

It has been stated that the existence of a saddle point singularity is a necessary condition for separation. Typically, a saddle point separates the skin friction lines issuing from adjacent nodes of attachment, as shown in Fig. 3.5, and used by Lighthill (1963) to illustrate the condition for separation. He defined the line emanating from the saddle point, into which the skin friction lines from both nodes converge but are prevented from crossing, as the *line of separation*.

However, this definition of separation is not unique. An example of a different type of separation is the experimentally observed line of separation emanating from a nodal point of attachment under certain flow conditions,

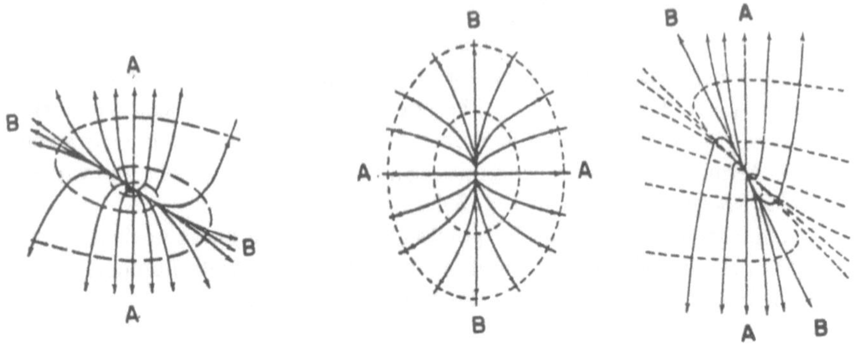

a. Various types of nodal points of attachment, $J > 0$, $\Omega > 0$

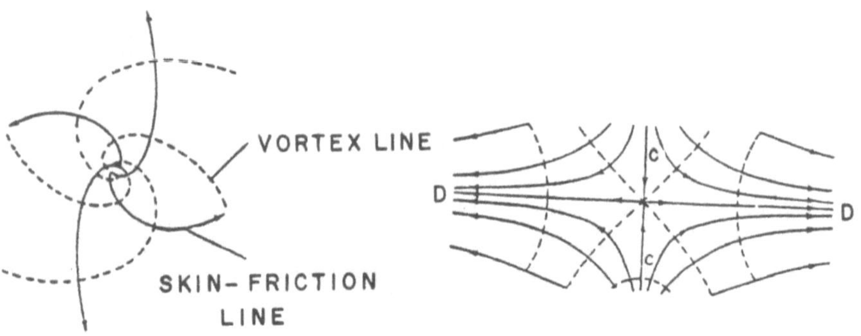

b. Foci points

c. Saddle points

Fig. 3.4. The pattern of skin friction lines and vortex lines near singularities (Lighthill 1963)

as can be seen in Fig. 2.16. There, a line of separation emerges from a focus, to be interpreted as a line of local separation. Indeed, this observation led Wang (1976) to distinguish between two types of separation: "open separation" where the skin-friction line on which the other lines converge does not emanate from a saddle point, and "closed separation" in which

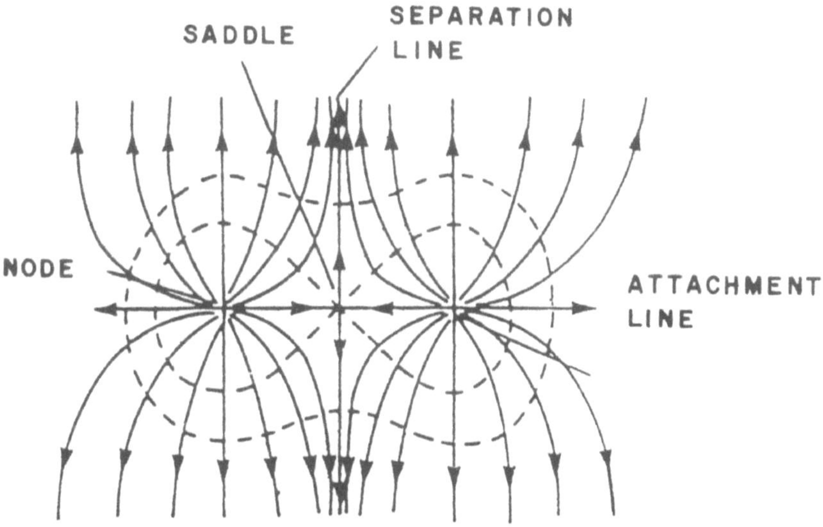

Fig. 3.5. Separation line of the saddle point between two nodes of attachment (Lighthill 1963)

this line does emanate from a saddle point, in keeping with Lighthill's definition. Generalizing the differences between the two types of separations, Tobak and Peake (1982) define a *global* line of separation as one which emanates from a saddle point. When the line on which other skin friction lines converge does not emanate from a saddle point, they define it as a *local* line of separation. It is then possible to define *global* and *local* flow separation, respectively.

The flow on both sides of the line of separation must be separated by a stream surface which emerges from the line of separation. In the case of global separation – separation stemming from a saddle point – the global line of separation is a smooth curve on the wall from which the *dividing surface* originates. Tobak and Peake (1982) define this *dividing surface* as the barrier separating the set of limiting streamlines that have risen from the surface on one side of the global line of separation from the set arisen from the other side. As it passes downstream, the dividing surface rolls-up to form the concentrated rolled-up vortex. While, the local separation is formed by a line of separation emerging from a focus type singular point.

3.2.5 Rules of Topology Applied to Three-Dimensional Separated Flows

It is becoming evident that there is a mathematical basis for the behavior of the elementary singular points and for the topological rules which they obey. Therefore, it is to be expected that a more general theory may be developed to enable the reconstruction of the three-dimensional flow field with separations and reattachments from the surface skin friction lines. The observation of skin friction lines make it evident that the rules governing their behavior can be adapted so as to yield rules governing the flow itself. Following analysis by Hunt et al. (1978), define the mean velocity vector in the plane of the surface $z = z_o =$ constant as:

$$\mathbf{V} = [u(x, y, z_o), v(x, y, z_o), w(x, y, z_o)] \tag{3.9}$$

where the components u and v are defined in the plane $z = z_o$ above a surface situated at $y = Y(x, z_o)$, as shown in Fig. 3.6. The mean streamlines in the plane are solutions to the equation

$$\frac{dx}{u} = \frac{dy}{v} \tag{3.10}$$

Equation 3.10 is a direct counterpart of the equation for the skin friction lines on the surface. It is important to note that these streamlines, defined by equation 3.10, are not necessarily the projections of three-dimensional streamlines on the plane $z = z_o$, nor are they particle paths, even for steady flows. Only for special planes, for example, a streamwise plane of symmetry (where $w(x, y, z_o) = 0$), are the streamlines defined by the equation identified with particle path lines in the steady flow plane, or with instantaneous streamlines when the flow is unsteady. In any case, since $[u(x, y)v(x, y)]$ is a continuous vector field $\mathbf{V}(x, y)$ with only a finite number of singular points in the interior of the flow at which $\mathbf{V} = 0$, it follows that nodes and saddle points can be defined in the plane just as they are defined for skin friction lines on the surface.

Nodes and saddle points within the flow, excluding the boundary $y = Y(x, z_o)$, are labeled N and S, respectively, and their typical streamline structure is indicated in Fig. 3.6. The new features required for this flow structure are the considerations of the singular points on the boundary $y = Y(x, z_o)$. As introduced by Hunt et al. (1978), the singular points on the boundary are defined as half-nodes N' and half-saddles S', indicated in Fig. 3.6. This addition to the collection of allowable singular points enables the use of all the previously discussed descriptions relevant to the analysis of the skin friction lines on surfaces to be carried over to the analysis of streamlines in the flow within a plane.

Therefore, just as the singular points in the skin friction lines on the surface obey topological rules, so must the singular points in any of the sectional views of three-dimensional flows obey these topological rules. The

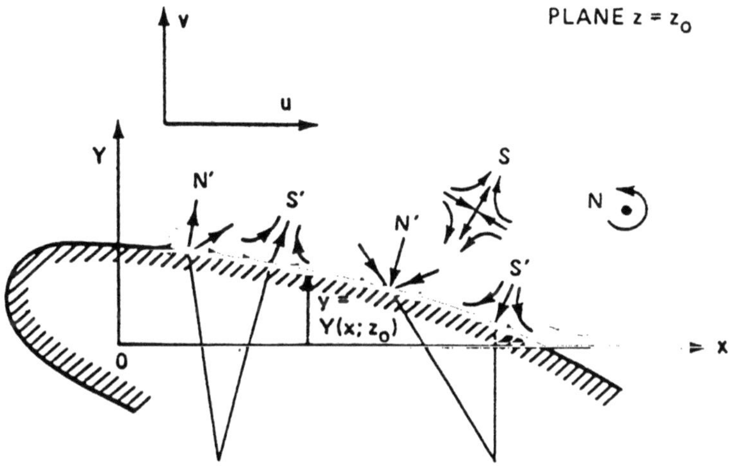

Fig. 3.6. Singularities in the cross plane $z = z_o$ for the flow over a surface and the free stream

practical rules useful for the study of flows around aerodynamic configurations at high angles of attack, can be represented by the five following topological formulas – assuming that the body is simply connected and immersed in a flow that is uniform far upstream.

1. For singular points in the skin friction lines on a three-dimensional body:

$$\sum_N - \sum_S = 2 \tag{3.11}$$

2. For singular points in the skin friction lines on a three-dimensional body B connected simply (without gaps) to a plane wall P that either extends to infinity both upstream and downstream or is the surface of a torus:

$$\left(\sum_N - \sum_S\right)_{P+B} = 0 \tag{3.12}$$

3. For singular points in the streamlines on a two-dimensional plane cutting a three-dimensional body:

$$\left(\sum_N + \frac{1}{2}\sum_{N'}\right) - \left(\sum_S + \frac{1}{2}\sum_{S'}\right) = -1 \tag{3.13}$$

4. For singular points in the streamlines on a vertical plane cutting a surface that extends to infinity both upstream and downstream:

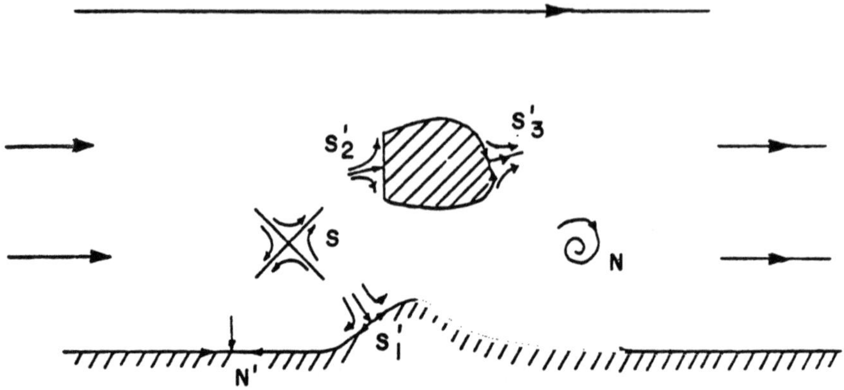

Fig. 3.7. The singularities for the flow over an infinite surface and a body placed in the flow, N, S denote nodes and saddle points in the flow and N', S' denote half nodes and half saddle points on the surfaces

$$\left(\sum_N + \frac{1}{2}\sum_{N'}\right) - \left(\sum_S + \frac{1}{2}\sum_{S'}\right) = 0 \tag{3.14}$$

5. For singular points in the streamlines on the projection onto a spherical surface of a conical flow past a three dimensional body:

$$\left(\sum_N + \frac{1}{2}\sum_{N'}\right) - \left(\sum_S + \frac{1}{2}\sum_{S'}\right) = 0 \tag{3.15}$$

A formulation generalizing these topological rules for defining singularities in the streamline pattern on a vertical section extending to infinity both downstream and upstream through the flow, including the presence of multiple bodies above the surface, was stated by Hunt et al. (1978):

$$\left(\sum_N + \frac{1}{2}\sum_{N'}\right) - \left(\sum_S + \frac{1}{2}\sum_{S'}\right) = 1 - n \tag{3.16}$$

For example, $n = 1$ for a singly connected region with no body, then Eq. 3.16 is identical to the rule stated in Equation 3.14. In the presence of one body $n = 2$, and Eq. 3.16 is equivalent to the rule stated in Eq. 3.13. An example of such a flow with singularities on the surfaces and in the free stream is shown in Fig. 3.7.

Knowledge of the variety of structures for the skin friction lines and the vortex lines near the singular points, those associated with the various types of *nodal* and *foci points* or *saddle points*, and the topological rules for these surface singularities as specified in rules 1 and 2 of Section 3.2.3, allow for the tracing of streamlines on the surface.

However, as noted by Dallmann (1983), a unique relationship does not exist between streamline patterns on the surface and the flow field above the surface. Various flow fields can be reconstructed from the same surface streamline pattern and the same topological singularities, as shown by Hornung and Perry (1982) and Bippes and Turk (1983). Dallmann (1983) suggests that analyzing the flow fields as dynamical systems may enable the explaination of the elementary topological structures of vortex flow separation. Tobak and Peake (1982) point to the possibilities of describing three-dimensional separated flows by use of Andronov's (et al. 1973) concepts of dynamical systems. Such a description leads to definitions of structural stability of the phase portrait of the surface shear stress vector (Andronov et al. 1971); asymptotic instability of the external flow leads to the use of the theory of bifurcation (Andronov et al. 1971, Benjamin 1978). Knowledge of the appropriate topological formulas as well as the surface streamline patterns and flow types near these singularities allows for expanded analysis to facilitate the reconstruction of the flow in the free stream away from the surface. In this way, the flow on and about surfaces with separations and reattachment can be studied, and the streamlines, skin friction lines and vortex lines can be drawn so as to describe the flow field qualitatively. It may then be possible to match the qualitative flow structure derived from the topological rules with the quantitative flow properties calculated by the Navier-Stokes equations. Some encouraging developments in such investigation are presented by Shirayama (1989) and Shirayama and Kuwahara (1990) on the structure of the vortical flow over delta wings and on the flow past a sphere, respectively.

The use of topological concepts for qualitative interpretation of experimentally measured skin friction lines on various configurations was included in the descriptions of the flows at high angles of attack in Chapter 2. It may be of interest to re-examine some of these interpretations for flow over the surface of a wing at high angles of attack, as in Fig. 2.28, flow over a slender body, as in Fig. 2.19, and the comparison of the measured surface skin friction lines with the calculated flow over a conical skirt at an angle of attack, in Figs. 2.26 a, b and c.

References

1. Andronov, A.A., Leontovich, E.A., Gordon, I.I. and Maier, A.G. (1971), "Theory of Bifurcations of Dynamical Systems on a Plane", NASA TT F-556.

2. Andronov, A.A., Leontovich, E.A., Gordon, I.I. and Maier, A.G. (1973), "*Qualitative Theory of Second Order Dynamical Systems*", Wiley New York.

3. Benjamin, T.B. (1978), "Bifurcation Phenomena in Steady Flows of a Viscous Fluid. I. Theory. II. Experiments", Proc. Royal Soc. London, Ser. A. Vol. 359, pp. 1–43.

4. Bippes, H. and Turk, M. (1983), "Oil Flow Patterns of Separated Flow on a Hemisphere Cylinder at Incidence", DFVLR-IB 222-83 A 07.

5. Chapman, G.T. (1986), "Topological Classification of Flow Separation on Three-Dimensional Bodies" AIAA Paper 86-0485.

6. Chong, M.S., Perry, A.E. and Cantwell, B.J. (1990), "A General Classification of Three-Dimensional Flow Fields", Physics of Fluids A, Vol. 2, No. 5, pp. 765–777.

7. Dallmann, U. (1983), "Topological Structures of Three-Dimensional Vortex Flow Separation", AIAA Paper 83-1735.

8. Flegg, G.C., (1974) *From Geometry to Topology,* English University Press.

9. Hesselink, L. and Helman, J. (1987), "Evaluation of Flow Topology from Numerical Data", AIAA Paper 87-1181.

10. Hornung, H. and Perry, A.E. (1982), "Streamsurface Bifurcation, Vortex Skeletons and Separation", DFVLR-IB 222-82 A 25.

11. Hunt, J.C.R., Abell, C.J., Peterka, J.A. and Woo, H. (1978), "Kinematical Studies of the Flows Around Free or Surface-Mounted Obstacles; Applying Topology to Flow Visualization", Journal of Fluid Mechanics, Vol. 86, Part 1, pp. 179–200.

12. Karamcheti, K. (1960), *Principles of Ideal-Fluid Aerodynamics*, J. Wiley, New York.

13. Legendre, R. (1956), "Séparation de l'écoulement laminaire tridimensionnel", ONERA Rech. Aéro., 54, pp. 3–8.

14. Legendre, R. (1977), "Lignes de Courant d'un Ecoulement Permanent: Découllement et Séparation", ONERA Rech. Aérosp., No. 6, pp. 327–335.

15. Lighthill, M.J. (1963), "Attachment and Separation in Three Dimensional Flow", in *Laminar Boundary Layers* Section 2.6, Rosenhead, L., ed. Oxford University Press, pp. 72–82.

16. Nash, C. and Sen, S. (1983), *Topology and Geometry for Physicists,* Academic Press.

17. Peake, D.J. and Tobak, M. (1980), "Three-Dimensional Interactions and Vortical Flows With Emphasis on High Speeds", NASA TM81169.

18. Poincaré, H., (1928), *Oeuvres de Henry Poincaré, Tôme 1.* Gauthier-Villars et Cie, Paris.

19. Sarpkaya, T. (1989), "Computational Methods With Vortices – The 1988 Freeman Scholar Lecture", Transactions of the ASME, Journal of Fluid Engineering, Vol.111, pp. 5–52.

20. Sears, W.R., (1948), "Boundary Layers on Yawed Cylinders", Journal of the Aeronautical Sciences, Vol. 15, No. 1, pp. 49–52.

21. Shirayama, S. (1989), "A Structure of Leading-Edge and Tip Vortices at a Delta Wing", AIAA Paper 89-1803.

22. Shirayama, S. and Kuwahara, K. (1990), "Flow Past a Sphere: Topological Transitions of the Vorticity Field", AIAA Paper 90-3105.

23. Smith, J.H.B., (1975), "A Review of Separation in Steady, Three-Dimensional Flow", AGARD CP-168.

24. Tobak, M. and Peake, T.J. (1979), "Topology of Two-Dimensional and Three-Dimensional Separated Flows", AIAA Paper 79-1480.

25. Wang, K.C. (1976), "Separation of Three-Dimensional Flow", In *Reviews in Viscous Flow*, Lockheed-Georgia Co. Symposium LG 77 E R 0044, pp. 341–414.

Chapter 4

Linear Aerodynamics of Wings and Bodies

The beginnings of modern mathematical analysis of Fluid Mechanics may be attributed to Euler (1707–1783). The differential equations representing the conservation rules of mass, momentum and energy for ***inviscid potential*** flows are known as Euler equations. Analytical solutions to Euler equations have been obtained for some two-dimensional and a few three-dimensional flow problems. For small disturbances in subsonic (below the critical Mach number) and supersonic flows, the equations for the perturbation velocities can be linearized, and exact or approximate analytical or numerical solutions can be developed. Small disturbance conditions require that aerodynamic configurations consist either of thin lifting surfaces and/or slender elongated bodies at small angles of attack. Linearized potential equations for subsonic and supersonic flows, the Laplace and the Wave equations, respectively, can be solved by superposition of elementary solutions of the corresponding subsonic or supersonic singularities – source, doublet and/or vortex. The full potential flow equations for transonic speeds, which are inherently nonlinear, can be solved by numerical methods.

The Euler equations can be solved using today's supercomputers. However, the various panel methods can produce useful results with much smaller computer resources. It will be shown in Chapters 6 and 7 that at high angles of attack, the nonlinearity of the aerodynamic characteristics is produced in the panel methods through the boundary conditions dependency on the rolled-up wake, shape and position. It is, therefore, justified to include here, a detailed introduction to the methods of calculation of the linear aerodynamic characteristics by the various panel methods. In these panel methods, distribution of singularities, sources and doublets or vortices, on the surfaces of the configurations are used for the calculations of the linear aerodynamic characteristics at small angles of attack. It will be shown that these methods, with the proper modifications for the shape and position of the rolled-up vortices, are the basic tools which can be used for the nonlinear aerodynamic calculation methods for the high angles of attack flows.

4.1 Equations for the Potential Subsonic Flows

Consider the flow of an incompressible inviscid fluid in a region (R'), exterior to (or interior to) a given boundary surface (S), shown as a single three-dimensional surface, as in Fig. 4.1. The velocity field ($\mathbf{V}$) can be expressed as a sum of two velocities:

$$\mathbf{V} = \mathbf{V}_\infty + \mathbf{v} \tag{4.1}$$

The velocity $\mathbf{V}_\infty$ is defined as the velocity field that would exist if all boundaries were simply transparent to fluid motion. It is assumed that $\mathbf{V}_\infty$ is known and represents a uniform parallel flow and is, therefore, a constant vector. The vector $\mathbf{v}$ is the disturbance velocity field, due to the boundary surface S.

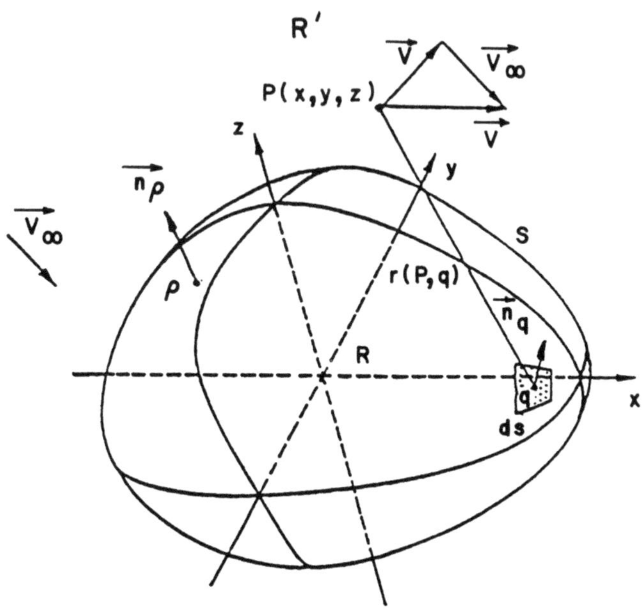

Fig. 4.1. Mathematical formulation of the flow (Hess 1975)

In an incompressible flow, both div $\mathbf{V}_\infty = 0$ and div $\mathbf{v} = 0$. Assuming also an irrotational flow, i.e. curl $\mathbf{v} = 0$, then a potential function ϕ is defined by the relation

$$\mathbf{v} = -\text{grad } \phi \tag{4.2}$$

and since div $\mathbf{v} = 0$, therefore

$$\nabla^2 \phi = 0 \quad \text{in } R' \tag{4.3}$$

The boundary condition on S is that the normal component of the velocity on the surface must vanish. Thus,

$$\frac{\partial \phi}{\partial \mathbf{n}} = \text{grad } \phi \cdot \mathbf{n} \quad \text{on } S \tag{4.4}$$

where $\mathbf{n}$ is the unit outward normal vector to S. Since the right side is known, equation 4.4 expresses a Neumann boundary condition for ϕ. In the usual exterior problem, a regularity condition in the far field is

$$\mid \text{grad } \phi \mid \Rightarrow 0 \quad \text{at infinity} \tag{4.5}$$

The Laplace equation 4.3 and the boundary conditions 4.4 and 4.5 comprise a well posed problem for the potential ϕ. Therefore, this problem can be solved under very general conditions by superposition of elementary solutions. Then, the elementary potential ϕ_e, which is a solution of the Laplace equation and satisfies the regularity conditions at infinity, must have singularities somewhere in space. Since ϕ_e is to be superimposed, there is no loss of generality in assuming that it is singular only at one point whose position is designated by q. The elementary potential at a point P is then denoted $\phi_e(P,q)$, to show its dependence on both the positions of P and q. The Laplace equation can be solved by superposition of the potentials of source, $\phi_s(P,q) = \log(1/r)$, and dipole $\phi_d(P,q) = \underset{q}{\text{grad}}(\phi_S) \cdot \mathbf{t}$, where the unit vector $\mathbf{t} = l\mathbf{i} + m\mathbf{j} + n\mathbf{k}$, and

$$r = [(x_P - x_q)^2 + (y_P - y_q)^2 + (z_P - z_q)^2]^{1/2}$$

The induced velocity due to a vortex element can be calculated by the Biot-Savart law.

For airfoils with sharp corners at the trailing edge, there is a unique value of γ that makes the potential flow surface velocity finite at this corner. This value of the circulation γ corresponds to a unique value of lift and also ensures that a streamline of the flow leaves the airfoil at the trailing edge with a direction along the bisector of the trailing edge. This condition is the well-known Kutta condition.

In order to calculate the lift on a three-dimensional wing, let us begin by defining γ_1 and γ_2 as the values of circulation about two sections of the wing. Since the region exterior to the closed three-dimensional body is simply connected, then for any closed path c in a potential flow without singularities, which implies $\gamma_1 = \gamma_2 = 0$, the line integral for the velocity vector is

$$\int_c \mathbf{v} \cdot ds = 0 \tag{4.6}$$

In order to obtain a nonzero section circulation, there must be some form of singularity in the exterior flow. The nature of this singularity can be evaluated by considering the path c as indicated in Fig. 4.2a. The line integral of the velocity around this path is

$$\int_c \mathbf{v} \cdot ds = \gamma_1 - \gamma_2 + \int_I (\mathbf{v}_+ - \mathbf{v}_-) \cdot ds \tag{4.7}$$

where I is the straight path joining the two section curves and $\mathbf{v}_+$ and $\mathbf{v}_-$ are the limiting velocities obtained by approaching I from two different directions on the surface (as seen in Fig. 4.2a). The condition expressed in equation 4.7 can be fulfilled by assuming a stream surface leaving the body along the path I. Since it is reasonable to assume that vorticity is introduced only to the fluid that passes by the body, therefore, the path I must lie along the trailing edge of the wing, as seen in Fig. 4.2b.

In this case, a vortex sheet issues from the trailing edge and, for steady flow, it proceeds to infinity. The average strength of the vortex sheet along I is proportional to the difference $\gamma_1 - \gamma_2$. Taking the limit as the two section curves approach each other, the result is that the local strength of the trailing vortex sheet is proportional to the spanwise derivative of the section circulation. It follows from this that the local strength of the trailing vorticity that issues from the wing trailing edge equals the spanwise derivative of the bound vorticity. The trailing vorticity is precisely of the right form so that the entire bound and trailing vorticity system may be thought of as being composed of constant strength vortex lines of infinitesimal strength, each of which proceeds spanwise along the wing and then turns and proceeds streamwise along the trailing vortex sheet to infinity – the familiar *"horseshoe vortices"*.

The trailing vortex sheet is assumed to be a stream surface of the flow; therefore, the pressure must be equal across the sheet. In principle, these conditions enable the calculation of the complete shape of the trailing vortex sheet. However, the basic flow problem is *nonlinear* in this case because the location of the trailing vortex sheet changes, in particular, as a function of conditions of the free stream with the variations of the wing's angle of attack relative to the free stream.

4.2 Equations for the Lifting Wing at Low Speeds

Using the representation of the wing and its wake by a vortex sheet of strength γ, the velocity induced at any point $\mathbf{r}$ on the sheet can be evaluated by the Biot-Savart law. As discussed in Part 4.1, both the distribution of vorticity on the bound and trailing vortex sheet and the shape of the trailing sheet can be calculated in principle, using the following conditions:

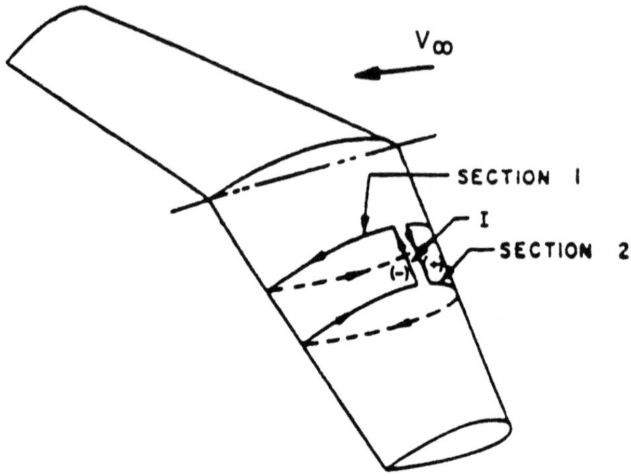

Fig. 4.2a. The path of integration for two sections on the wing

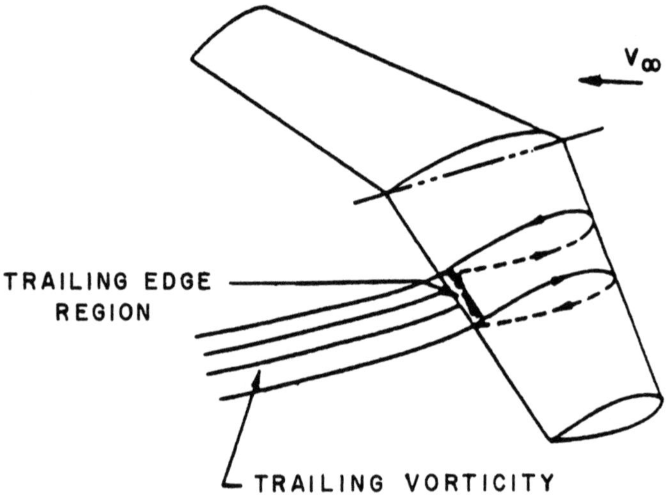

Fig. 4.2b. The path of integration for two sections on the wing with vorticity generation line at the trailing edge

1. The trailing vortex sheet is a stream surface; therefore, the velocity normal to any point on the surface is zero. 2. The pressure difference, Δp, is zero everywhere on the trailing vortex sheet. 3. The Kutta condition requires that $\Delta p = 0$ at the trailing edge of the wing. Since there is an interdependence between the distribution of the strength of the vortices, the shape of the free trailing vortex sheet and the flow conditions, namely, the angle of attack of the wing, the problem is *nonlinear* and the possibility of general solutions are obviously remote.

The developments of the linearized approximations due to Prandtl, are presented in some of the classical texts, e.g. Thwaites (1960), so that only a brief outline is contained here. It is shown that by assuming that the induced velocity, $\mathbf{v}$, is small relative to the free stream velocity V_∞ and that both the wing and the trailing vortex sheet lie in a plane parallel to the free stream, the equation for the downwash velocity at any point $P(x, y, 0)$ in the plane of the wing is

$$\frac{w(x,y)}{V_\infty} = \frac{1}{8\pi}\frac{\partial}{\partial y}\iint_S \frac{l(x',y')}{y-y'}\left[1+\frac{\sqrt{(x-x')^2+(y-y')^2}}{x-x'}\right]dx'dy' \tag{4.8a}$$

and an equivalent form of the equation for the downwash velocity is

$$\frac{w(x,y)}{V_\infty} = \frac{1}{8\pi}\iint_S \frac{l(x',y')}{(y-y')^2}\left[1+\frac{x-x'}{\sqrt{(x-x')^2+(y-y')^2}}\right]dx'dy' \tag{4.8b}$$

where the integration is over the wing surface S, and $l(x', y')$ is the local aerodynamic loading, given by

$$l(x',y') = 2\frac{\gamma(x',y')}{V_\infty} = \frac{\Delta p(x',y')}{\frac{1}{2}\rho V_\infty^2} \tag{4.9}$$

Equations 4.8a and 4.8b are known as the ***downwash integral equation*** and are the fundamental equations of the linearized lifting wing theory. Hence, the section lift coefficient at a given spanwise station y is

$$C_1(y) = \frac{1}{c(y)}\int_{x_{LE}}^{x_{TE}} l(x,y)\,dx \tag{4.10}$$

The overall lift coefficient for the wing is, therefore,

$$C_L = \frac{AR}{2b}\int_{-b/2}^{+b/2} C_1(y)c(y)\,d(y/b) \tag{4.11}$$

For wings with no sweep and of large Aspect Ratio, AR, the integral equation for the spanwise lift distribution becomes

$$\frac{C_1(y)}{2\pi} = \alpha(y) - \frac{1}{8\pi}\int_{-b/2}^{+b/2} \frac{d[C_1(y)c(y)]}{dy'}\frac{dy'}{y-y'} \tag{4.12}$$

For wings of small AR, the spanwise lift distribution is calculated to be

$$C_1(y) = 2\alpha\frac{b}{c(y)}\left[1-\left(\frac{y}{b/2}\right)^2\right]^{1/2} \tag{4.13}$$

the overall lift coefficient is then

$$C_L = \frac{1}{2}\pi\mathrm{AR}\alpha \tag{4.14}$$

and the induced drag, for the low AR wing, is calculated to be $C_{Di} = \frac{1}{2}C_L\alpha$.

4.3 Linear Panel Methods for the Calculation of the Subsonic Aerodynamic Coefficients for Wings and Bodies

4.3.1 The Multhopp Method

The theoretical method of Multhopp (1950) has received wide acceptance as an accurate method for calculating aerodynamic load distributions on wings of various shapes in low speed flows. The method was modified and formulated for complex planform shapes, which are used in modern high performance aircraft, and adapted into a form suitable for programming for high speed digital computers by Lamar (1968). The method uses the acceleration potential approach in conjunction with the linearized Euler equations, to relate the pressure difference across the wing to the downwash field on the wing surface. The appropriate equation for the downwash for the subsonic flow case (based on Eq. 4.8b) is given by Lamar (1968).

$$\left.\frac{w}{V_\infty}\right|_{z=0} = \frac{1}{8\pi q_\infty}\int_{-1}^{1}\int_{x'_{LE}}^{x'_{TE}} \frac{\Delta p(x',y')}{\eta}[1+\xi(\xi^2+\eta^2)^{-1/2}]\,dx'dy' \tag{4.15}$$

where $\xi = (x-x')/\beta$ and $\eta = y - y'$.

The solution is obtained by replacing $\Delta p(x',y')$ with as many as 10 pressure modes of unknown or known values, in the form of

$$\Delta p(x',y') = \frac{q_0(y')}{c(y')}\cot\frac{\vartheta}{2} + \sum_{j=1}^{9}\frac{q_j(y')}{c(y')}\sin j\vartheta \tag{4.16}$$

where

$$x' = -c(y')\cos\vartheta + d(y') \tag{4.17}$$

and then, employing the Multhopp quadrature formula to approximate the spanwise integration of the pressure doublets. A logarithmic singularity correction term is used to account for the integration across the second-order singularity. Applying these approximations to equation 4.15 and reforming it into a matrix equation, the result is

$$\left\{\frac{-4w(\phi_S, y_\nu)}{V_\infty}\right\} = \left[\bar{L}^j_{\nu n}(\phi_S)\right]\left\{\frac{q_j(y'_n)}{q_\infty}\right\} \tag{4.18}$$

The formulation of the influence coefficients $\bar{L}^j_{\nu n}(\phi_S)$ is accomplished by integrating the chordal influence that the pressure loading functions exert on each control point from every other spanwise station. The details of this evaluation are given in the paper by Lamar (1968). Control points are defined at the positions, determined from ϕ_S and y_ν, which are selected to specify either the tangential flow boundary condition or the downwash ratio. These control points are located in both spanwise and chordwise directions by the use of semicircles, which are laid along the span and along the chord as indicated in Fig. 4.3.

The matrix equation 4.18 can be solved for the values of the array of $q_j(y')/q_\infty$, then the local pressure loading distribution is evaluated by equation 4.17, from which the local and the total aerodynamic coefficients can be calculated. The Multhopp method can be also used to evaluate the required mean chamber line for a specified planform and a specified desired loading – the inverse problem.

The Multhopp method and its modifications give reasonable results for wings of varying planform shapes. Comparison of the calculated aerodynamic characteristics for a high aspect ratio swept back wing with the experimental data is shown in Fig. 4.4. Good agreement is found for the local lift coefficient distribution, Fig. 4.4a, and for the center of pressure, x_{cp}, Fig. 4.4b, but there is an appreciable difference between the calculated and measured local pressure distributions, Fig. 4.4c. In general, the accuracy of the calculations of the loading and lift coefficients is relatively good, while the calculations of the pressure distributions is much less accurate. It is also important to note that this method is restricted to wings alone and is not applicable either to bodies alone or to wing-body configuration.

4.3.2 The Surface-Source Distribution Method

The potential equation, the Laplace equation, and the appropriate boundary conditions for incompressible flows, as specified in Section 4.1, can all be solved under very general conditions by superposition of elementary solutions. A method based on distribution of sources over the surfaces of the

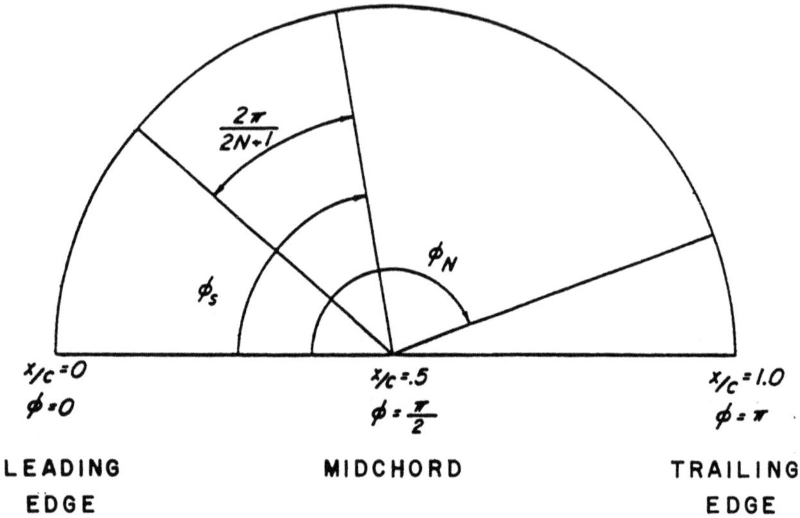

a. Spanwise divisions $\theta = \frac{\pi}{2} - \frac{\nu\pi}{m+1}$, where $-\frac{m-1}{2} < \nu < \frac{m-1}{2}$

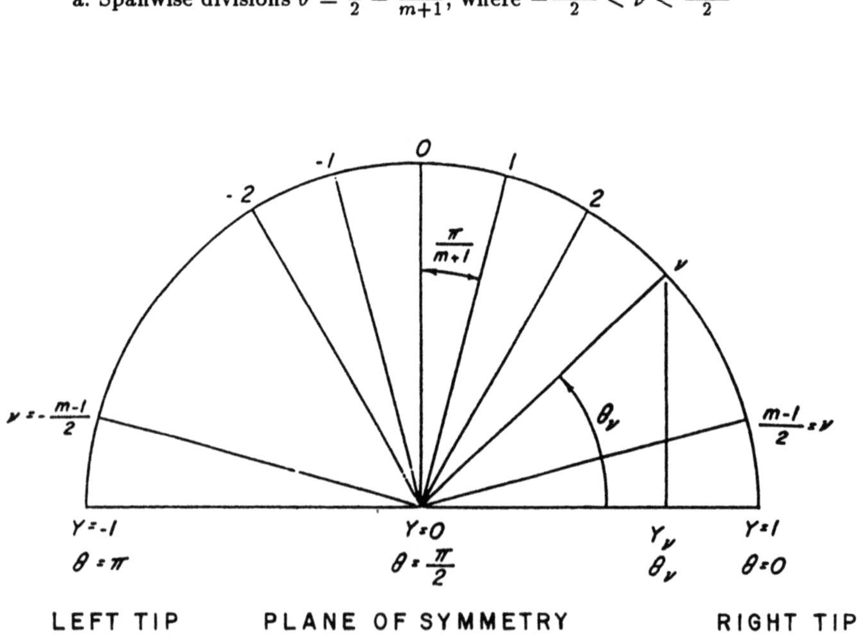

b. Chordwise divisions, $\Phi_s = \frac{2s\pi}{2N+1}$, where s ranges from 1 to N

Fig. 4.3. The spanwise and chordwise divisions of the wing (Lamar 1968)

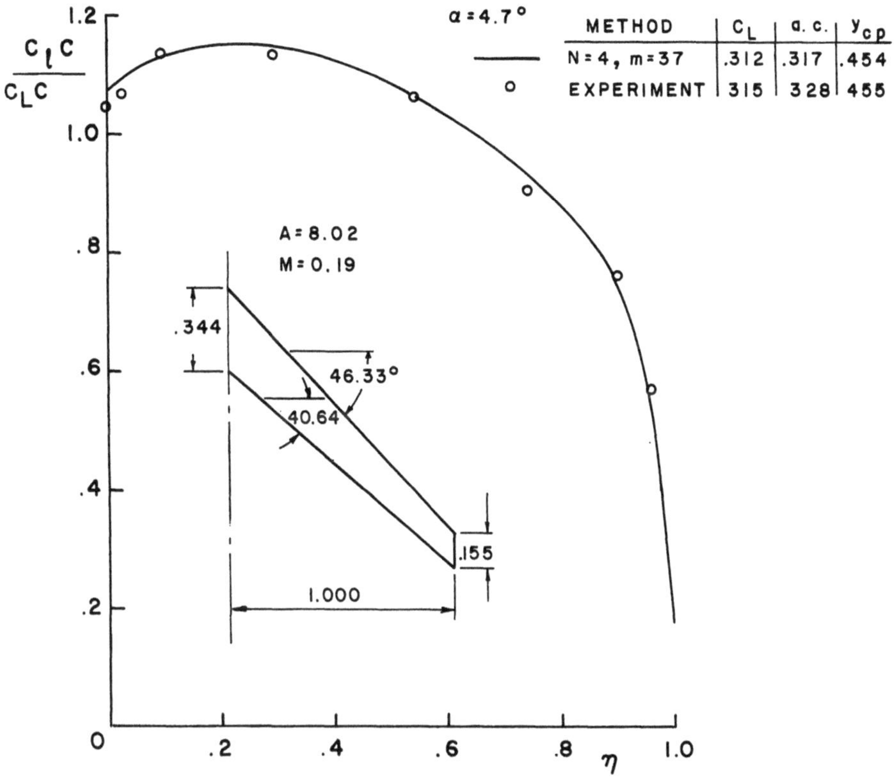

Fig. 4.4a. Spanwise loading distribution calculated by the modified Multhopp method as compared with experimental data (Lamar 1968)

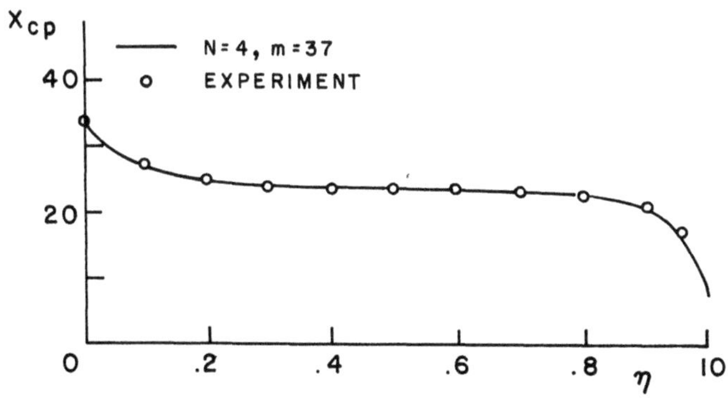

Fig. 4.4b. Centers of pressure calculated by the modified Multhopp method as compared with experimental data (Lamar 1968)

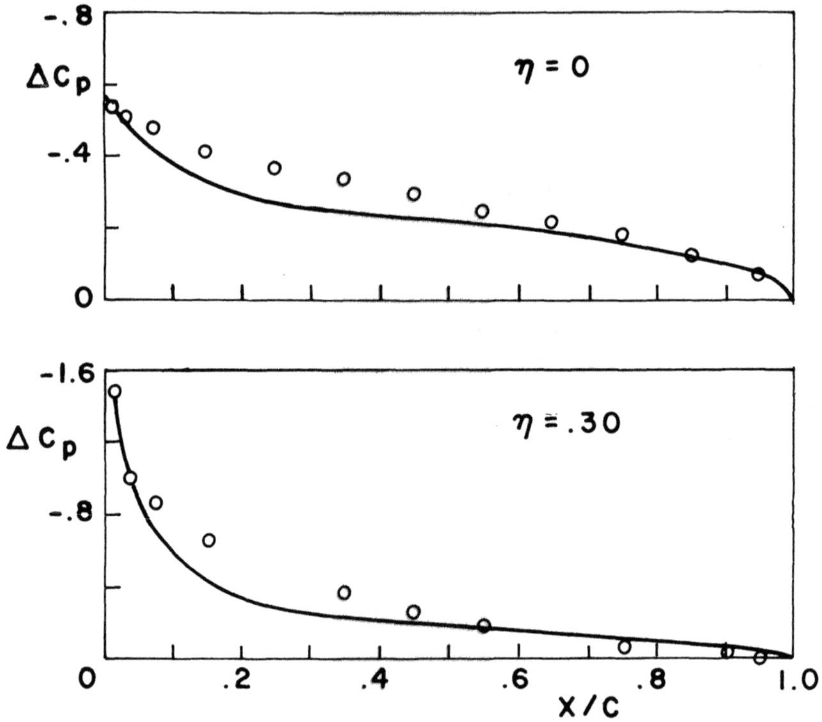

Fig. 4.4c. Chordwise loading distributions calculated by the modified Multhopp method as compared with experimental data (Lamar 1968)

configuration was developed at the Douglas Aircraft Company, and a general review of this method is presented by Hess and A.M.O. Smith (1966) and by Hess (1975).

The method for solving potential flow is based on the distribution of singularities on the boundary surface S (Fig. 4.1) which are obtained as solutions of integral equations. Using Green's Theorem, the potential at a point P exterior to S is

$$\phi(P) = \frac{1}{4\pi}\iint_S \phi(q)\frac{\partial}{\partial n_q}\left(\frac{1}{r(P,q)}\right) dS - \frac{1}{4\pi}\iint_S \frac{1}{r(P,q)}\frac{\partial \phi}{\partial n_q}(q)\, dS \tag{4.19}$$

while the potential at a point p on the surface S is

$$\phi(p) = \frac{1}{2\pi}\iint_S \phi(q)\frac{\partial}{\partial n_q}\left(\frac{1}{r(p,q)}\right) dS - \frac{1}{2\pi}\iint_S \frac{1}{r(p,q)}\frac{\partial \phi}{\partial n_q}(q)\, dS \tag{4.20}$$

For the source distribution method, the disturbance potential due to the boundary S may be represented by means of a distribution of sources alone on the surface, in the form

$$\phi(P) = \iint_S \frac{1}{r(P,q)} \sigma(q)\, dS \tag{4.21}$$

where the function $\sigma(q)$ denotes the source strength over S. The integral equation for the source density distribution σ becomes

$$2\pi\sigma(p) - \iint_S \frac{\partial}{\partial n_p}\left(\frac{1}{r(p,q)}\right) \sigma(q)\, dS = -\mathbf{n}_p \cdot \mathbf{V}_\infty \tag{4.22}$$

This is a Fredholm integral equation of the second kind whose kernel is

$$K(p,q) = -\frac{\partial}{\partial n_p}\left(\frac{1}{r(p,q)}\right) \tag{4.23}$$

which is the outward normal velocity at p due to a unit point source at q. After determining σ , the perturbation velocity of the flow over the surface S is obtained from

$$v(p) = -\iint_S \underset{p}{\text{grad}}\left(\frac{1}{r(p,q)}\right) \sigma(q)\, dS \tag{4.24}$$

The velocity in the space away from the boundary surface S can be calculated from equation 4.24, replacing p by P.

The above equations comprise the basic formulation of the surface source density method for solving the potential flow problem. Usually, only the surface velocity, as given in equation 4.24, is of interest. Equations 4.22 and 4.24 are for three-dimensional cases; for two-dimensional cases, the source potential $1/r$ is replaced by $\log(1/R^2)$.

These integral equations are solved by an approximate numerical method. The first step is to divide the surface S into small surface elements, each of which has a simple geometric form, which is generally planar. The surface source density on each element is assumed to be finite and continuous. It is usually taken as constant on the element, although it may vary discontinuously from element to element. Some schematics of the division of two-dimensional and axisymmetric bodies are presented in Fig. 4.5a and for three dimensional configuration in Fig. 4.5b.

The essence of the surface source density method is that the source distribution is determined to satisfy the normal velocity boundary condition, expressed in equation 4.22. This condition is applied to one selected point in each panel element, defined as the control point. The control point is selected to be at the mid-point of the line segment in two-dimensional and axisymmetric cases and at the centroid of the quadrilateral in three-dimensional cases.

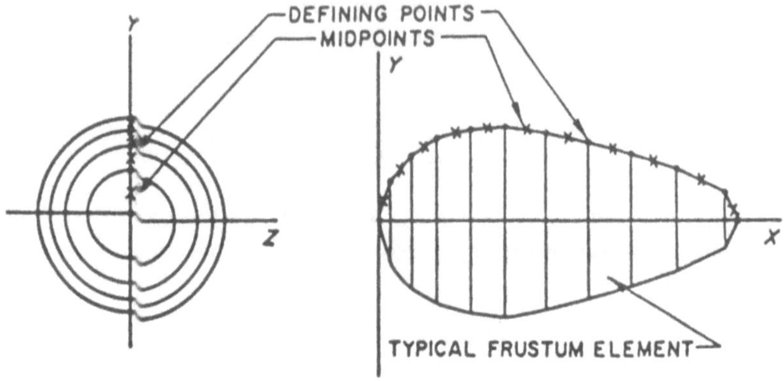

Fig. 4.5a. Schematic of division of two-dimensional and axisymmetric configurations (Hess 1975)

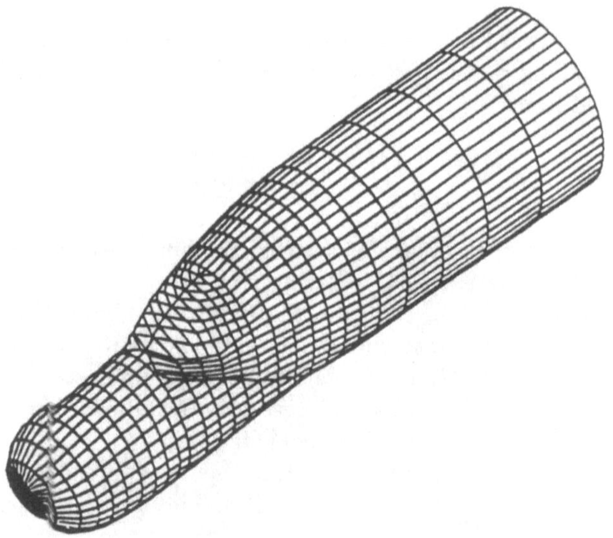

Fig. 4.5b. View of division of the surface of a three-dimensional body to panels (Hess 1975)

The surface S is divided into N elements, each one assumed to have a constant source density. Therefore, the problem of determining the continuous distribution function σ is reduced to determining the values of σ_i in each of the N element panels. As a result, the application of the boundary condition at each control point for all the elements gives a set of N linear algebraic equations for the N values of σ_i.

It is possible to define an influence matrix A_{ij} whose elements are given by $A_{ij} = \mathbf{n}_i \cdot \mathbf{V}_{ij}$, i.e. the normal velocity at the control point of the i-th element due to a unit source density on the j-th element, where V_{ij} is defined as the velocity at the control point of the i-th element due to the unit value of source density on the j-th element and $\mathbf{n}_i$ is the unit normal vector to the i-th element. Applying the boundary conditions at all control points gives, then, the following set of linear equations

$$\sum_{j=1}^{N} A_{ij}\sigma_j = -\mathbf{n}_i \cdot \mathbf{V}_\infty, \qquad i = 1, 2, \ldots, N \tag{4.25}$$

This is the numerical approximation for the evaluation of the integral equation 4.22. Equation 4.25 is solved numerically for the sources strength σ_i. Thus determining σ_i enables us to determine the perturbation velocities at all control points by the relation

$$v_i = \sum_{j=1}^{N} \mathbf{V}_{ij}\sigma_j \tag{4.26}$$

This is the numerical approximation of equation 4.24. The numerical methods for solving these linear algebraic equations and the applications to two-dimensional, axisymmetric and three-dimensional configurations are reviewed by Hess (1975). It is shown that the matrix equation can be solved either directly or by iterative procedures.

The accuracy of this calculation is determined by the method of division and the number of surface elements used to approximate the body surface. It is important to point out that surface singularity methods, like the source density method, are based on the numerical solution of an "exact" integral equation (in terms of the potential flow analysis). The numerical solution can be made as accurate as desired by increasing the number of surface elements – panel divisions. The practical accuracy is limited by the physics of the potential theory, where the effects of viscosity are neglected. It is found that for flow over smooth two-dimensional and axisymmetric bodies, 30 to 60 surface panel elements are sufficient. For complex configurations, the panel number may be increased to a few hundred on two-dimensional and axisymmetric bodies and up to a few thousand for three-dimensional bodies.

The surface source density method is particularly suited for the calculation of the velocity field and the pressure distributions over various *nonlifting* configurations in *subsonic* flows. An example of the excellent agreement between the calculated results by this method with the experimental measurements is presented in Fig. 4.6, where the pressure distributions on a three-dimensional body (a radome mounted on a fuselage, shown in Fig. 4.5b) are indicated.

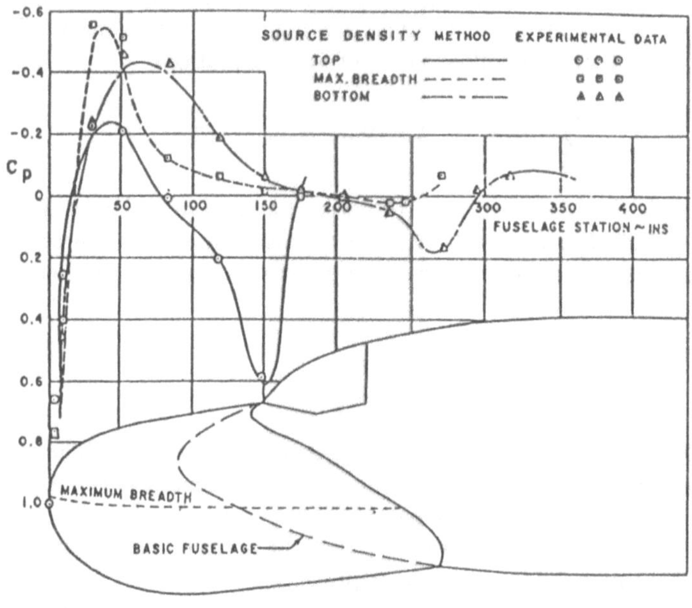

Fig. 4.6. Comparison of the surface source density method calculations with the experimentally measured pressure distributions on a radome mounted on a fuselage (Hess 1975)

As stated, the surface-source method is most suitable for the calculations of pressure distributions over two and three-dimensional bodies in incompressible flows. It was modified to include the effects of compressibility and was applied to various flow problems such as: inlets at prescribed mass flow ratio (Hess and AMO Smith 1965), multi-element airfoils (Callaghan and Beatty 1972), unsteady flows with wakes (Giesing 1970), first-order two-dimensional free surface effects (Giesing and AMO Smith 1967) and axisymmetric bodies at angles of attack (Hess 1969).

It was shown that the surface-source method can be modified for flows with lift, by calculating the effects of the superposition of distribution of vortices, in addition to the effects of the distribution of sources over the surfaces. In order to define the total circulation uniquely, the strength of these vortices is determined so as to fulfill the Kutta condition at the sharp trailing edges of the lifting airfoils in two dimensions and at the trailing edges of the wings for aircraft configurations. The modified method for the calculations of two- and three-dimensional lifting cases is presented by Hess (1972, 1975). In the two-dimensional lifting airfoil case, the calculation also includes a viscous boundary layer correction, resulting in reasonable agreement with experimental data, as seen in Fig. 4.7.

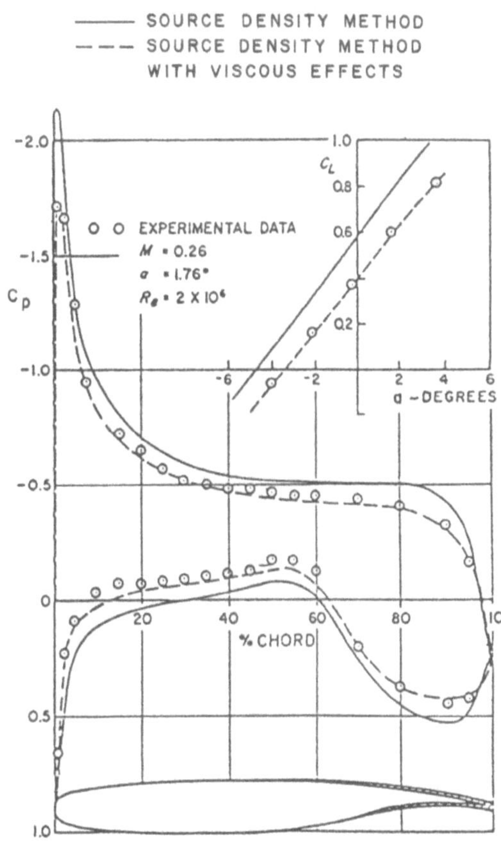

Fig. 4.7. Comparison of experimental surface pressure distributions and lift coefficients with the calculations of the modified surface-source method (Hess 1975)

The application of the surface-source distribution method to calculate the flow about a lifting aircraft configuration is much more complicated and much less successful then the previously discussed cases. The distribution of vortices over the finite span lifting surfaces require their extension as a trailing vortex sheet into the wake. In general the position of the trailing vortex sheet must be determined by the induced flow from the complete configuration and the boundary conditions: the existence of the Kutta condition at the sharp trailing edges and the fact that the vortex sheet cannot carry any load, i.e. $\Delta p = 0$ across the free vortex sheet. As in all linear methods, it is assumed that the free vortex sheet remains flat in the plane of the wing and that the free vortices can be visualized as straight lines trailing from the wings' trailing edges. The velocities, induced by the complete vortex structure on the wings and the wake, can be calculated, assuming that elements with bound vortices as strips on the wing surfaces are followed by the corresponding strips in the trailing vortex wake, as seen in Fig. 4.8.

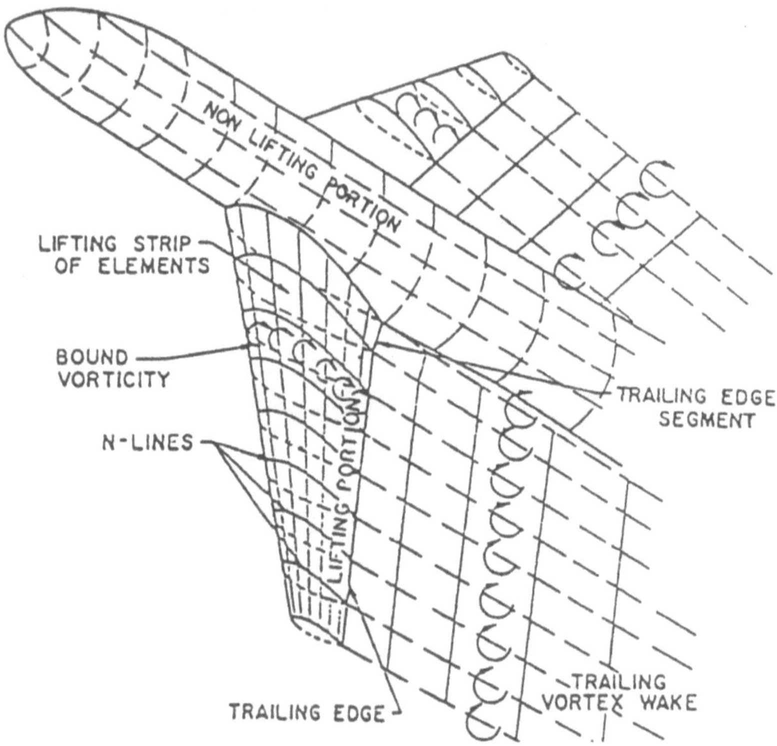

Fig. 4.8. Distribution of vortices on a typical lifting configuration

Reasonable results are obtained at low angles of attack, where the linear method can be applied, as can be seen in the results presented in Fig. 4.9.

The surface-source distribution method can be applied successfully in the subsonic speed range. The results are very accurate for the nonlifting cases, and the calculated pressure distributions are found to be in excellent agreement with the corresponding experimental data. The calculations of

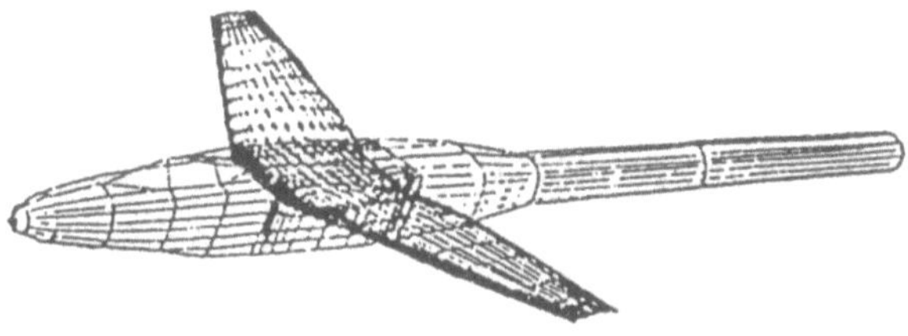

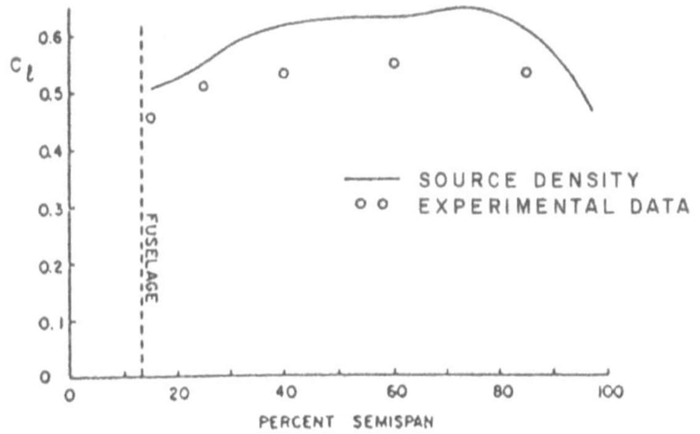

a. Spanwise distribution of section lift coefficient

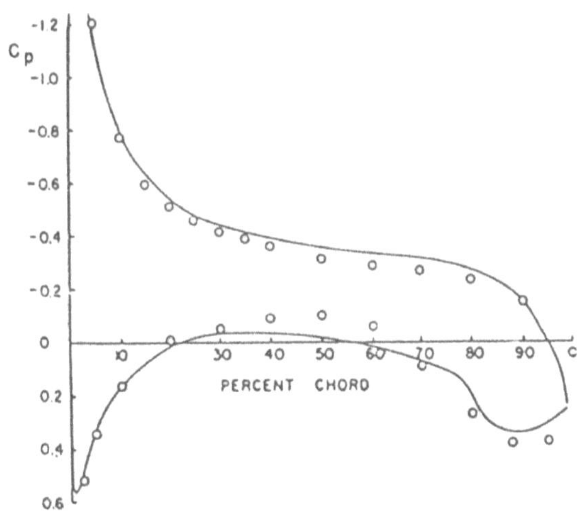

b. Chordwise pressure distribution at 25% span station

Fig. 4.9. Comparison of calculated and experimental results on a wing-fuselage configuration at $\alpha = 7^o$ (Hess 1975) the lift on aircraft configurations in subsonic flow are reasonable but cannot be extended to supersonic flows.

4.3.3 The Linear Vortex Lattice Method for Subsonic Flow

The lifting line and the lifting surface theories (based on the concepts of Prandtl and Glauert) have been used very successfully to determine the linear aerodynamic characteristics of flat, thin wings of medium and high aspect ratios. It may be, therefore, expected that a good representation of the lifting characteristics of a wing with a complex planform can be obtained by dividing the wing planform into many quadrilateral panel elements, placing a horseshoe vortex in each one of these panels. A method for calculating the lift of wings, based on this concept of *vortex lattice*, was proposed by Falkner (1947, 48). A computer program for the vortex lattice calculations for evaluating the lifting characteristics of the wings of complex planforms in subsonic flows was presented by Margason and Lamar (1971).

The calculation first requires the determination of the geometrical division of the planform into the panel elements and the placement of horseshoe vortex systems into each of these panels. An example of such a division is presented in Fig. 4.10. The planform is divided by a series of straight lines which are projected onto the reference plane $\bar{X}-\bar{Y}$ from the actual deflected wing planform, double delta wing with dihedral deflections, as shown in the Fig. 4.10. The program is capable of handling wings with camber and twist and as well as variable sweep and movable control surfaces in both symmetric and asymmetric deflections.

The planform is thus divided chordwise and spanwise into trapezoidally shaped elemental panels. A horseshoe vortex is placed in each panel, as shown in Fig. 4.11. Each horseshoe vortex is composed of a bound vortex which is positioned to coincide with the quarter chord line of the panel in the plane of the wing and two trailing vortices which extend chordwise as straight lines parallel to the free stream to infinity behind the wing, as indicated in Fig. 4.10. A *control point* is defined for each panel at the midspan three-quarter panel-chord line, as seen in Fig. 4.11.

This method specifies that the boundary condition of "no flow" through the wing surface is fulfilled at these control points. The downwash velocity for a particular horseshoe vortex is evaluated by the Biot-Savart law

$$w(x,y,z) = \frac{\Gamma}{4\pi} F_w(x',y,z,s,\psi',\phi) \tag{4.27}$$

and the sidewash velocity is

$$v(x,y,z) = \frac{\Gamma}{4\pi} F_v(x',y,z,s,\psi',\phi) \tag{4.28}$$

where F_w and F_v are the downwash influence coefficient and the sidewash influence coefficient, respectively. The "no flow" condition requires that

$w - v \tan\phi = V_\infty \alpha$. In terms of the influence coefficients, the boundary condition for a vortex lattice of N elements is

$$\sum_{n=1}^{N} (F_{w,n} - F_{v,n} \tan\phi_n) \frac{\Gamma_n}{V_\infty} = 4\pi\alpha \tag{4.29}$$

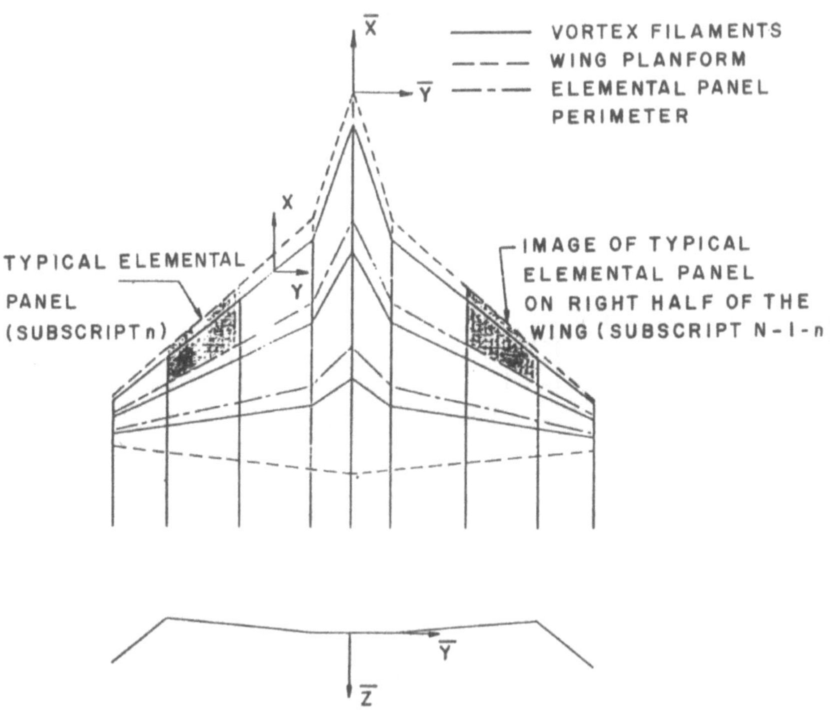

Fig. 4.10. Layout of wing planform for VLM calculations – axis system, panel elements and horseshoe vortices (Margason and Lamar 1971)

Therefore, for this vortex lattice of N elemental panels, the strength of the vortex in each panel Γ_n can be calculated by solving the following matrix equation which is obtained from the application of equation 4.29 to all the panels

$$\left[F_{w,n,k} - F_{v,n,k} \tan\phi_n \right] \left\{ \frac{\Gamma_n}{V_\infty} \right\} = 4\pi \left\{ \alpha_k \right\} \tag{30}$$

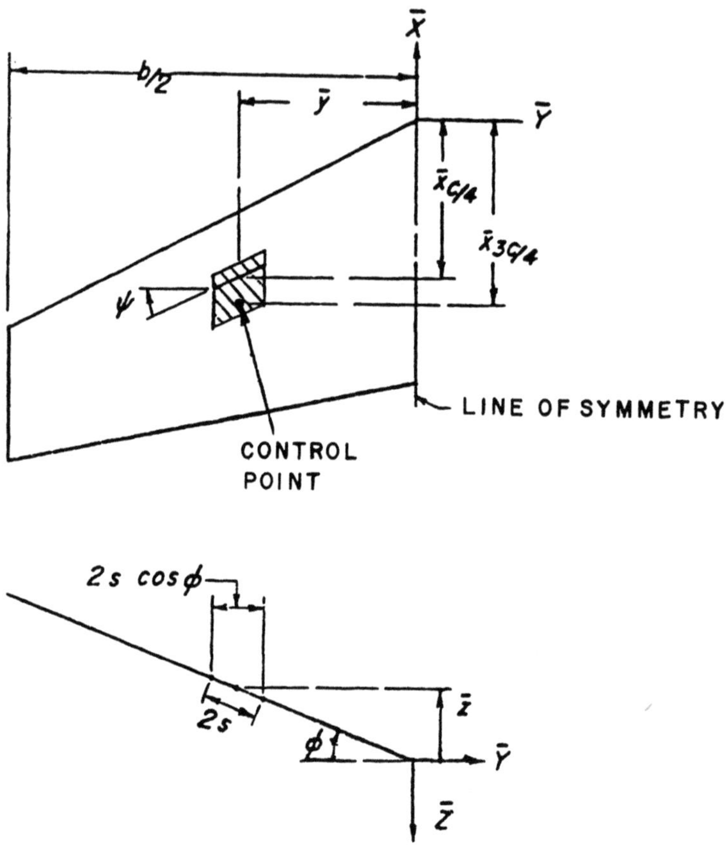

Fig. 4.11. The definitions of the geometry of a panel (Margason and Lamar 1971)

where α_k is the local angle of attack at the control point. The detailed specifications of the downwash and sidewash influence coefficients are presented by Margason and Lamar (1971). Once this matrix equation is solved for the strength of the vortices in each panel it is possible to compute the aerodynamic coefficients

$$C_L, \quad C_D, \quad C_m, \quad C_L, \quad \frac{\partial C_m}{\partial C_L}, \quad C_{m_o}$$

as well as the chordwise and spanwise loading and the spanwise center of pressure y_{cp}. It is also possible to calculate the rotary stability derivatives

$$C_{lp}, \quad C_{mq} \quad \text{and} \; C_{Lq}$$

as discussed by Margason and Lamar (1971).

This version of the Vortex Lattice Method is restricted to subsonic flows. It is a relatively simple calculation method with a very short execution time using presently available high speed computers. The basic method does not include thickness effects, so it cannot be applied to wing-body configurations. The results for wings alone, including multi-surface cases, are generally in good agreement with the experimental data.

4.4 Linear Panel Methods for the Calculation of the Subsonic and Supersonic Aerodynamic Coefficients for Wings and Bodies

The concept of representing a configuration by dividing its surfaces into many panels and placing in each panel a superposition of singularities, which represent the elementary solutions to the Laplace equation, is the basis for the various subsonic panel methods. The surface-source distribution method and the linear vortex lattice method, discussed in 4.3.2 and 4.3.3, represent the first applications of these methods in subsonic flows. Further developments of panel methods resulted in operational computer codes, such as the Woodward (1968) programs, which are used in many aerodynamic design offices as the main calculation tool for calculating the aerodynamic characteristics of practical aircraft configurations at low angles of attack for both subsonic and supersonic flows.

Panel methods, which use flat planar panels on which the singularities are distributed uniformly, are defined as low-order methods. It is possible to introduce curved panels which better describe the actual surface shape and to distribute on them singularities with varying strength, this being defined as a high-order method. In addition to the linear VLM and the surface-source (Hess) methods, two more methods exemplifying the low-order surface singularity panel methods which use quadrilateral panels to represent arbitrary three-dimensional configurations are the VSAERO (Vortex Separation Aerodynamic Program) and the QUADPAN (Quadrilateral Panel Aerodynamics Program). The VSAERO program is presented by Maskew (1982a,b) and the QUADPAN is presented by Coppersmith et al. (1983). High-order panel methods are the MCAERO, presented by Bristow and Hawk (1983), and the PAN AIR, presented by Carmichael and Erickson (1981) and appearing in NASA publications PAN AIR (1980). A panel method for computing the subsonic aerodynamic characteristics of complex configurations is the three-dimensional ECOPAN code developed at ONERA. The use of this code in the European programs of Ariane and Airbus is presented by Roggero and Larguier (1990).

The panel methods applications can also be extended to supersonic potential flows. A program for subsonic and supersonic linear potential

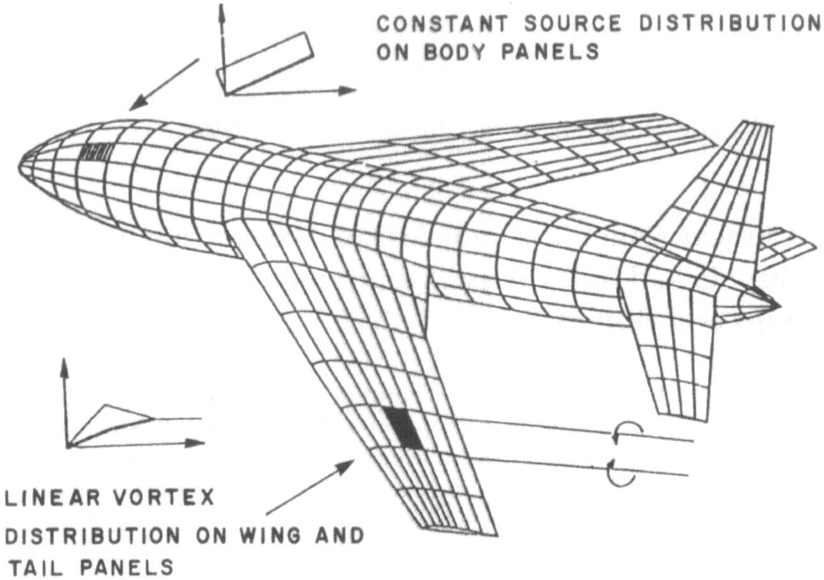

Fig. 4.12. Schematic of the division of a wing-body-tail configuration

flows is the USSAERO, presented by Woodward (1968, 1973 and 1980). An improved high order modification of this program was developed at MBB, designated as the HISSS program, presented by Woodward and Fornasier (1984) and Fornasier (1984). A supersonic linear Vortex Lattice Method, the VORLAX program, was presented by Miranda et al. (1977). A high-order supersonic program is included in the PAN AIR code.

The main features of three of the panel programs – USSAERO, Supersonic VLM and PAN AIR – are presented in the following sections to illustrate the essential characteristics of the many panel programs. The first two are low-order methods while the last represents a high-order method.

4.4.1 Low Order Subsonic/Supersonic Method – USSAERO

In order to apply this method to a typical aircraft, comprising of a wing, body, tail, the configuration is divided into a large number of planar panels, and into each one is introduced an aerodynamic singularity distribution. A constant source distribution is used on the body panels and a combination of source and vortex distributions is used on the lifting surfaces, wing and tail, panels. A typical configuration panel subdivision is presented in Fig. 4.12.

The perturbation velocity field, induced by each panel singularity distribution, is evaluated and is used to calculate the coefficients of a system of linear algebraic equations, relating the magnitude of the normal velocities at the panel control points to the unknown singularity strengths. The singularity strengths, which satisfy the boundary condition of "no flow" through the control points for a given Mach number and angle of attack, are determined by solving this system of equations using an iterative solution of a matrix equation. The pressure coefficients at the control points are then evaluated from the perturbation velocity components, and the aerodynamic forces and moments are evaluated by numerical integration.

The expressions for the perturbation velocity components, induced by constant and linearly varying source and vortex singularities in compressible subsonic and supersonic flows, are presented by Woodward (1973). Both planar and nonplanar boundary conditions are used in versions B and C (Woodward 1980), and are given in the following equations.

The nonplanar boundary conditions, with linearly varying vortex singularities and constant source singularities, version B, sketched in Fig. 4.13a and 4.13b, are

$$w_{B_i} = \sum_{j=1}^{NB} a_{BB_{ij}}\sigma_{B_j} + \sum_{k=1}^{NC}\left(\sum_{j=1}^{NR} a_{BW_{ij}}\sigma_{W_j} + \sum_{j=NR+1}^{2NR} b_{BW_{ij}}\gamma_{W_j}\right) \tag{4.31}$$

$$w_{W_i} = \sum_{j=1}^{NB} a_{WB_{ij}}\sigma_{B_j} + \sum_{k=1}^{NC}\left(\sum_{j=1}^{NR} a_{WW_{ij}}\sigma_{W_j} + \sum_{j=NR+1}^{2NR} b_{WW_{ij}}\gamma_{W_j}\right) \tag{4.32}$$

and the constant vortex distribution boundary conditions, version C, shown in Fig. 4.13c, are

$$w_{B_i} = \sum_{j=1}^{NB} a_{BB_{ij}}\sigma_{B_j} + \sum_{k=1}^{NC}\left(\sum_{j=1}^{2NR} a_{BW_{ij}}\sigma_{W_j} + \gamma\sum_{j=1}^{2NR} b_{BW_{ij}}\right) \tag{4.33}$$

$$w_{B_i} = \sum_{j=1}^{NW} a_{WB_{ij}}\sigma_{B_j} + \sum_{k=1}^{NC}\left(\sum_{j=1}^{2NR} a_{WW_{ij}}\sigma_{W_j} + \gamma_k\sum_{j=1}^{2NR} b_{WW_{ij}}\right) \tag{4.34}$$

The planar boundary conditions are applied to the lifting surfaces only and have identical formulations in versions B and C. Wing thickness effects are represented by linearly varying source distributions, and wing lifting effects are represented by linearly varying vortex distributions. Both singularities are assumed to lie in the plane of the wing. The source-strength distribution is set equal to the known chordwise slope of the thickness distribution, while the "no flow" boundary condition at the control points determine the strengths of the vortex distribution. The equations for the boundary conditions in the planar case are

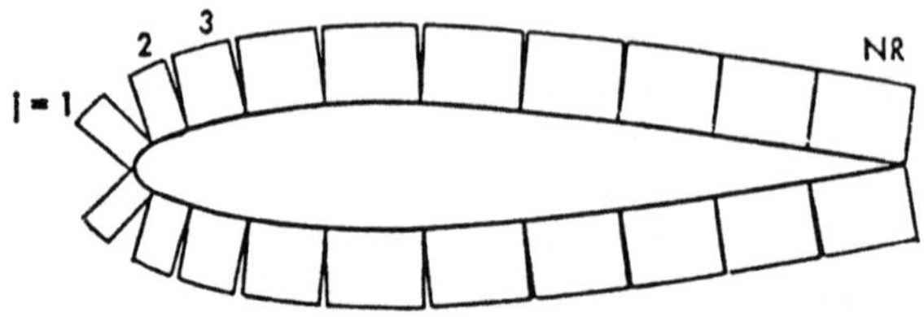

SYMMETRICAL DISTRIBUTION OF CONSTANT SOURCE SINGULARITIES ON UPPER AND LOWER SURFACES OF AIRFOIL

a. Constant source singularities

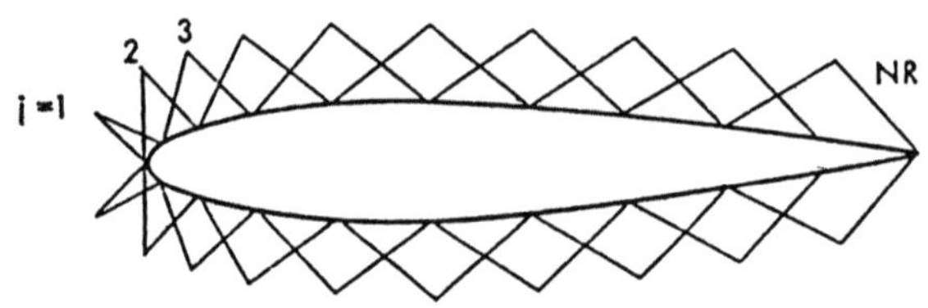

SYMMETRICAL DISTRIBUTION OF LINEARLY VARYING VORTEX SINGULARITIES ON UPPER AND LOWER SURFACES OF AIRFOIL

b. Linearly varying vortex singularities

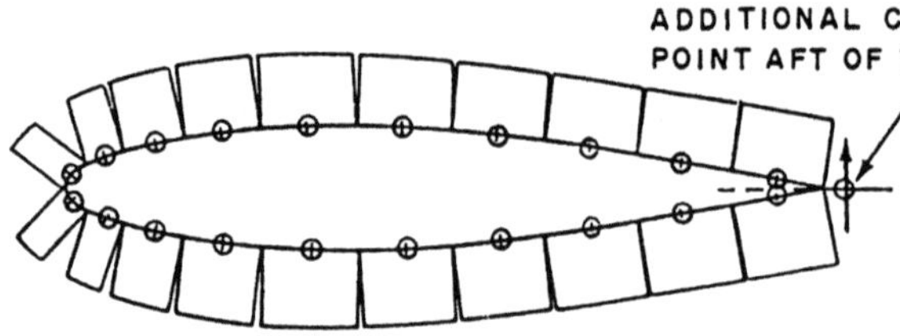

CONSTANT VORTEX DISTRIBUTION ON AIRFOIL SURFACE

c. Constant vortex singularities

Fig. 4.13. Distributions of singularities on airfoil section

$$w_{B_i} = \sum_{j=1}^{NB} a_{BB_{ij}} \sigma_j + \sum_{j=1}^{NW} b_{BW_{ij}} \gamma_j \tag{4.35}$$

$$w_{W_i} = \sum_{j=1}^{NB} a_{WB_{ij}} \sigma_j + \sum_{j=1}^{NW} b_{WW_{ij}} \gamma_j \tag{4.36}$$

These boundary conditions can be expressed in a matrix equation that can be solved for the singularities distributions. The matrix equation can be subdivided into four partitions as follows:

$$\begin{bmatrix} A_{BB} & A_{BW} \\ A_{WB} & A_{WW} \end{bmatrix} \begin{Bmatrix} \sigma \\ \gamma \end{Bmatrix} = \begin{Bmatrix} w_B \\ w_W \end{Bmatrix} \tag{4.37}$$

where A_{BB} represents the influence of the body source panels on the body control points; A_{BW} represents the influence of the wing source and vortex panels on the body control points; A_{WB} represents the influence of the body source panels on the wing control points and A_{WW} represent the influence of the wing source and vortex panels on the wing control points. These influence coefficients are calculated by a procedure presented by Woodward (1973) for singularities in subsonic flows (including compressibility corrections) and for singularities in supersonic flows.

The matrix equation is solved by iteration procedure, the program presented by Woodward (1980) being limited to the maximum order of each of the four partitions to 600. Once the singularities' strength is known, it is possible to calculate the local perturbation velocities and the local pressure coefficient at each control point. Then, the aerodynamic forces and moments are obtained by integrating these pressures over all panels on the configuration.

Results of calculations of the pressure distributions over a swept back wing of AR=3.0 at $M = 0.80$ and $\alpha = 4^o$ are shown in Fig. 4.14, including comparison with experimental measurements.

An example of calculations of C_P, using USSAERO B and C for supersonic flow over an uncambered arrow wing of AR=2.24 at $M = 2.01$ and $\alpha = 4^o$ including comparison with experimental data, is presented in Fig. 4.15.

The agreement of the calculated results with the experimental results is rather good in the subsonic case, where version B seems to be in even better agreement, and is reasonable in the supersonic case except near the leading edge of the sharp outboard tip of the arrow wing.

4.4.2 Low Order Supersonic Linear VLM – VORLAX

It can be argued intuitively, that in analogy to the methods of solution for potential subsonic flows by superposition of elementary solutions to the

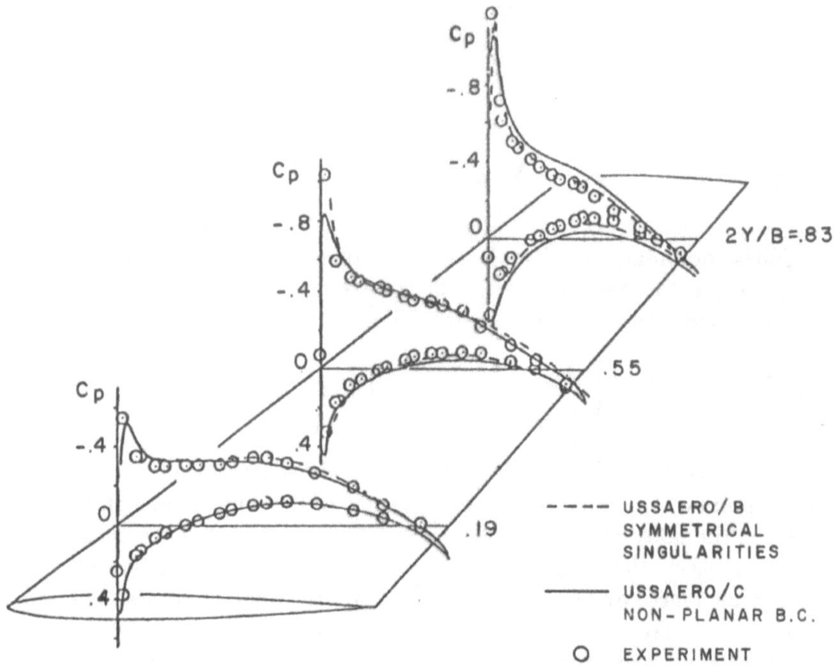

Fig. 4.14. Results of calculations of C_p with USSAERO B and C programs and comparison with experimental data for swept uncambered wing at $M = 0.80$ and $\alpha = 4^o$ (Woodward 1980)

Laplace equation, it should be possible to construct analogous solutions based on superposition of elementary solutions to the potential supersonic flow equation, which is the hyperbolic potential wave equation for $M > 1$ and $\beta^2 = 1 - M^2$.

$$\beta^2 \phi_{xx} + \phi_{yy} + \phi_{zz} = 0 \tag{4.38}$$

The mathematical formulation of these elementary solutions and the definitions of the equivalent formulation of source, doublet and vortex elements in supersonic flows are presented by Robinson (1948), Sears (1954) and Ward (1955). By analogy to the incompressible potential elementary solutions, the supersonic elementary solutions to the hyperbolic potential equation 4.38 can be found by introducing the definitions of hyperbolic parameters, the hyperbolic length

$$R_H = \{(x - x_o)^2 - \beta^2[(y - y_o)^2 + (z - z_o)^2]\}^{1/2}$$

and the hyperbolic operator $\nabla h\beta$ and the notation $\Delta h\beta$ were introduced by Robinson (1948), so that

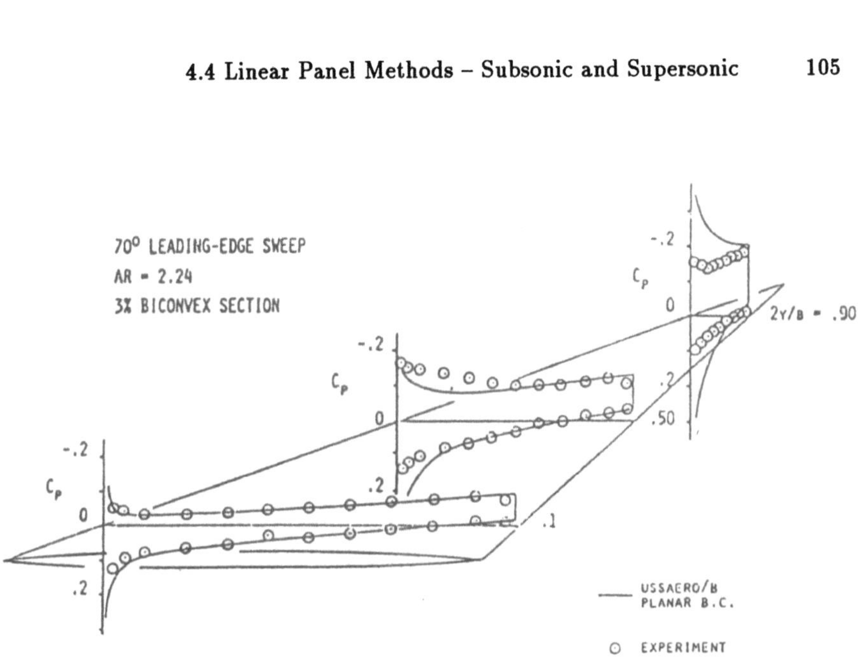

Fig. 4.15. Results of the calculation of C_P using USSAERO B and C programs, and comparison with experimental data for uncambered arrow wing at $M = 2.01$ and $\alpha = 4^o$ (Woodward 1980)

$$\nabla h\beta = \left(-\beta^2 \frac{\partial}{\partial x}, \frac{\partial}{\partial y}, \frac{\partial}{\partial z}\right)$$

and $\Delta h\beta$ =div gradh β =divh β grad.

Source in Supersonic Flow. The potential for *source* of strength m is

$$\phi_s = \frac{m}{2\pi} \frac{1}{[x^2 - \beta^2(y^2 + z^2)]^{1/2}} = \frac{m}{2\pi} \frac{1}{R_H} \tag{4.39}$$

where $R_H^2 = x^2 - \beta^2(y^2 + z^2)$, and the solution exists when $R_H > 1$. The velocity components are

$$u = \frac{\partial \phi_s}{\partial x} = \frac{m}{2\pi} \frac{x}{R_H^{3/2}}; \qquad v = \frac{\partial \phi_s}{\partial y} = \frac{m}{2\pi} \frac{y}{R_H^{3/2}};$$
$$w = \frac{\partial \phi_s}{\partial z} = \frac{m}{2\pi} \frac{z}{R_H^{3/2}} \tag{4.40}$$

The streamlines are shown schematically in Fig. 4.16.

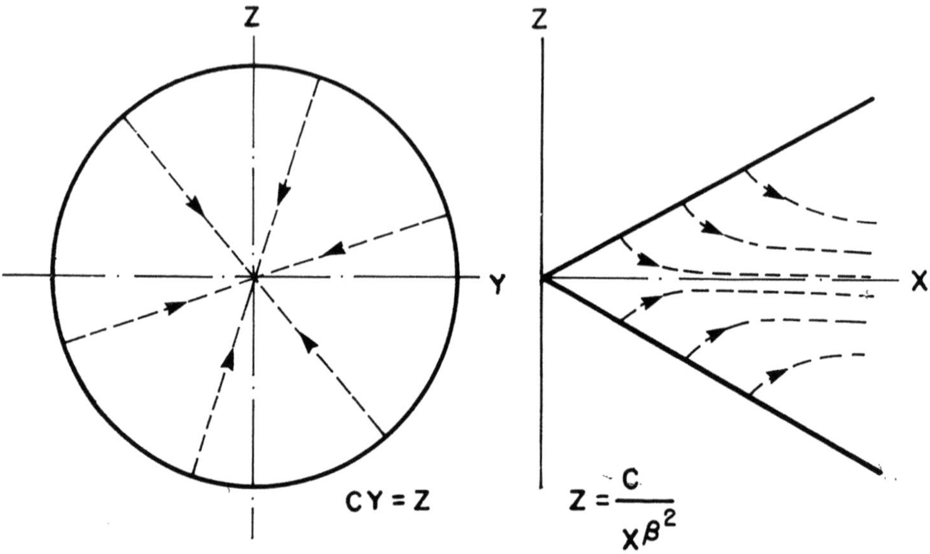

Fig. 4.16. The streamlines for a supersonic source

Doublet in supersonic flow. The potential for a doublet with axis normal to the flow direction

$$\phi_D = -\frac{m}{2\pi}\frac{\beta^2 z}{R_H^{3/2}} \qquad R_H > 1 \tag{4.41}$$

and the velocity components are

$$\begin{aligned} u &= \frac{\partial \phi_D}{\partial x} = \frac{3m\beta^2 xz}{R_H^{5/2}}, \qquad v = \frac{\partial \phi_D}{\partial y} = -\frac{3m\beta^4 yz}{R_H^{5/2}}, \\ w &= \frac{\partial \phi_D}{\partial z} = -\frac{m\beta^2}{2\pi}\frac{x^2 - \beta^2(y^2+2z^2)}{R_H^{5/2}} \end{aligned} \tag{4.42}$$

Note that on the plane $z = 0$, only the w component is not zero. The three-dimensional streamlines are projected in Fig. 4.17.

The potential for a doublet with axis parallel to the velocity direction

$$\phi_D = \frac{m}{2\pi}\frac{x}{R_H^{1/2}} \qquad R_H > 1 \tag{4.43}$$

and the velocity components are

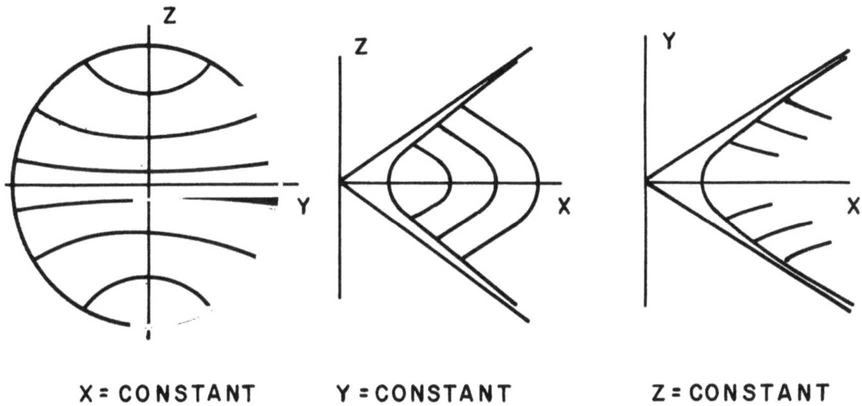

Fig. 4.17. The streamlines for a doublet with axis normal to the velocity direction

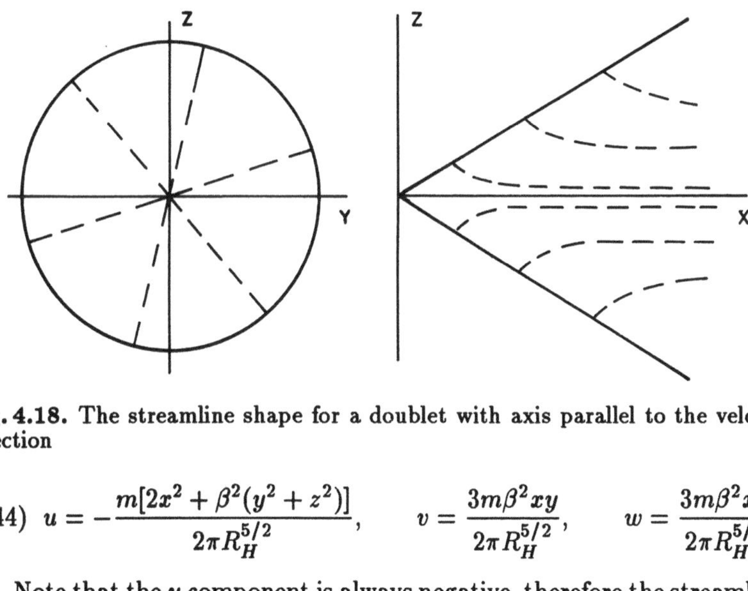

Fig. 4.18. The streamline shape for a doublet with axis parallel to the velocity direction

$$(4.44)\quad u = -\frac{m[2x^2 + \beta^2(y^2 + z^2)]}{2\pi R_H^{5/2}}, \qquad v = \frac{3m\beta^2 xy}{2\pi R_H^{5/2}}, \qquad w = \frac{3m\beta^2 xz}{2\pi R_H^{5/2}}.$$

Note that the u component is always negative, therefore the streamlines resemble those of a sink. The streamline shape is shown in Fig. 4.18.

The potential for a doublet with its axis at an angle θ from the velocity direction

$$(4.45)\qquad \phi_D = \frac{m}{2\pi}\frac{x\cos\theta - \beta^2 z \sin\theta}{R_H^{3/2}}$$

and the velocity components are

$$
(4.46)\quad \begin{cases} u = (2\pi R_H^{5/2})^{-1} m[-2x^2\cos\theta + 3\beta^2 xz\sin\theta - \beta^2\cos\theta(y^2+z^2)] \\ v = 3(2\pi R_H^{5/2})^{-1}\beta^2 my(x\cos\theta - \beta^2 z\sin\theta) \\ w = (2\pi R_H^{5/2})^{-1} m\beta^2[3xz\cos\theta + (\beta^2y^2 - x^2 - 2\beta^2z^2)\sin\theta] \end{cases}
$$

The relations for parallel and normal doublet orientations are obtained for $\theta = 0$ and $\pi/2$, respectively.

Vortex in Supersonic Flow. The induced velocity due to vortices distributed on the surface of a body S_B and on the surface of its wake S_W is obtained by a relation which is analogous to the potential Biot-Savart Law

$$
(4.47)\qquad \mathbf{V}(\mathbf{R}_1) = \frac{\beta^2}{2\pi}\int_{S_B+S_W} \gamma(R)\times\frac{\mathbf{R}-\mathbf{R}_1}{R_H^3}\,dS
$$

Dividing the surfaces to N panels τ_i(where i has values from 1 to N), and let $\bar{\gamma}_i$ denote the average value of γ in the panel i, then equation 4.47 can be expressed as a sum

$$
(4.48)\qquad \mathbf{V}(\mathbf{R}_1) = \frac{\beta^2}{2\pi}\sum_{j=1}^{N}\int_{\tau_j}\bar{\gamma}_j\times\frac{\mathbf{R}-\mathbf{R}_1}{R_H^3}\,dS
$$

When $\mathbf{R}_1$ is outside the surface elements τ_j, the integral over this discrete area can be approximated by the mean value theorem

$$
\int_{\tau_j}\bar{\gamma}_j\times\frac{\mathbf{R}-\mathbf{R}_1}{R_H^3}\,dS = \bar{\gamma}_j\delta_j\times\int_{C_j}\frac{\mathbf{R}-\mathbf{R}_1}{R_H^3}\,dl
$$

where C_j is a line in τ_j parallel to the average direction of $\bar{\gamma}_j$ in τ_j, δ_j is a distance normal to C_j, and dl is the arc length element along C_j. Therefore, the velocity components at a point x_o, y_o, z_o outside the surface due to vortex line of strength Γ per unit length positioned at point x, y, z is

$$
(4.49)\qquad \begin{aligned} u &= \frac{\beta\Gamma z_o^2}{2\pi}\int_C\frac{dy}{R_H^{3/2}}, \qquad v = -\frac{\beta\Gamma z_o^2}{2\pi}\int_C\frac{dx}{R_H^{3/2}} \\ w &= \frac{\beta^2\Gamma}{2\pi}\int_C\frac{(x-x_o)\,dy-(y-y_o)\,dx}{R_H^{3/2}} \end{aligned}
$$

The integrations are to be carried out along that part of C which satisfies the conditions $R_H > 0$ and $x - x_o < 0$ when $M > 1$.

The velocity induced by a skewed horseshoe vortex, where the notation is defined in Fig. 4.19, is calculated by the expressions presented by Miranda et al. (1977) and by Tennenbaum (1990),

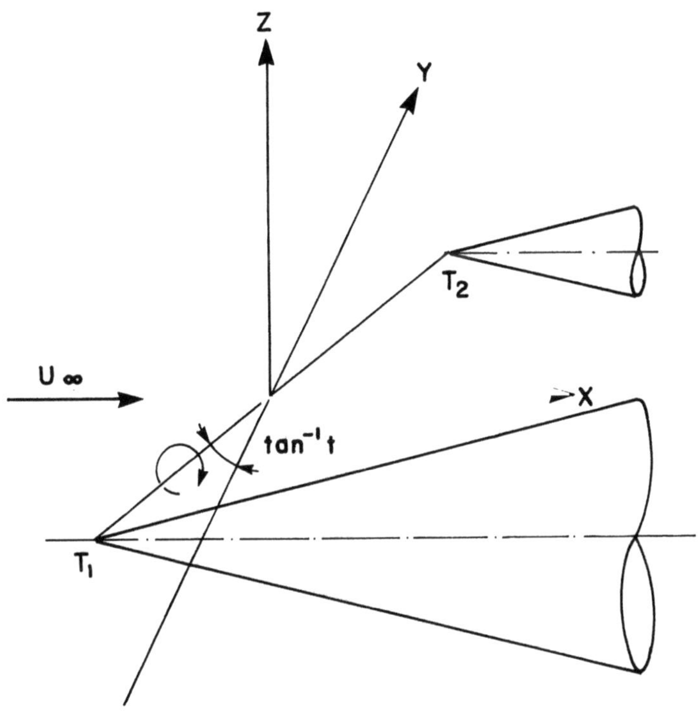

Fig. 4.19. The skewed horseshoe vortex in supersonic flow

$$u(x_o, y_o, z_o) = \frac{\Gamma}{2\pi} \frac{z_o}{x^{*2} + (t^2 - \beta^2) z_o^2} \left[\frac{t x_1 - \beta^2 y_1}{R_{H_1}^{1/2}} - \frac{t x_2 - \beta^2 y_2}{R_{H_2}^{1/2}} \right], \tag{4.50a}$$

$$v(x_o, y_o, z_o) = \frac{\Gamma}{2\pi} \frac{t z_o}{x^{*2} + (t^2 - \beta^2) z_o^2} \left[\frac{t x_1 - \beta^2 y_1}{R_{H_1}^{1/2}} - \frac{t x_2 - \beta^2 y_2}{R_{H_2}^{1/2}} \right], \tag{4.50b}$$

$$w(x_o, y_o, z_o) = \frac{\Gamma}{2\pi} \frac{x^*}{x^{*2} + (t^2 - \beta^2) z_o^2} \left[\frac{t x_1 - \beta^2 y_1}{R_{H_1}^{1/2}} - \frac{t x_2 - \beta^2 y_2}{R_{H_2}^{1/2}} \right], \tag{4.50c}$$

where

$$x_1 = x + ts, \quad x_2 = x - ts, \quad x^* = x - ty, \quad y_1 = y + s, \quad y_2 = y - s,$$

and $R_{H_1} = [x_1^2 - \beta^2(y_1^2 + z^2)]^{1/2}$, $\quad R_{H_2} = [x_2^2 - \beta^2(y_2^2 + z^2)]^{1/2}$.

In supersonic flow, in the plane of the horseshoe vortex, where $z_o = 0$, then $u = v = 0$ and the upwash velocity is found by using the equation (Miranda et. al, 1977)

$$w(x_o, y_o, 0) = -\frac{\Gamma}{2\pi}\frac{1}{x^*}\left(\frac{[x_1^2 + \beta^2 y_1^2]^{1/2}}{y_1} - \frac{[x_2^2 + \beta^2 y_2^2]^{1/2}}{y_2}\right) \tag{4.51}$$

The regions of influence of the supersonic vortex have been determined by Eqs. 4.50a to 4.50c to be limited to the "Mach cone" issued from the tip of the trailing vortices, as indicated in Fig. 4.19. Note the important difference between the limited region of influence of this supersonic vortex and the subsonic potential vortex which induces velocities over the complete volume according to the Biot-Savart law.

When the field point $(x_o, y_o, 0)$ is within the distributed vorticity in the panel e.g., the control point associated with the horseshoe, the upwash must be complemented by additional velocity distribution, due to the generalized principal part of the upwash integral

$$w^* = -\frac{\gamma \cos \Lambda}{2}[\beta^2 - t^2]^{1/2} \tag{4.52}$$

where Λ is the angle between the y-axis and the direction of the vorticity in the panel τ and in supersonic flow $\beta^2 > 0$. It has been shown by Miranda et al. (1977) that the relationship between γ of equation 4.52 and the Γ in equation 4.51 is

$$\Gamma = \gamma \cos \Lambda \delta x \tag{4.53}$$

where δx is the distance, measured in the x-direction, occupied by the distributed vorticity γ which has been lumped into the discrete transverse leg of the circulation Γ.

Method of Calculations. The region of influence of the singularities in supersonic flow is limited to the space within or on the surfaces of the Mach cone. In the case of the supersonic vortex the influence is limited to the Mach cones issued from the edges of the bound vortex; these surfaces are singular, in analogy to the singularity on the line of the subsonic vortex.

The calculation procedure starts by dividing the surfaces into small panels and imbedding a horseshoe vortex into each one. In order to be consistent with the subsonic panel method, it may be advisable to position the bound vortex on the quarter chord line of the panel and to place the control point at the midchord of the three-quarter chord line (Tennenbaum 1990). Induced velocity at any point outside the surfaces is evaluated by Eqs. 4.50 to 4.50c, while at the control point, the additional upwash, presented by equation 4.52, must be added. A schematic drawing of the division of a lifting surface and the positions of the horseshoe vortices and control points used by Miranda et al. (1987) is shown in Fig. 4.20.

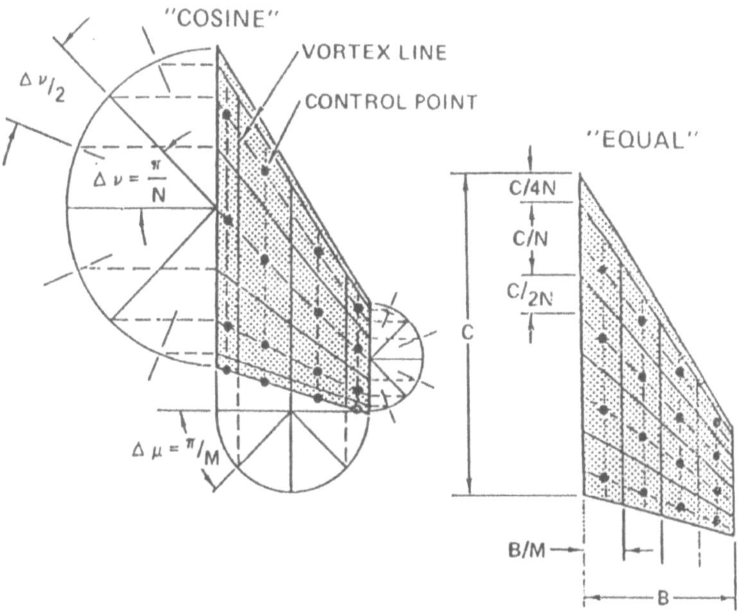

Fig. 4.20. Various panel divisions and position of vortices for the supersonic VLM (Miranda et al. 1987)

Imposing the "no flow" boundary condition at the control points of each of the N panels results in a set of N algebraic linear equations which must be solved in order to determine the strength of the vortices in each panel. This matrix equation is in the form of

$$\left[A_{ij}\right] \cdot \left\{\Gamma_i\right\} = \left\{\mathbf{V}_i\right\}, \quad i, j = 1, 2, \ldots, N \tag{4.54}$$

where the matrix A_{ij} is the influence matrix, which is evaluated by Eqs. 4.50a to 4.50c when $i \neq j$ and on the diagonal terms, when $i = j$, the upwash is evaluated by adding equation 4.52. In the linear VLM, the trailing vortices remain straight lines in the plane of the wake, while the wake may be directed at various pitch and yaw angles to simulate sideslip, Miranda et al. (1977), as indicated in Fig. 4.21.

Results of the calculations of aerodynamic coefficients of Arrow wings using the VORLAX program of Miranda et al. (1977) are shown in Fig. 4.22 and the calculated pressure distribution for a swept back rectangular wing at $M = 1.92$ is shown in Fig. 4.23. It can be seen that the estimation of the lift, aerodynamic center and induced drag are in good agreement with the conical flow theory. Pressure distribution calculated by the VORLAX program is also in close agreement with the conical flow theory as far as the pressure levels and the pressure variations are concerned but is less accurate in following the jumps in the pressures across the Mach lines, as is indicated in Fig. 4.23

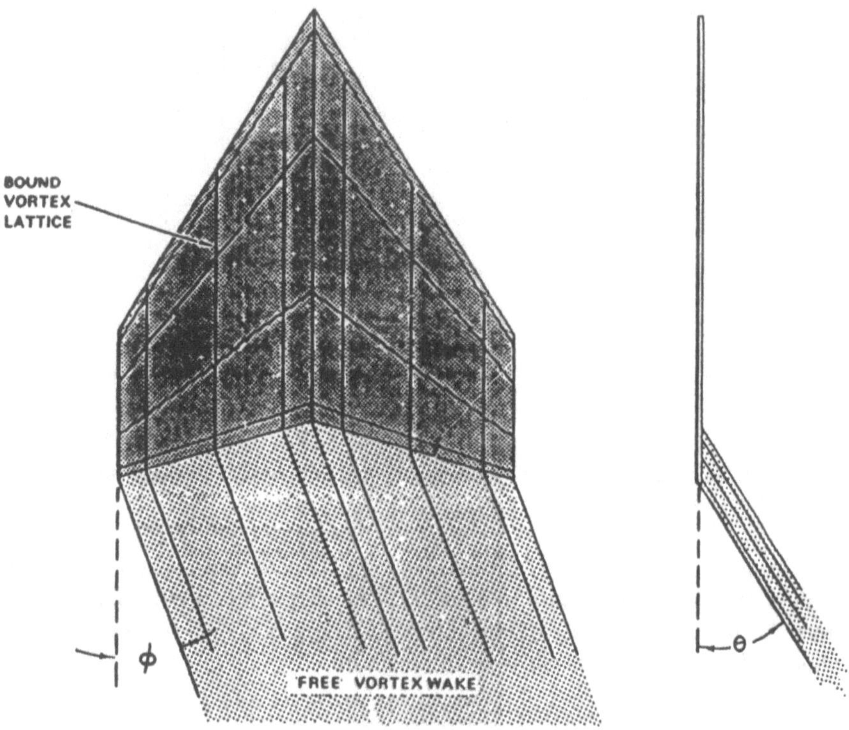

Fig. 4.21. The plane vortex wake with yaw and pitch for the supersonic linear VLM calculations (Miranda et al. 1987)

4.4.3 High Order Subsonic/Supersonic Method – PAN AIR

The PAN AIR is a higher order panel method for calculating subsonic or supersonic linear potential flows on arbitrary three-dimensional configurations. This program uses higher order singularities, linear source and quadratic doublet distributions, in order to improve the accuracy and ensure convergence for arbitrary configurations. It is important to eliminate discontinuous jumps in the doublet strengths at panel boundaries; therefore, the program incorporates a continuous quadratic doublet strength distributed over a piecewise flat panel. This, combined with the linear source strength, facilitates numerical stable solutions at both supersonic and subsonic speeds for arbitrary configurations incorporating panels inclined to

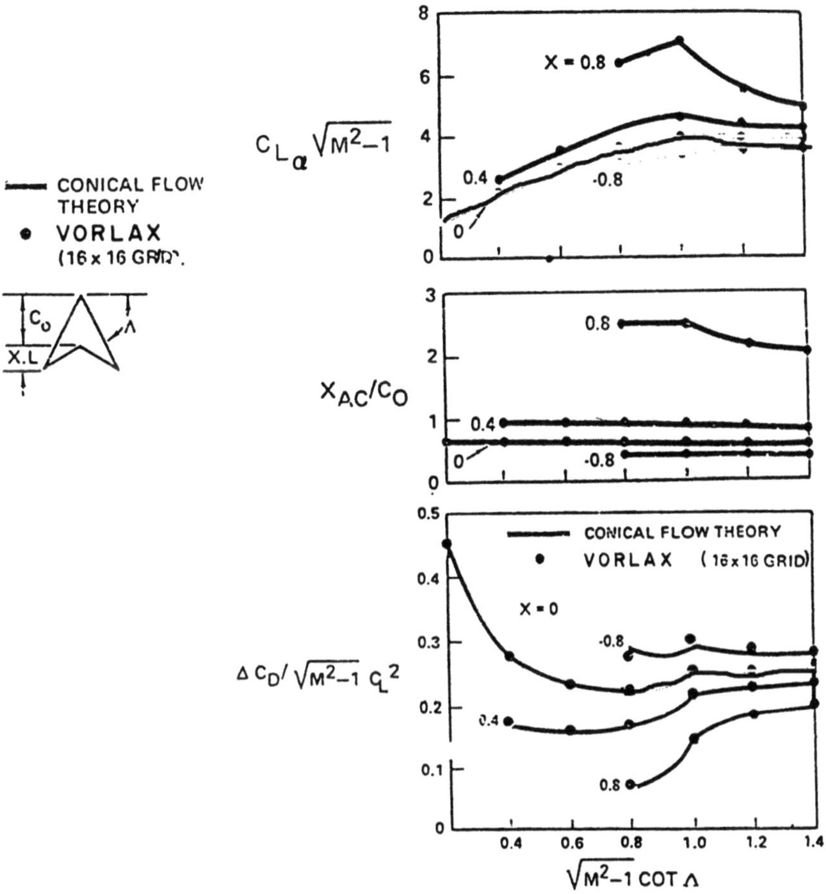

Fig. 4.22. The aerodynamic coefficients of arrow wings calculated by the supersonic program and by the conical flow theory (Miranda et al. 1977)

the incoming flow and boundary conditions simulating intake and exhaust flows. The wide range of options for the input and output of the program indicate that the PAN AIR method is probably the most advanced panel program for calculation of the *linear aerodynamic coefficients* of arbitrary three-dimensional configurations in subsonic or supersonic flows. A review of the program and its capabilities is presented by Carmichael and Erickson (1981) and details of the computer program are presented in the PAN AIR NASA reports (1980).

The PAN AIR computer program. As in most panel methods, the first step is the establishment of the grid geometry. An example of the partitioning of the surface of an aircraft configuration into several networks of surface grid

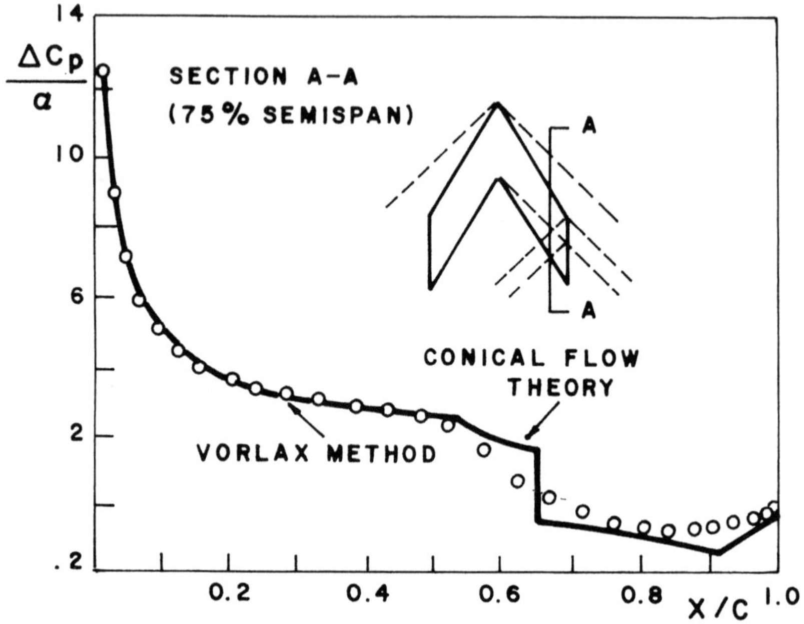

Fig. 4.23. The pressure distribution on a swept back rectangular wing by the supersonic VLM and by the conical flow theory (Miranda et al. 1977)

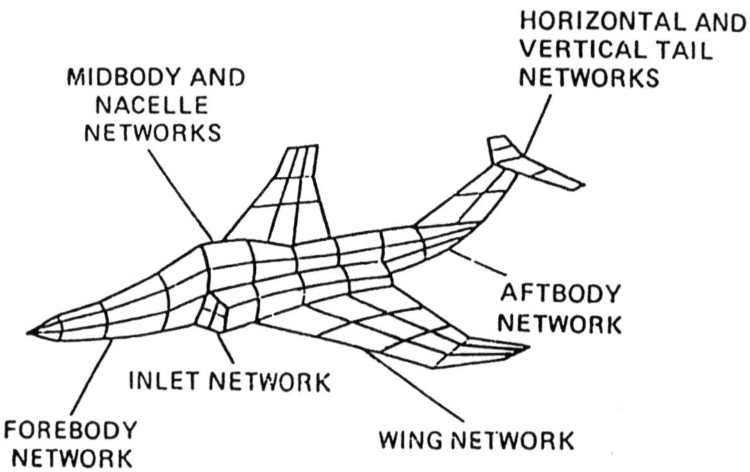

Fig. 4.24. Networks of surface panels for an aircraft configuration (Carmichael and Erickson)

points is presented in Fig. 4.24. The surface grid points, must be computed by the user and supplied as input to the program on a network-by-network basis; the PAN AIR connects the grid points with piecewise flat panels. The input includes also the wake surfaces panel network (not shown in Fig. 4.24).

In subsonic flow, the panels can be oriented at any arbitrary inclination on the configuration. In supersonic flow, the panels representing the solid boundaries must be inclined at an angle less than the local Mach cone angle, "subinclined panels", as indicated in Fig. 4.25. This precludes the use of the PAN AIR program for configurations with blunt supersonic leading edges. However, "superinclined panels" can be used in supersonic flow, provided the boundary conditions on such panels are allowed only on their downstream side – that is, simulating intake and exhaust flows.

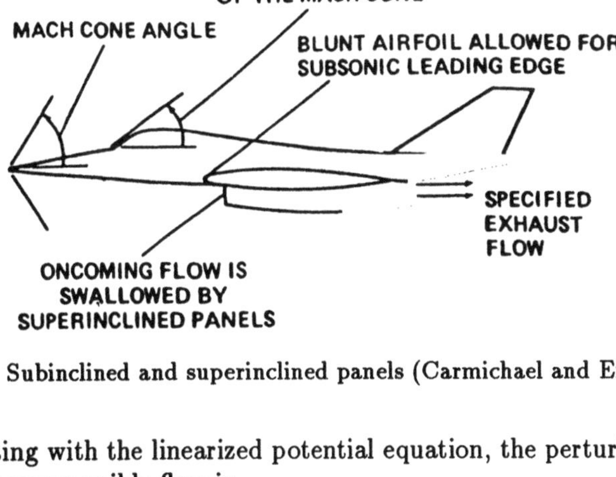

Fig. 4.25. Subinclined and superinclined panels (Carmichael and Erickson 1981)

Starting with the linearized potential equation, the perturbation velocity for incompressible flow is

$$\mathbf{v}(x,y,z) = \frac{1}{4\pi} \iint_S [-\sigma(\xi,\eta)\nabla(R^{-1}) + \mu(\xi,\eta)\nabla\{\mathbf{n}\cdot\nabla_Q(R^{-1})\}]\, d\xi\, d\eta \tag{4.55}$$

where $\sigma(\xi,\eta)$ and $\mu(\xi,\eta)$ are the source and doublet strengths distributions, respectively, R is the distance between the point $P(x,y,z)$ and the point $Q(\xi,\eta)$ on the panel surface, as indicated in Fig. 4.26, and the terms R^{-1} and $\mathbf{n}\cdot\nabla_Q(R^{-1})$ are the source and doublet solutions, respectively, of the potential equation. The boundary conditions on the surface of the configuration and on the wake are determined according to the formulation of Morino (1973). The perturbation velocities at the control points can be calculated from Equation 4.55 for distribution of sources and doublets on the surface S and its wake by imposing the boundary conditions at these control points. Similar equations for the perturbation velocities are available for compressible subsonic and supersonic potential flows. In the PAN AIR method, the surface S is divided into N panels. Linearly varying (with ξ and η) source strength and quadratic varying strength of doublets are imbedded into each panel, so that Equation 4.55 can be integrated over each single panel and the perturbation velocity can be expressed in terms of the unknown singularity parameters. If the point P is located at the control point of the panel, then, summing up the effects from all N panels results in a matrix equation for strength of the singularities, λ_i

$$\left[A_{ij}\right]\left\{\lambda_i\right\} = \left\{\alpha_i\right\} \quad \text{for } i,j = 1,2,\ldots N \tag{4.56}$$

where α_i is determined by the known boundary conditions at the control points of each panel and the A_{ij} terms are the coefficients of the influence matrix.

The PAN AIR computer program package was developed by the Boeing Military Airplane Company under the joint sponsorship of NASA, USAF and Navy. The PAN AIR consists of separate modules, which are defined by the system architecture, shown in Fig. 4.27, where the definition of the data structure and the movements of data from discs to core is achieved by a Scientific Data Management System (SDMS).

The input modules include the Module Execution Control (MEC) for user input control directives and the Data Input Processor (DIP) which reads the input records defining the aerodynamic conditions, network grid points, boundary conditions and the quantities to be computed. The Defining Quantities Generator (DQG) calculates quantities relating to panel geometry, control points, boundary conditions, network abutments and source and doublets spline matrices. The Matrix Generator module (MAG) computes the influence coefficients matrix. The Real Matrix Solve (RMS) decomposes the influence coefficient matrix into upper and lower triangular matrices. The Right Hand Side generator (RHS) creates the right hand side of equation 4.56 and solves for the unknown singularity parameters. The Minimal Data Generator (MDG) collects the basic information (generated by the upstream modules) needed for the post processing modules. The post processing modules are: 1. the Point Data Processor (PDP) which computes the local flow quantities such as pressures and velocities at points on the vehicle surface or at points in the flow field; 2. the Configuration Data Processor (CDP) computes the force and moment coefficients; and 3. the Print-Plot Processor (PPP) creates the output files of geometry data from DQG, results from point data from PDP and results from configuration data from CDP for subsequent processing by user supplied plot programs. There is an additional library of programs (PALIB) of frequently used routines which can be accessed by the various modules. The PAN AIR can be operated with any number of panels, limited by the computer system. A

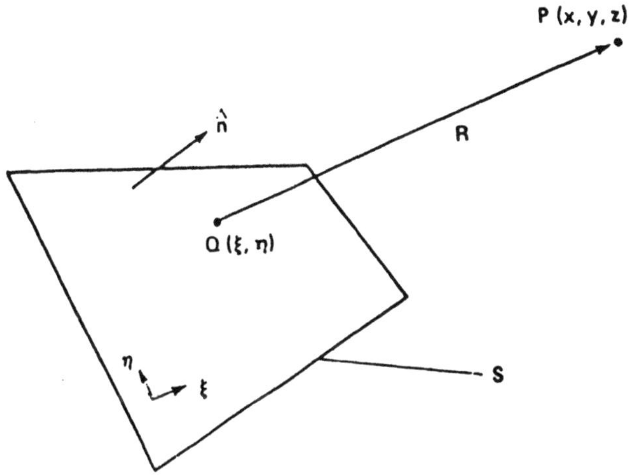

Fig. 4.26. Geometry for calculation of the perturbation velocity

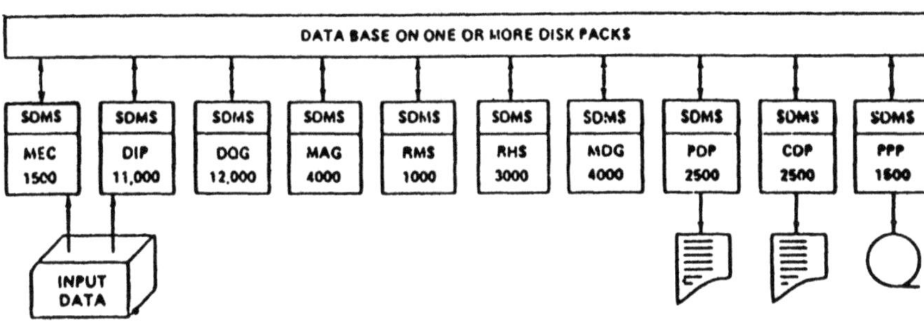

Fig. 4.27. Schematic of the PAN AIR architecture (Carmichael and Erickson 1981)

1000 panel program requires 8 million words of storage for permanent data bases and several millions of additional words for temporary data bases. The PAN AIR has been run on the CDC 7600 and the CYBER 74 and CYBER 175 computers. CPU run time for an aircraft configuration with 700 panels is about 1 hour on the CDC 7600.

The results of these calculations indicate good agreement with exact solutions to the small velocity perturbations potential flow equations both in the subsonic and supersonic speed ranges. The use of linear sources distribution with continuous quadratic doublets distribution proved able to eliminate many of the numerical stability problems of earlier panel methods. The panel networks used in this program are capable of simulating the surface shapes of real aircraft type configurations. Results of calculations of the pressure distributions on test cases like conical and cone-cylinder-cone bodies showed good agreement with exact theoretical results. By calculating the momentum balance over a control volume it was possible to evaluate the leading edge thrust force and to evaluate the vortex and wave drag forces on wings and on complete configurations for small angles of attack, presented in Fig. 4.28.

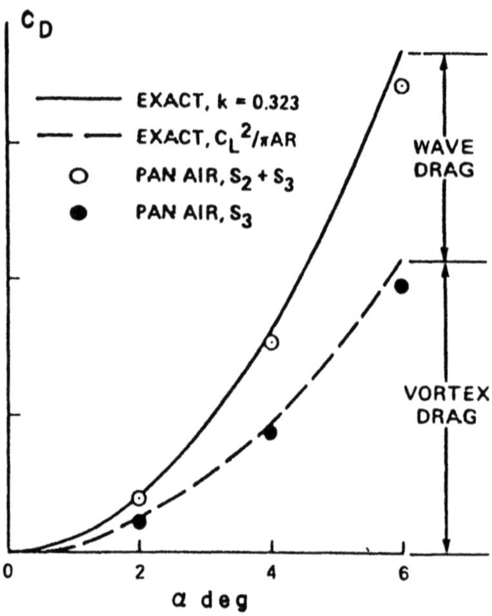

Fig. 4.28. Calculated drag coefficients for a flat-plate delta wing of AR=2 at $M_\infty = .53$; S_1 is wing area, S_2 and S_3 are control volume surface areas

4.5 Comparison of the Various Panel Methods

The definition of the various panel methods present these as methods for the numerical solution for the differential-integral equation of the potential function and the perturbation velocities resulting from singularities distributed on the surface of the aerodynamic configuration. The numerical calculation begins by dividing the surface of the configuration and its wake into small panels and by imbedding into each panel distributions of singularities, sources and/or doublets or vortices. A control point is defined at an arbitrary position in each panel and the condition of "no flow" through the boundary is applied on solid boundaries at that point. For panels on the wake and in regions of engine intakes and exhausts, bleed or injection of fluids, the appropriate boundary conditions must be applied, again, at the corresponding control points of these panels. In this way, the evaluation of the differential-integral equation for the perturbation velocity vector reduces to the calculation of the N unknown values of the singularities by the solution of N algebraic equations fulfilling the boundary conditions at the control points of the N panels. In this case, the following matrix equation is obtained

$$\left[A_{ij}\right]\left\{q_i\right\} = \left\{\alpha_i\right\} \tag{4.57}$$

where A_{ij} are the terms of the influence coefficient matrix which represent the induced velocity at the control point of panel i due to the singularities of panel j; the q_i are the strengths of the singularities in panel i and the α_i are local induced angles of attack which must equal the local surface inclination at the control point of panel i.

The various panel methods differ mainly in the types of singularities and the specific distributions of these singularities in the panels. The available methods for linear aerodynamic characteristics are presented in Table 4.1.

The various programs can be compared, first, on their capabilities to simulate correctly complex geometries and on the convergence of the calculation of the unknown singularities for complex configurations. Furthermore, these programs can be compared with experimental and more exact theoretical calculations, if available.

The linear VLM program of Margason and Lamar (1971) does not in-clude source distribution. Therefore, it is not capable of including effects of thickness. Some improvements can be realized by distributing the vortices over the panels on the upper and lower surfaces or by the double sheet of vortex distribution suggested by Miranda et al. (1977) for supersonic flows. Representing potential function by the distribution of only one type of singularity means either neglecting the effect of thickness (without source distribution) or neglecting the lifting effect (without vortex distribution). Trying to simulate effect of thickness by distribution of only vortices

Table 4.1. Distinguishing features of the linear panel method

Method	Panel Geometry	Doublet/Vortex chordwise	Doublet/Vortex spanwise	Source Variation chordwise	Source Variation spanwise	Reference
VLM	Planar Quadrilateral	constant	constant	——— *	——— *	Lamar 1968
VORLAX	Planar Quadrilateral	constant	constant	——— *	——— *	Miranda 1977
Source Panel	Planar Quadrilateral	pres-cribed	constant	constant	constant	Hess 1975
VSAERO	Planar Quadrilateral	constant	constant	constant	constant	Maskew 1982a,b
USSAERO	Planar Quadrilateral	linear	linear	constant	constant	Woodward 1973
MARCAP	Planar Triangular	linear	linear	constant	constant	Maruyama 1986
SOUSSA	Hyperboloidal Quadrilateral	constant	constant	constant	constant	Morino 1980
QUADPAN	Planar Quadrilateral	constant	constant	constant	constant	Cooper-smith 1983
MCAERO	Quadrilateral 8 subpanels	quadratic	quadratic	constant	constant	Bristow 1983
PAN AIR	Quadrilateral 8 subpanels	quadratic	quadratic	linear	linear	PAN AIR 1980
ECOPAN						Roggero, Larguier 1990
TRANAIR	Transonic full potential equation solution program fitted to the panelling scheme of PAN AIR					Samant et al. 1987

* source distribution can be included, not available in the linear VLM programs.

on the separated surfaces cannot fulfill correctly the boundary conditions on a thick wing, for example. It will be shown, however, that it is possible to include source distributions in the VLM programs as was done in the nonlinear programs by Rom et al. (1981) for subsonic flows, and by Tennenbaum (1990) for supersonic flows. Solutions of the matrix equation associated with the linear VLM generally converge, except for multi-surface configurations where special care must be taken when the trailing vortices shed from the trailing edges of one surface may intersect a second surface or its wake. In such cases, the concept of minimum core distance between approaching vortices as a criterion for joining these vortices, which was developed for the Nonlinear VLM, can be utilized to eliminate numerical problems in such calculations.

In the methods using the doublet singularity distribution on the panels it has been found that numerical problems may arise if the continuity of the doublet strengths at the edges of matching panels is not ensured. This is particularly important in the supersonic speed range. The most advanced method in this class is the PAN AIR which uses a linear source distribution and a quadratic doublet distribution on each panel. It then spline fits the doublet strength between neighboring panels to ensure complete doublet strength continuity. Therefore, the PAN AIR is probably the most versatile method that can be applied to complex configurations with many options for boundary conditions (for example, simulating engine intake and exhaust flows). However, the PAN AIR is also the most expensive method in terms of computer resources and CPU time, as discussed by Margason et al. (1985) and presented in Fig. 4.29.

The comparison of the calculated aerodynamic characteristics of various aerodynamic configurations (swept wings alone and wing-body combinations) using the various linear panel methods and the comparison with the corresponding experimental results at subsonic speeds are presented by Margason et al. (1985). Illustrative results are included to enable evaluation of these methods. The comparison of the lift coefficients calculated for two wing-body configurations are shown in Figs. 4.30 and 4.31.

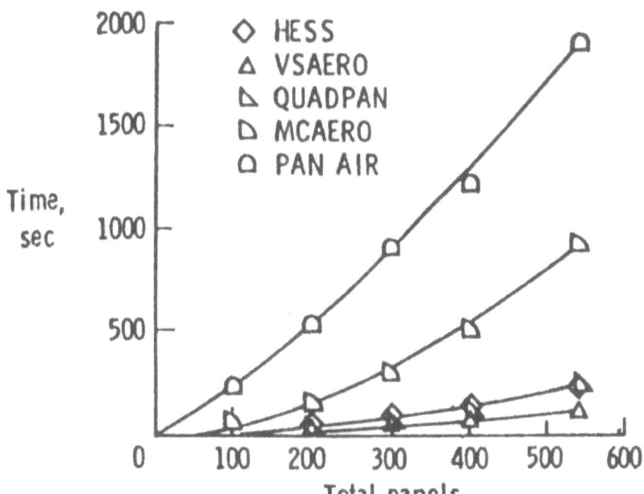

Fig. 4.29. Comparison of CPU execution time for various panel methods on Cyber 175 computer (Margason et al. 1985).

The calculations of chordwise pressure distributions on a swept back wing profile and on the body near a wing-body junction are shown in Figs. 4.32 and 4.33 for low order and high order methods, respectively.

The comparison of the calculated spanwise load distributions for unswept and swept wings are shown in Figs. 4.34 and 4.35, respectively.

The effect of Mach number on the lift curve slope was calculated using the various panel methods, compared with experimental data by Margason et al. (1985) and presented in Fig. 4.36. The VLM, QUADPAN and

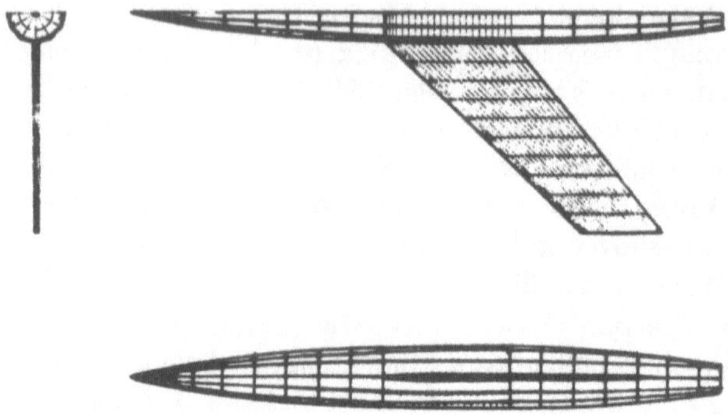

a. Panel representation of wing-body configuration

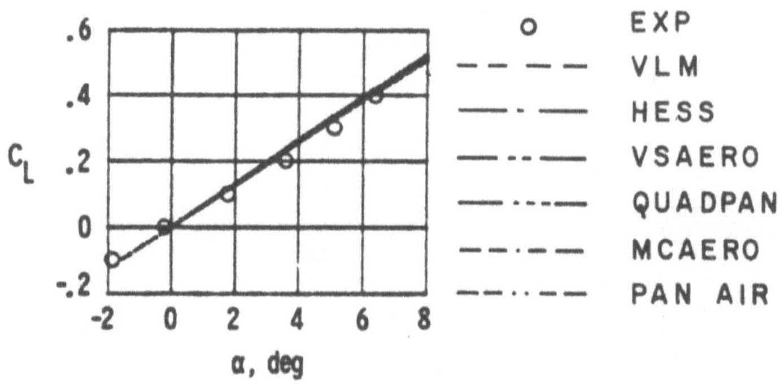

b. Lift coefficient variation as a function of α

Fig. 4.30. Comparison of the calculated lift coefficients for various linear panel methods with experimental data for symmetrical wing-body configuration (Margason et al. 1985)

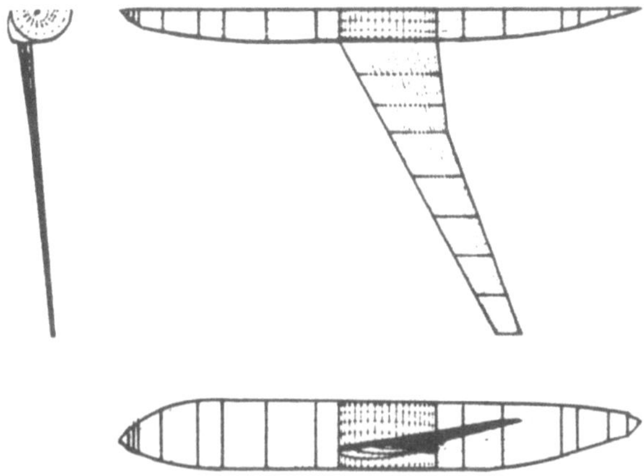

a. Panel representation of the EET configuration

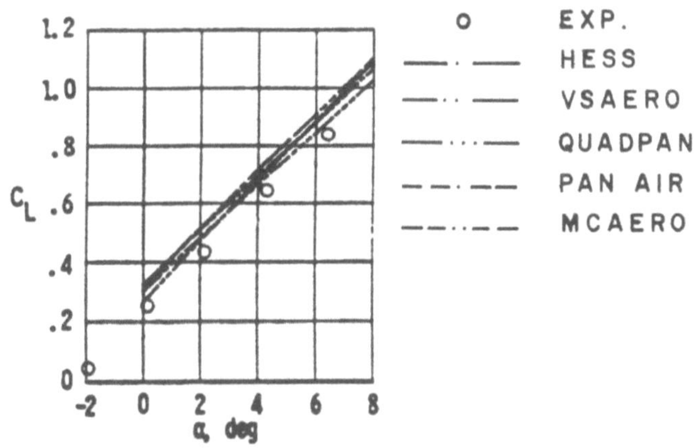

b. Lift coefficient variation as a function of α

Fig. 4.31. Comparison of the calculated lift coefficients for various linear panel methods with experimental data for the Energy-Efficient Transport (EET) aircraft configuration (Margason et al. 1985)

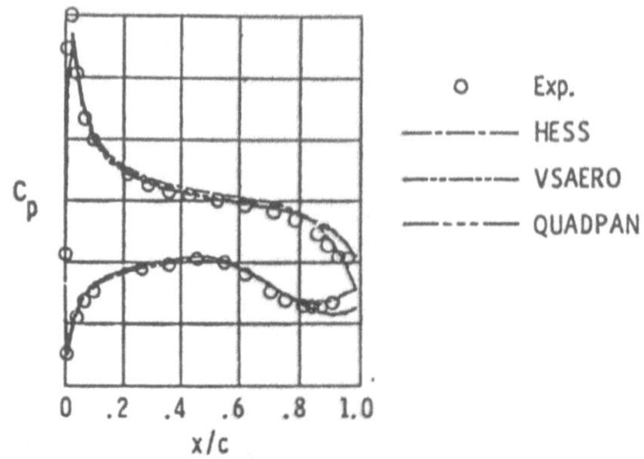

a. Low order methods

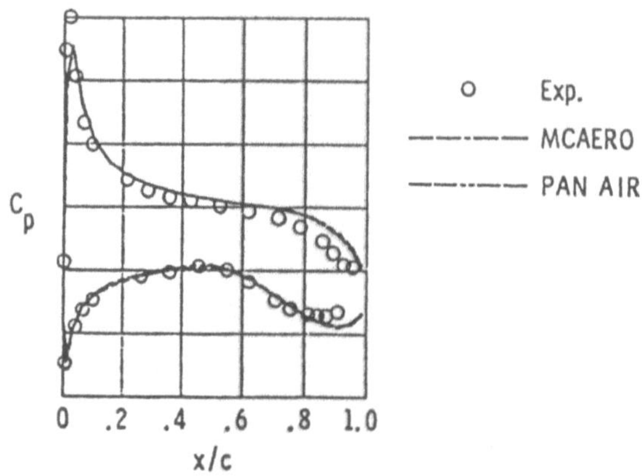

b. High order methods

Fig. 4.32. Chordwise pressure distributions on a swept back wing profile of the EET configuration at $\eta = 0.55$ and $\alpha = 4.31°$ (Margason et al. 1985)

MCAERO use second order expansions to calculate the pressure relations at increasing Mach numbers; their results are in good qualitative agreement with the experimental data. The Hess, VSAERO and PAN AIR use the isentropic pressure relation and under-predict the lift increase at increasing Mach number. Therefore, it seems that it is more important to use the correct compressibility relation than the high or low order panel methods in order to obtain good agreement with the experimental Mach number effects.

Calculations presented by Margason et al. (1985) indicate that for sub-

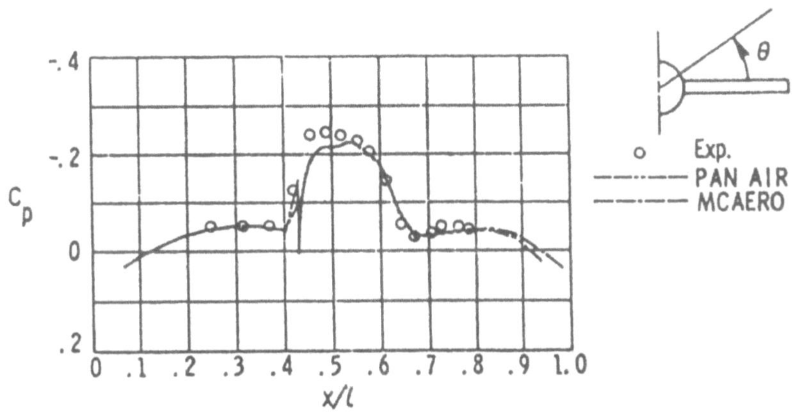

a. Low order methods

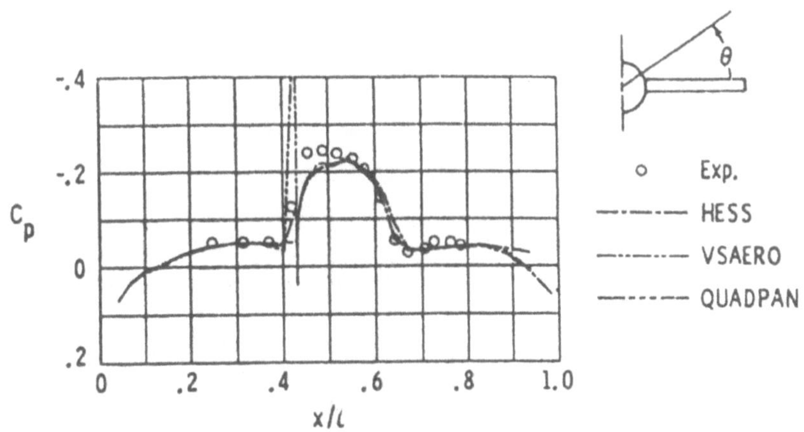

b. High order methods

Fig. 4.33. Pressure distributions on the body near the wing junction of a wing-body configuration at $M = 0.6$, $\theta = 15°$ and $\alpha = 4°$ (Margason et al. 1985)

sonic flows there is no particular advantage to the high order linear panel methods for accuracy of the evaluation of the aerodynamic characteristics. This finding accords with similar conclusions presented by Maskew (1982) and Maruyama et al. (1986). Observation supports the conclusion that when the matrix calculations for the linear panel methods converge, the panel loading values and the associated aerodynamic coefficients are about the same for all reasonable panel methods. The high order me-

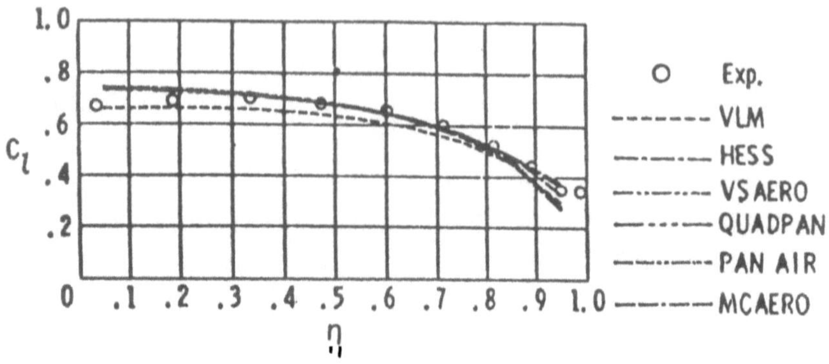

Fig. 4.34. Spanwise load distribution for the unswept semi-span wing at $\alpha = 8^0$ (Margason et al. 1985)

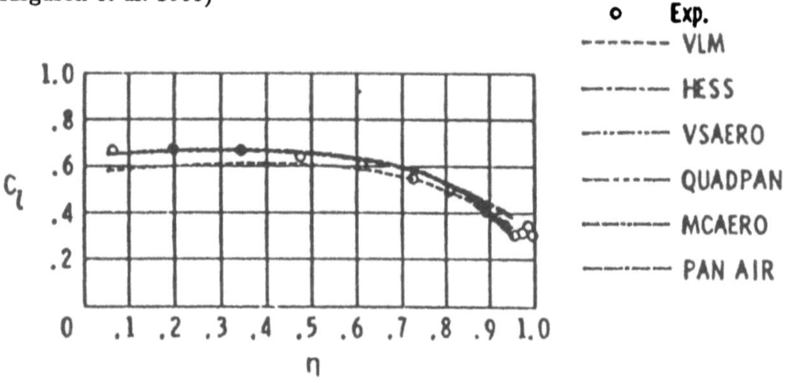

Fig. 4.35. Spanwise load distribution for the 20° semi-span swept wing at $\alpha = 8°$ (Margason et al. 1985)

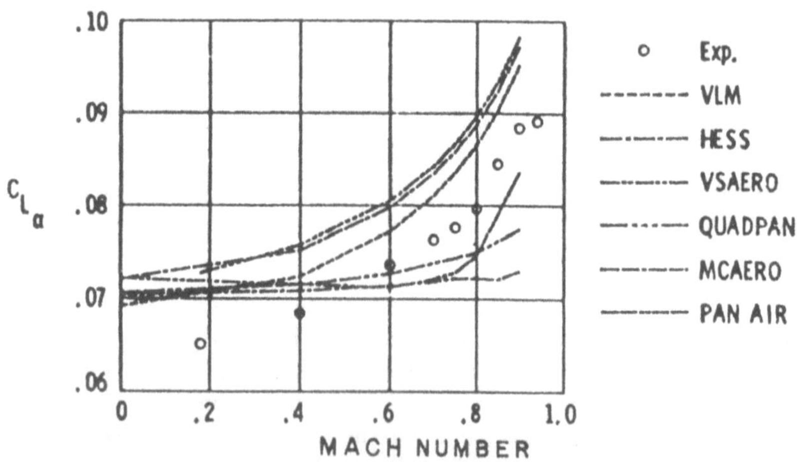

Fig. 4.36. Comparison of calculated lift curve slope by various panel methods with experimental data for 37° swept back wing (Margason et al. 1985)

thods may have advantages in ensuring convergent solutions of the matrix equations and eliminating spurious jumps at panel interfaces that may be encountered in some panel methods in supersonic flows, as observed in the PAN AIR (1980) papers. The use of low order panel methods reduces the computation periods to a few minutes on a large computer, thus making these methods an excellent aerodynamic tool for preliminary design.

The utility of the panel methods, even the more complex ones with a large number of panels, can be improved by the use of parallel computing methods on a parallel computer. An example of the application of the Connecting Machine for the solution of the low order panel method of Hess and Smith (Section 4.3.2) is presented by Chriske and Bogucz (1991). It is demonstrated that the Connecting Machine Model CM-2 operating with a 32K processor can outperform the Cray code in computing the influence coefficients and is somewhat slower in solving the linear equations needed for the evaluation of the aerodynamic coefficients. Overall, it is shown that the Connecting Machine CM FORTRAN provides a convenient programming environment for achieving a high level of computational performance on dense matrix calculations. Therefore, using such parallel computer codes may further accelerate the more complex panel methods, making them even more applicable for preliminary design calculations.

References

1. Bristow, D.R. and Hawk, J.D. (1983), "Subsonic Panel Method for Designing Wing Surfaces from Pressure Distribution", NASA CR 3713.

2. Callaghan, J.G. and Beatty, T.D. (1972), "A Theoretical Method for the Analysis and Design of Multi-Element Airfoils", Douglas Aircraft Co. Engineering Paper No. 5967.

3. Carmichael, R.L. and Woodward, F.A. (1966), "An Integrated Approach to the Analysis and Design of Wings and Wing-Body Combinations in Supersonic Flow", NASA TN D-3685.

4. Carmichael, R.L. and Erickson, L.L. (1981), "PAN AIR - A Higher Order Panel Method for Predicting Subsonic or Supersonic Linear Potential Lows About Arbitrary Configurations", AIAA Paper 81-1255.

5. Cenko, A., Tinoco, E.N., Dyer, R.D. and DeJongh, J. (1981), "PAN AIR Applications to Weapon Carriage and Separation", Journal of Aircraft, Vol. 18, No. 2, pp. 128–134.

6. Coppersmith, R.M., Youngren, H.H. and Bouchard, E.E. (1983), "Quadrilateral Element Panel Method (QUADPAN)", Theoretical Report (Version 3), Lockheed-California Co. LR-30500.

7. Derbyshire, T. and Sidewell, K.W. (1982), "PAN AIR Summary Document (Version 1.0)", NASA CR-3250.

8. Ehlers, F.E., Johnson, F.T. and Rubbert, P.E. (1976), "A Higher Order Panel Method for Linearized Supersonic Flow", AIAA Paper 76-381.

9. Ehlers, F.E., Epton, M.A., Johnson, F.T., Magnus, A.E. and Rubbert, P.E. (1979), "A Higher Order Panel Method for Linearized Supersonic Flow", NASA CR-3062.

10. Erickson, L.L. and Strande, S.M. (1983), "PAN AIR: Evolution and New Directions", AIAA Paper 83-1831.

11. Falkner, V.M. (1947), "The Solution of the Lifting Plane Problems by Vortex Lattice Theory", British ARC R&M 2591.

12. Falkner, V.M. (1948), "Calculated Loadings Due to Incidence of a Number of Straight and Swept Back Wings", British ARC R&*M* 2596.

13. Fornasier, L. (1984), "Treatment of Supersonic Configurations by an Updated Low-Order Panel Method" Journal of Aircraft, Vol. 21, No. 4, pp. 227–237.

24. Fornasier, L. (1984), "HISSS – A Higher-Order Subsonic/Supersonic Singularity Method for Calculating Linearized Potential Flow" AIAA 84-1646.

15. Geising, J.P. and Smith, A.M.O. (1967), "Potential Flow About Two-Dimensional Hydrofoils", Journal of Fluid Mechanics, Vol. 28, pp. 113–129.

16. Geising, J.P. (1970), "Two-Dimensional Potential Flow Theory for Multiple Bodies in Small Amplitude Motion", AIAA Journal, Vol. 8, pp. 1944–1953.

17. Goodsell, A., Madson, M.D. and Melton, J. (1989), "TRANAIR and Euler Computations of a Generic Fighter Including Comparisons with Experiments", AIAA Paper 89-0263.

18. Hess, J.L. and Smith, A.M.O. (1965), "A General Method for Calculating Low Speed Flows About Inlets", AGARDograph 103, Aerodynamics of Power Plant Installation.

19. Hess, J.L. and Smith, A.M.O. (1966), "Calculation of Potential Flow About Arbitrary Bodies", in Progress in Aeronautical Sciences, Vol. 8, pp. 1–138.

20. Hess, J.L. (1972), "Calculation of Potential Flow About Arbitrary Three-Dimensional Lifting Bodies", Final Technical Report, McDonald-Douglas Co., Report No. MDC J5679-01, (AD 755 480).

21. Hess, J.L. (1975), "Review of Integral Equation Techniques for Solving Potential Flow Problems with Emphasis on the Surface-Source Method", in Computer Methods in Applied Mechanics and Engineering, Vol. 5, pp. 145–196.

22. Hunt, B. (1978), "The Panel Method for Subsonic Aerodynamic Flows: A Survey of Mathematical Formulations and Numerical Models and an Outline of the New British Aerospace Scheme", Von-Karman Institute for Fluid Dynamics, Lecture Series 1978-4.

23. Johnson, F.T. and Rubbert, P.E. (1975), "Advanced Panel-Type Influence Coefficient Methods Applied to Subsonic Flows", AIAA Paper 75-50.

24. Johnson, F.T. (1980), "A General Panel Method for the Analysis and Design of Arbitrary Configurations in Incompressible Flow", NASA CR-3079.

25. Kraus, W. (1978), "Panel Methods in Aerodynamics", in *Numerical Methods in Fluid Dynamics*, Wirz, H.J. and Smolderen, J.J. (eds.), McGraw-Hill Book Co., N.Y., pp. 237–297.

26. Lamar, J.E. (1968), "A Modified Multhopp Approach for Predicting Lifting Pressure and Camber Shape for Composite Planforms in Subsonic Flow", NASA TN D-4427.

27. Landrum, E.J. and Miller, D.S. (1981), "Assessment of Analytic Methods for the Prediction of Supersonic Flow Over Bodies", AIAA Journal, Vol. 19, No. 2, pp. 160–164.

28. Madson, M.D. (1987), "Transonic Analysis of the F-16A with Under-Wing Fuel Tanks: An Application of the TRANAIR Full-Potential Code" AIAA Paper 87-1198.

29. Margason, R.J. and Lamar, J.E. (1971), "Vortex Lattice Fortran Program for Estimating Subsonic Aerodynamics Characteristics of Complex Planforms", NASA TN D 6142.

30. Margason, R.J., Kjelgaard, S.O., Sellers, W.L. III, Morris, C.E.K. Jr., Walkey, K.B. and Shields, E.W. (1985), "Subsonic Panel Methods – A Comparison of Several Production Codes", AIAA Paper 85-0280.

31. Maruyama, Y. and Akishita, S. (1986), "Numerical Simulations of Supersonic Flows About Arbitrary Configurations Using A New Panel Method Program", AIAA Paper 86-2181.

32. Maskew, B. (1982a), "Prediction of Subsonic Aerodynamic Characteristics: A Case for Low-Order Panel Methods", Journal of Aircraft, Vol. 19, No. 2, pp. 157–163.

33. Maskew, B. (1982b), "Program VSAERO, A Computer Program for Calculating the Nonlinear Aerodynamic Characteristics of Arbitrary Configurations", NASA CR-166476.

34. Miller, S.G. and Youngblood, D.B. (1983), Applications of USSAERO-B and the PAN AIR Production Code to the CDAF Model a Canard/Wing Configuration" AIAA Paper 83-1829.

35. Miranda, L.R., Elliot, R.D. and Baker, W.M. (1977), "A Generalized Vortex Lattice Method for Subsonic and Supersonic Flow Applications", NASA CR 2865.

36. Morino, L. (1974), "A General Theory of Unsteady Compressible Potential Aerodynamics", NASA CR 2464.

37. Morino, L. and Kuo, C.C. (1974), "Subsonic Potential Aerodynamics for Complex Configurations: A General Theory", AIAA Journal, Vol. 12, No. 2.

38. Morino, L., Chen, L.T. and Suciu, E.O. (1975), "Steady and Oscillatory Subsonic and Supersonic Aerodynamics Around Complex Configurations", AIAA Journal, Vol. 13, No. 3, pp. 368-374.

39. Morino, L. (1980), "Steady, Oscillatory and Unsteady Subsonic and Supersonic Aerodynamics – Production Version (SOUSSA - P1.1), Vol. I. Theoretical Manual", NASA CR-159130.

40. Multhopp, H. (1950), "Methods for Calculating the Lift Distribution of Wings (Subsonic Lifting-Surface Theory)", British ARC R&M No. 2884.

41. "PAN AIR – A Computer Program for Predicting Subsonic or Supersonic Linear Potential Flows About Arbitrary Configurations Using a Higher Order Panel Method (Version 1).
Vol. I – Theory Document, NASA CR-3251, 1980.
Vol. II – Users Manual, NASA CR-3252, 1980.
Vol. III – Case Manual, NASA CR-3253, 1980.
Vol. IV – Maintenance Manual, CR-3254, 1980.

42. Robinson, A. (1948), "On Source and Vortex Distributions in Linearized Theory of Steady Supersonic Flow", Quart. J. Mech. Appl. Math. I, pp. 408-432, also in *Selected Papers of Abraham Robinson*, Yale University Press, 1979, pp. 126–150.

43. Roggero, F. and Larguier, R. (1990), "Aerodynamic Calculation of Complex Three-Dimensional Configurations", ICAS Paper 90-6.4.3. pp. 770–781.

44. Rom, J., Almosnino, D. and Gordon, R. (1981), "Method for the Calculation of the Non-Linear Aerodynamic Characteristics of Thick Wings and Bodies Including Symmetric Vortex Separation in Subsonic Flow", TAE Report No. 437, Dept. of Aerospace Engineering, Technion, I.I.T., Haifa, Israel.

45. Rubbert, P.E. and Saaris, G.R. (1972), "Review and Evaluation of a Three-Dimensional Lifting Potential Flow Analysis Method for Arbitrary Configurations", AIAA Paper 72-188.

46. Samant S.S., Bussoletti, J.E., Johnson, F.T., Burkhart, R.H., Everson, B.L., Melvin, R.G., Young, D.P., Erickson, L.L., Madson, M.D. and Woo, A.C. (1987), "TRANAIR: A Computer Code for Transonic Analysis of Arbitrary Configurations", AIAA Paper 87-0034.

47. Sears, W.R. (ed), (1954), *High Speed Aerodynamics and Jet Propulsion; Vol. VI: General Theory of High Speed Aerodynamics*, Princeton University Press.

48. Tennenbaum, M.G. (1991), "Calculation of the Interaction between Close-Coupled Surfaces in Supersonic Flow", M.Sc. Thesis, Faculty of Aerospace Engineering, Technion, I.I.T., Haifa, Israel.

49. Thomas, J.L. and Miller, D.S. (1979), "Numerical Comparisons of Panel Methods at Subsonic and Supersonic Speeds", AIAA Paper 79-0404.

50. Thomas, J.L., Lurcking, J.L. and Sellers, W.L. III (1983), "Evaluation of Factors Determining the Accuracy of Linearized Subsonic Panel Methods", AIAA Paper 83-1826.

51. Tseng, W., Feinberg, E. and Cenko, A. (1989), "TRANAIR Applications to Fighter Configurations", AIAA Paper 89-2220 CP.

52. Tinoco, E.N., Johnson, F.T. and Freeman, L.M. (1979), "The Application of a Higher Order Panel Method to Realistic Supersonic Configurations", AIAA Paper 79-0274. Also, Journal of Aircraft, Vol. 17, No. 1, 1980, pp. 38–44.

53. Ward, G.N. (1955), *Linearized Theory of Steady High-Speed Flow*, Cambridge University Press.

54. Woodward, F.A. (1968), "Analysis and Design of Wing-Body Combinations at Subsonic and Supersonic Speeds", Journal of Aircraft, Vol. 5, No. 6, pp. 528–534.

55. Woodward, F.A. (1973), "An Improved Method for the Aerodynamic Analysis of Wing-Body-Tail Configurations in Subsonic and Supersonic Flow, Part I – Theory and Application", NASA CR-2228.

56. Woodward, F.A. (1980), "USSAERO Computer Program Development, Versions B and C", NASA CR 3227.

57. Woodward, F.A. and Landrum, E.J. (1980), "The Supersonic Triplet – A New Aerodynamic Panel Singularity with Directional Properties", AIAA Journal, Vol 18, No. 2, pp. 138–142.

58. Woodward, F.A. and Fornasier, L. (1984), "Investigation of the Triplet Concept Using a Higher-Order Supersonic Panel Method", ICAS Paper 84-1.1.1. pp. 31–40.

Chapter 5

Vortex Flows and the Rolled Up Vortex Wake

The trailing vortex wake behind the lifting wing is an excellent example of the natural tendency of free vortical shear layers to roll up into well-defined concentrated vortices with the essentially inviscid outer flow rotating around an inner viscous core. Such is the case when the well known tip vortices are established in the wake of high AR lifting wings. High speed computers provide the means for calculating the characteristics of the complete flow field over the wing as well as its wake and the rolled-up tip vortices. These calculations will be presented and discussed in the following chapters. However, before diving into the details of the calculations, it is of particular interest and importance to gain some insight into the flow field over wings and their wakes. This can best be done by examining analytical solutions, such as those developed by Sprieter and Sacks (1951), to the problems of simulating the main features of the rolled-up vortex flow.

Some early investigations of the characteristics of flow over wings and their wakes were presented by Kaden (1931), Rosenhead (1931), Betz (1933) and Westwater (1935). Introducing the model of discreet line vortices to represent the vortex wake behind the trailing edge of the wing, Westwater (1935) calculated the trajectories of these line vortices and showed that induced velocities cause the deformation of these line vortices into helical trajectories which cause these vortices to roll into two concentrated tip vortices.

Interest in the rolling-up effect of the vortex wake was further stimulated by practical problems associated with the tip vortices trailing behind flying aircraft. It is well known that the vortex layer, formed by the union of upper and lower wing surface boundary layers, behind the trailing edge of a flying aircraft, rolls up into two concentrated vortex cores. In the case of large heavy aircraft, Garodz (1971) reports tangential velocities of up to 300 ft/sec near the core of the tip vortices. Such velocities are high enough to become a serious hazard for flight safety which must be considered when determining the separation distance between aircraft during takeoff and landing on busy airport runways and in flight in busy flight corridors.

The vorticity in the cores of the tip vortices intensifies with increasing aircraft weight. This is due to the fact that the overall circulation around the vortex cores must be equal to the total circulation on the wing; this circulation is proportional to the total lift on the aircraft wing. However, while the total circulation is determined by the aircraft lift, the diameter of the tip vortex cores and, hence, their maximum tangential velocities may be amenable to variation by proper aerodynamic design. However, such design requires a better understanding of the parameters (such as wing geometry, tip blowing, winglets) affecting vortex wake structure and the rolling-up process.

Another possible technique for reducing long range effects of the trailing vortices consists of attempts to increase the dissipation in the cores. The viscous dissipation at the Reynolds numbers of flight is relatively weak, so that, barring any other effects, the trailing tip vortices with the high tangential velocities may persist for very long distances, measured in thousands of wing span lengths. This means a distance of a few miles behind a large aircraft. One mechanism which accelerates the breakup and dissipation of these tip vortices is the instability which causes the amplification of disturbances and oscillatory motions of the vortex cores, discussed by Crow (1970). It may be possible to trigger such oscillations earlier by control of time-dependent variations of the lift and/or the lift distribution on the wing.

All of these applications require better understanding of the rolling-up process of vortex wakes behind trailing edges of aircraft wings. The work of Kaden (1931) marks the earliest step in the study of the roll-up process. Kaden succeeded to calculate analytically the evolution with time of the shape of the spiral roll-up of the edges of a semi-infinite, initially flat two-dimensional vortex sheet of zero thickness. This solution can be interpreted as evaluating the shape of the spiral roll-up of the edges of the vortex wake as a function of distance behind the wing's trailing edge. An important result of this analysis shows that, from the very onset of the rolling-up process, a spiral near axisymmetric form with an infinite number of turns, is established at the sheet edge, caused by infinite velocity at the edge (a consequence of the assumption of zero sheet thickness).

The Westwater's (1935) calculation considers a finite span, zero thickness wake formed by an elliptically loaded wing and assumes that the roll-up can be treated as a two-dimensional, time-dependent process. The configurations at successive stages in time represent successively downstream sections of the wake, as fixed by the forward speed of the wing multiplied by the time. Westwater further simplified his method by replacing the continuous vortex sheet with a row of infinite line vortices. This method is suitable for use on wakes which roll up relatively slowly, as is the case far behind wings of high Aspect Ratio. Clements and Maull (1973) extended the line vortex method to cover nonelliptic loadings. Further developments of methods of calculations utilizing the line vortex model to include three-dimensional effects due to bound vortices and effects of curvature of the line vortices by the use of vortex lattice procedures were presented by Hackett and Evans (1971), Hancock and Butter (1971), Labrujere (1972), Maskew (1973) and Rom, Zorea and Gordon (1974).

The roll-up of a finite thickness wake was studied by Portnoy and Rom (1971) and Portnoy (1975). The analysis attempts to remove the unrealistic feature of models using zero wake thickness and vorticity concentrated on lines. It is therefore assumed that the wake vorticity is contained in a layer of finite thickness with some plausible cross-sectional shape. The wake flow is assumed to be two-dimensional and rolling-up is analyzed via the time dependent development of this model using the Westwater calculation. This calculation also defines a slow rolling-up wake behind high aspect ratio wings. The introduction of vorticity distributed continuously throughout the finite thickness wake cross-section enables a more reasonable picture of the initial roll-up phase than that of the infinite spiral obtained by Kaden (1931).

Some of the main features of the methods proposed to facilitate analysis of the rolling-up process and the characteristics of the two- and three-dimensional vortex wakes will be presented in this chapter.

5.1 Vortex Core of the Rolled-Up Wake

A highly illuminating theoretical analysis for the flow field in the wake of lifting wings has been developed by Sprieter and Sacks (1951). It was discussed previously, that the paths of the trailing vortices cannot be prescribed a priori but must coincide with the stream lines, which are in turn influenced by the induced velocities of these vortices. Therefore, Sprieter and Sacks start from considerations of similarity relations that are applicable to the rolled-up vortex wake. Let us examine two wings of similar span loading but differing in span b and total circulation Γ_o. The symbols representing the reference wing are denoted by an asterisk; second wing symbols are plain marked. The ratio of induced velocities q at corresponding stations behind the wings, neglecting the small influence of the difference in chordwise loadings, is given by

$$q/q^* = (\Gamma_o/b)/(\Gamma_o^*/b^*) \tag{5.1}$$

The ratio of the distances d, in terms of the wing spans, measured from the trailing edge to stations having similar degrees of roll-up of trailing vortex sheets is

$$\frac{d/b}{d^*/b^*} = \frac{V_o/q}{V_o^*/q^*} = \frac{V_o b \Gamma_o^*}{V_o^* b^* \Gamma_o} = \frac{\mathrm{AR}/C_L}{\mathrm{AR}^*/C_L^*} \tag{5.2}$$

From this equation, it can be demonstrated that the distance e required for the trailing vortex sheet to become essentially rolled-up is

$$e/c = K(\mathrm{AR}/C_L)(b/c) \tag{5.3}$$

It is assumed that a reasonable approximation of the value of K was calculated by Kaden (1931) for wings with elliptical span loading, where $K = 0.28$. The above equation 5.3, indicates that the rolling-up distance e/c depends on the wing's geometry; thus, for a rectangular wing of AR=6, the rolling-up distance is $(e/c)_{\text{rec}} = 0.28(36/C_L)$ while for a delta wing of AR=2, the rolling-up distance is $(e/c)_{\text{del}} = 0.28(2/C_L)$. The distance for the rolling-up of the trailing vortices for a low Aspect Ratio delta wing is in the order of 18 times shorter than that for the high Aspect Ratio rectangular wing. Thus, for wings of low Aspect Ratio, the vortex sheet may roll-up into two trailing vortices within less than one chord length, while this distance is considerably longer in the case of the high Aspect Ratio wings.

It is expected that the rolled-up trailing vortices far behind the wing are not idealized line vortices but must have cores of finite diameter. An estimation of the core diameter, which is a good approximation at either subsonic or supersonic speeds, is presented by Sprieter and Sacks (1951). Their argument is based on energy considerations, assuming that the vortex cores are circular in shape, rotating as solid bodies, with a radius of a_c. Equating the total kinetic energy of the fluid outside of the vortex cores and inside these cores to the induced drag D_i of the wing, and denoting the distance between the rolled-up vortices as $2b'$ and assuming that $a_c/b' << 1$ it is then shown that

$$D_i = \frac{\rho_o \Gamma_o^2}{8\pi}\left(1 + 4\log\frac{2b' - a_c}{a_c}\right) \tag{5.4}$$

Therefore, the ratio of the core radius a_c to the semi span of the rolled-up vortices b' is then

$$\frac{a_c}{b'} = 2\left\{1 + \exp\left[\frac{4\pi\text{AR}\cdot C_{Di}}{C_L^2}\left(\frac{b'}{b}\right)^2 - \frac{1}{4}\right]\right\}^{-1} \tag{5.5}$$

For elliptic span loading, $C_{Di} = (C_L^2/\pi)\text{AR}$ and $b'/b = \pi/4$, the corresponding core radius is

$$a_c = 0.197b' = 0.155b \tag{5.6}$$

5.2 The Rolled Up Tip Vortices

The approximation that simplifies an estimate of the shape of the trailing vortex sheet is based on the assumption of slow variation of the motion of trailing vortices in the direction of the stream. In such a case, the axial variations are neglected and the shape of the trailing vortex sheet can be estimated by the calculation of the motion of free vortices in the cross flow, i.e. the motion of free vortices in a two-dimensional, nonsteady, incompressible flow. It is possible to calculate the motion of the center of gravity of the fully rolled-up vortices far behind the wing using the Betz method (1933). The velocity of the center of gravity of the vortices behind elliptically loaded wings is calculated by Spreiter and Sacks (1951) to be, at the trailing edge

$$w(x_{TE}) = -0.43 V_\infty (C_L/\pi)\mathrm{AR} \tag{5.7}$$

and far behind the wing where the vortices are fully rolled-up

$$w(x_{TE}) = -0.41 V_\infty (C_L/\pi)\mathrm{AR} \tag{5.8}$$

A detailed analysis of the behavior of the vortex system can be made by replacing the continuous sheet of vortices with a finite number of discrete line vortices and calculating, step by step, the motion of each vortex line. Such a calculation was made by Westwater (1935) for elliptically loaded wings. In this calculation, the vortex sheet was replaced by 20 vortices of equal strength, and the results represent the positions of each of the vortices at various distances behind the wing, as shown in Fig. 5.1. The positions of the vortex center of gravity and that of the center of the vortex sheet are presented in Fig. 5.2, where the corresponding positions evaluated by Eq. 5.7 and Eq. 5.8 are indicated. The good agreement between the calculated vortex position by the "exact" numerical evaluation of the vortex lines trajectories with the positions evaluated by the cross flow analysis as indicated in Fig. 5.2 can be accepted as support for using these methods of calculation approximations of potential flow.

5.3 The Rolling-Up of the Vortex Wake Behind Wings

In linear aerodynamic calculations, the vortex wake, which is shed at the trailing edge, is assumed to be a rigid sheet of zero thickness with a fixed geometry (generally a flat planar sheet), independent of the wing' lifting force. It is well known that in the actual flow the edges of the vortex sheet roll-up into concentrated tip vortices. Therefore, it is necessary to evaluate the shape of the vortex wake as the angle of attack is varied and to calculate the effects of the rolled-up vortex wake on the wing's aerodynamic characteristics. Methods for calculations of rolled-up wake are presented in Section 5.3.1. (for zero thickness vortex sheet) and in Section 5.3.2. (for finite thickness sheet).

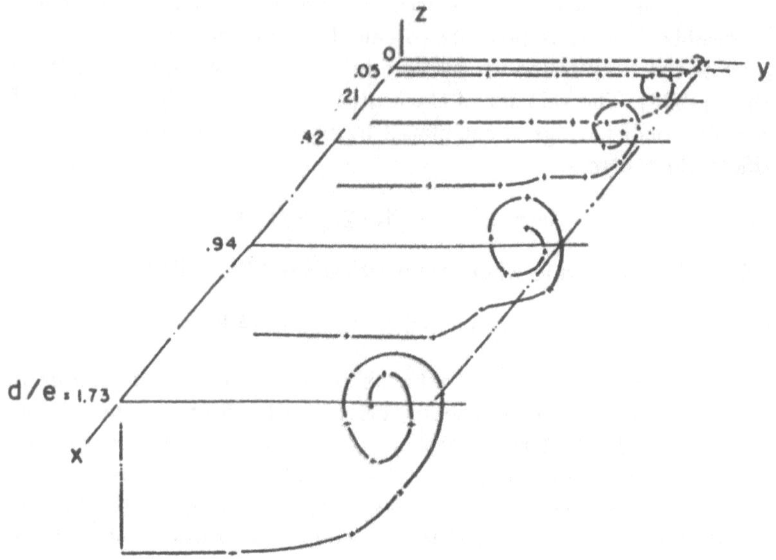

Fig. 5.1. Shape of vortex sheet for elliptic loading (Westwater 1935)

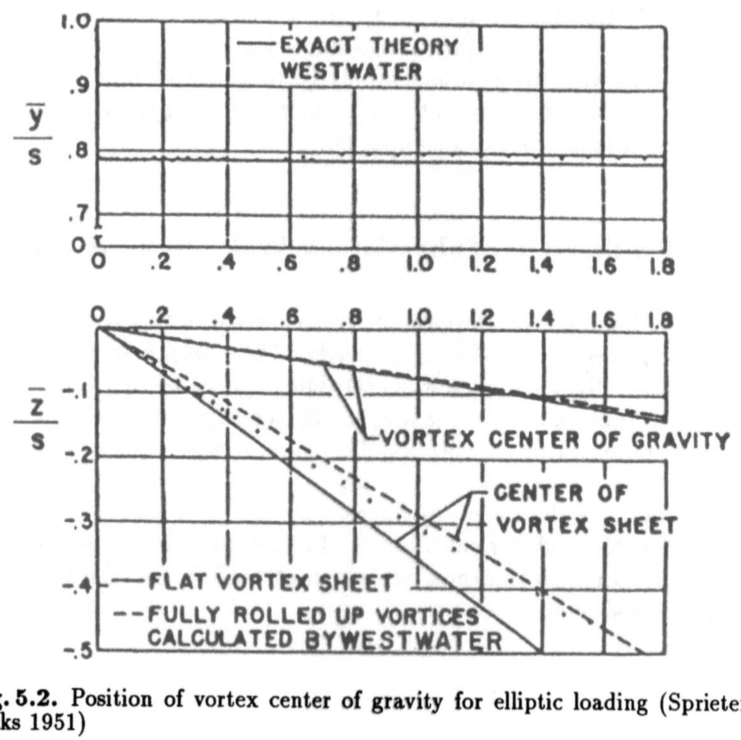

Fig. 5.2. Position of vortex center of gravity for elliptic loading (Sprieter and Sacks 1951)

5.3.1 The Rolling-Up of the Vortex Lines of the Zero Thickness Vortex Sheet

In the linear aerodynamic analysis, it is assumed that the vortex wake which is shed from the trailing edges has a rigid shape (generally, a flat sheet in the plane of the wing). It is well known that the edges of the vortex sheet roll-up into a spiral. Kaden (1931) calculated the shape and position of the roll-up of the edge of a two-dimensional plane vortex sheet into an infinite spiral, as shown in Fig. 5.3.

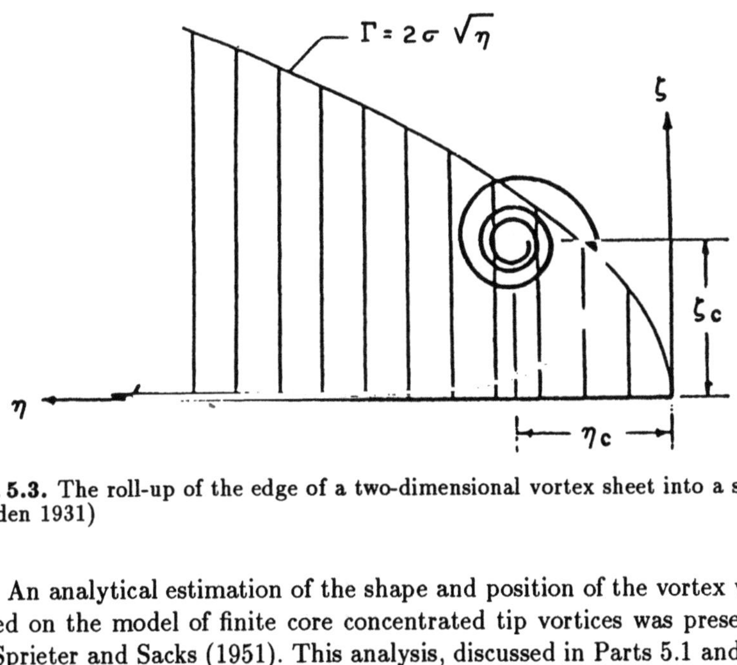

Fig. 5.3. The roll-up of the edge of a two-dimensional vortex sheet into a spiral (Kaden 1931)

An analytical estimation of the shape and position of the vortex wake based on the model of finite core concentrated tip vortices was presented by Sprieter and Sacks (1951). This analysis, discussed in Parts 5.1 and 5.2, allows for reasonable estimates of the vortex wake characteristics, particularly for lightly loaded high AR wings; the results are shown in Fig. 5.2.

Other methods for calculations of vortex wake characteristics are based on representation of the vortex sheet by discrete line vortices. The first calculation that demonstrated the utility of the discretization model was Westwater's (1935). In the Westwater calculation, the elliptic load distribution of the wing is represented by a discrete number of equal strength infinite line vortices. This solution is equivalent to the problem of calculating the unsteady motion of a two-dimensional array of point vortices moving under their own mutual influence. Results of the calculation of the vortex wake behind a wing are presented in Fig. 5.1. Since in this discretization, the vortices are of equal strength, the spacing between the vortices becomes smaller as one moves towards the wing tips; the mutual induced velocities do not result in the infinite spiral calculated by Kaden (1931).

With the use of high speed digital computers, it has been possible to improve and enlarge the scope of these calculations. A more realistic model was devised by Hackett and Evans (1971), by Butter and Hancock (1971) and by Clements and Maull (1973). The wing is represented by a bound vortex at the 1/4 chord line of the wing. Semi-infinite discrete vortices are trailing from the wing. These calculations enable the evaluation of the three-dimensional vortex wake. The schematic of the calculation procedures for the two-dimensional and three-dimensional vortex wakes for elliptically loaded wings are presented in Figs. 5.4a and 5.4b, respectively.

The representation of the continuous vorticity of the vortex sheet by discrete line vortices raises many problems because of the singularity as the distance between line vortices tend to zero. These numerical problems were discussed by Hackett and Evans (1971), Moore (1971, 1974) and Rom and Zorea (1973). In particular, Moore (1971) questioned the justification for representing the vortex sheet with continuous distributed vorticity using discrete vortex lines. Various investigators have indicated that certain difficulties result from numerical instabilities generated by the discretization, so that it should be expected that improved numerical procedures may provide more accurate calculations. Even then, however, there will be a limit on the number of discrete line vortices to be used in such calculations.

A number of improvements in the calculation methods have been reported by Hackett and Evans (1971), Rom and Zorea (1973), Nielsen and Schwind (1971), Chorin and Bernard (1973), Moore (1974) and Kuwahara and Takami (1973). Additional numerical problems encountered in the calculations of the rolling-up near the tip are manifested by a variety of phenomena including: crossover of vortex lines, shooting out of a vortex line from within the ordered sheet of the rolled-up vortex, and irregular oscillation problems in the positions of vortices in the rolled-up sheet. These problems are particularily evident when large numbers of equal-strength line vortices are used in the representation of the sheet, (resulting in very close spacing of vortices near the tips of the wing). Hackett and Evans (1971) determined that some of these difficulties can be alleviated by the use of vortices that are equally-spaced but of varied strength. Further improvements that enable calculations of complex flow fields including flows generated by interactions between number of lifting surfaces have been introduced by Rom and Zorea (1973) and Rom, Zorea and Gordon (1974) where the use of finite core vortex model was investigated. The replacement of the finite core vortex structure to model the line vortices by a definition of a "cut-off" distance for amalgamation of vortex lines into a single line, when these vortices approach each other towards a limiting distance, was introduced by Almosnino, Zorea and Rom (1978).

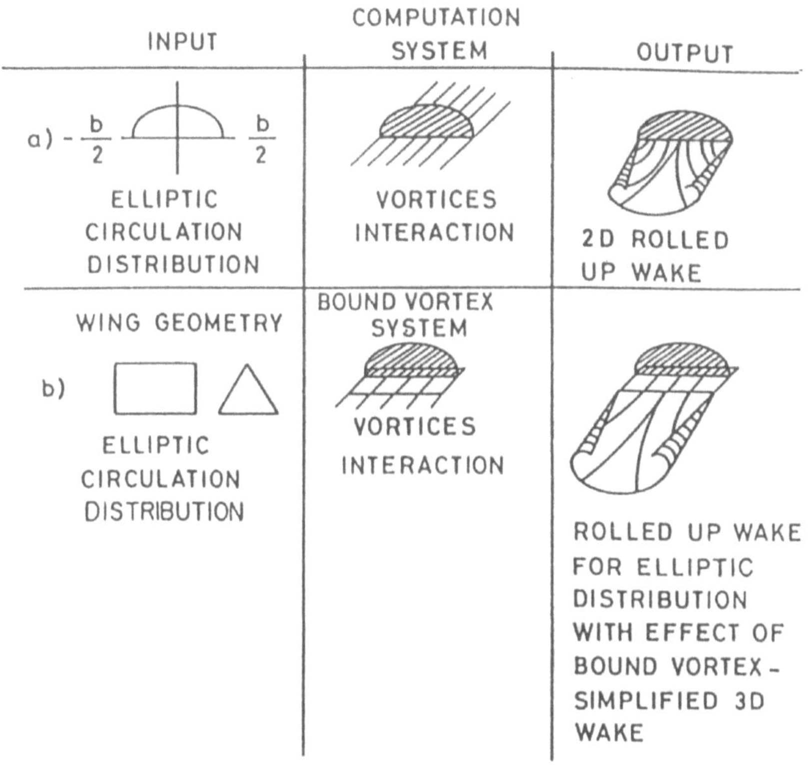

a. Two-dimensional make (Westwater model)

b. Three-dimensional wake model

Fig. 5.4. The schematic for the calculations of the vortex wake of elliptically loaded wings

The free vortex line trajectories are constructed out of a number of straight, finite length segments between designated points. The calculations, following the Rom and Zorea (1973) procedure, begin with an arbitrarily selected initial position; the final position is obtained through an iterative process. For that purpose, the downstream point of each segment is moved so that the segment is aligned approximately with the local streamline direction. The new coordinates, $y_j^{(i)}$ and $z_j^{(i)}$, of the downstream end of segment j are calculated using the downstream end point of segment $j-1$ (known from the previous step), by equations 5.9 and 5.10, with the superscript i denoting the iteration step number

(5.9)
$$y_j^{(i)} = y_{j-1}^{(i)} + \left(\frac{v_{j-1}^{(i)}}{V_\infty + u_{j-1}} \right) (x_j - x_{j-1})$$

(5.10)
$$z_j^{(i)} = z_{j-1}^{(i)} + \left(\frac{w_{j-1}^{(i)}}{V_\infty + u_{j-1}^{(i)}} \right) (x_j - x_{j-1})$$

To prevent the generation of exaggerated induced velocities due to the singularity of the potential flow vortex elements, a "cutoff distance" was introduced by Almosnino, Zorea and Rom (1978) and is used in various discrete vortex line methods. A relation for the determination of the "cutoff distance" was presented by Almosnino (1985). Here, the "cutoff distance" is used to determine the range from the vortex segment, within which a solid body rotation can be assumed for the influencing vortex. This "cutoff distance", r^*, is determined by the equation

(5.11)
$$r^* = \left[\frac{\Gamma \, \Delta x \, (\cos \beta + \cos \gamma)}{4\pi V_\infty} \right]^{1/2}$$

where the geometric notations in Eq. 5.11 are defined in Fig. 5.5.

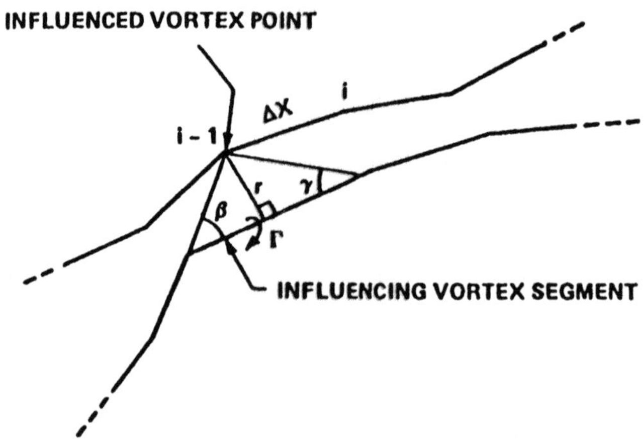

Fig. 5.5. Geometric notations used for the cutoff distance evaluation

The cutoff distance, r^*, defined by Eq. 5.11, is dependent on two parameters characterizing the vortex model, namely the strength of the influencing vortex, Γ, and the lengths of both segments (i.e. the integration step length). This makes the cutoff distance criterion sensitive to local conditions at each step of the iterative process, rather than an arbitrary choice of a single valued cutoff distance used over the whole flow field. The size

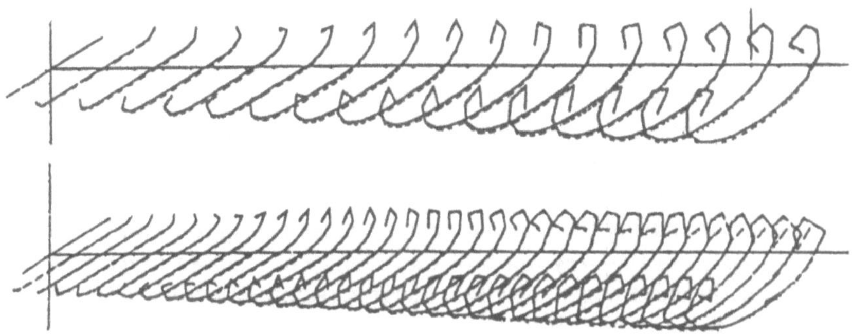

Fig. 5.6. The rolled-up vortex wake calculated by the modified vortex lattice method

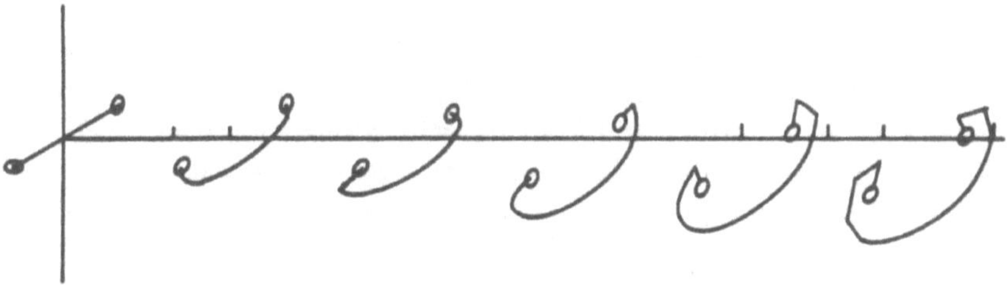

Fig. 5.7. The rolled-up vortex wake calculated by the modified vortex lattice method with finite core vortices

of this cutoff distance influences the rate of roll-up of the vortex lines; the proper values are evaluated by numerical experimentation.

The vortex wake with the rolled-up edges, calculated by the modified vortex lattice method of Rom and Zorea utilizing line vortices, is presented in Fig. 5.6. A calculation of the vortex wake using the finite core vortices, applying the criterion of amalgamation to the rolled-up edges, is shown in Fig. 5.7.

5.3.2 The Rolling-Up of a Finite Thickness Vortex Sheet

Most of the investigations of the rolling-up process of the vortex wake behind wings of finite span assume that this vortex sheet is of zero thickness. Kaden's (1931) first study introduced the solution for the rolling-up in time of a semi-infinite, straight, two-dimensional vortex sheet of zero thickness in an analytical form. Focusing on conditions very close to the edge of the zero thickness sheet during the initial stages of the rolling-up process, Kaden indicates that from the very onset of the rolling-up – caused by the infinite velocity at the sheet edge – a spiral of near axisymmetric form with infinite number of turns is established at the edge. It remains to be determined as to whether the roll-up of the edge of the vortex sheet into the spiral form with infinite number of turns is due to the effects of the assumed zero thickness of the vortex sheet. Experimental evidence clearly indicates that the vortex wake behind wings has some thickness and the concentrated tip vortices have finite radius cores.

An attempt to remove the unrealistic assumption of zero wake thickness and of vorticity concentrated on lines, by assuming that the wake vorticity is contained in a layer of finite thickness with some plausible cross-section, was presented by Portnoy and Rom (1971) and Portnoy (1975). Here, the wake flow is assumed to be two-dimensional and the rolling-up is studied via the time dependent development of this model. As in Westwater's research, Portnoy and Rom deal with a slow rolling-up which takes place far behind the wing. Assuming that a reasonable wake cross-section can be selected, an elongated elliptical section was used in the calculations, as presented in Fig. 5.8. The wake cross-section area was divided into triangular elements; within each element, the vorticity is assumed to be constant. Furthermore, it is assumed that the periphery of the wake is moving downwards with a velocity determined by the spanwise position and wing loading, exactly as in ordinary wing-wake theory. The triangular division is shown in Fig. 5.9. The velocity field and the strength of the vorticity within each triangle have been calculated by Portnoy (1975).

Examples of the initial roll-up edge shapes calculated for an elliptic thickness wake subject to uniform downwash (corresponding to elliptic spanwise loading) is presented in Figs. 5.10a, 5.10b, 5.10c and 5.10d. A view of the shape of the half wake with the rolled-up edge is shown in Fig. 5.11. In this calculation, the dimensionless time t^* is $t^* = 2C_L \cdot x/\pi \mathrm{AR} \cdot s$.

It should be noted that in order to clarify the inner detail of the roll-up spiral, the vertical scale in Figs. 5.10 and 5.11 has been exaggerated. Few of the rolled-up vortex characteristics vary with dimensionless time: the percentage of vorticity within the core, the ratio of mean core diameter to wing semi-span, the height-to-width ratio of the core and the number of turns of the spiral are calculated for elliptic wakes of thickness of 4%, 5% and 6% and presented in Figs. 5.12 to 5.15, respectively.

It can be clearly seen that the thick wake procedure enables calcula-

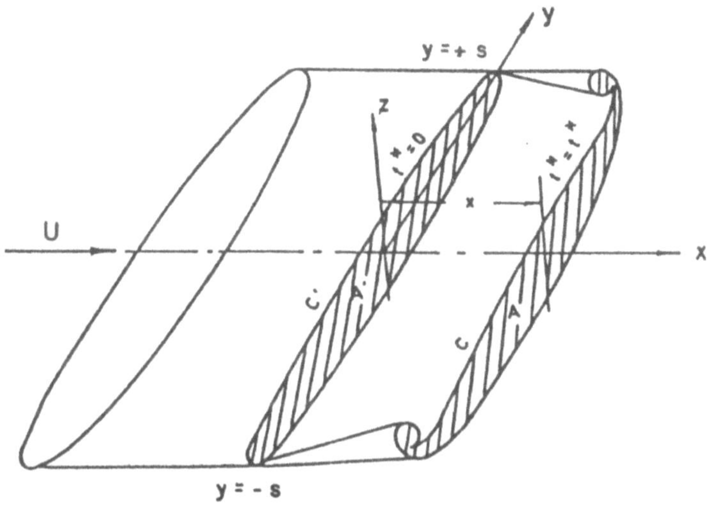

Fig. 5.8. Thick wake cross-sections and system of coordinates (Portnoy 1975)

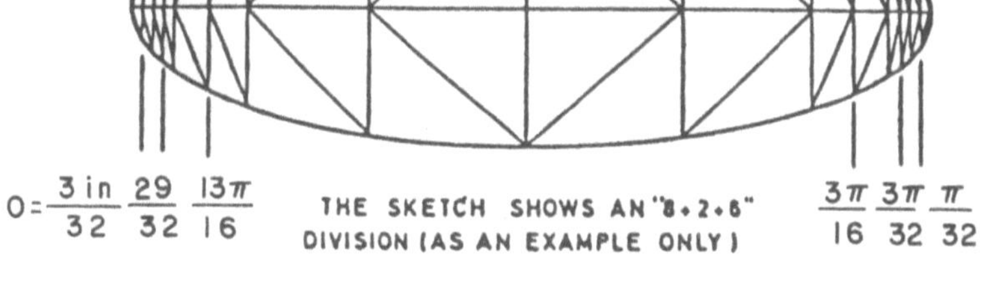

Fig. 5.9. Triangular elements in the elliptic cross-section wake (Portnoy 1975)

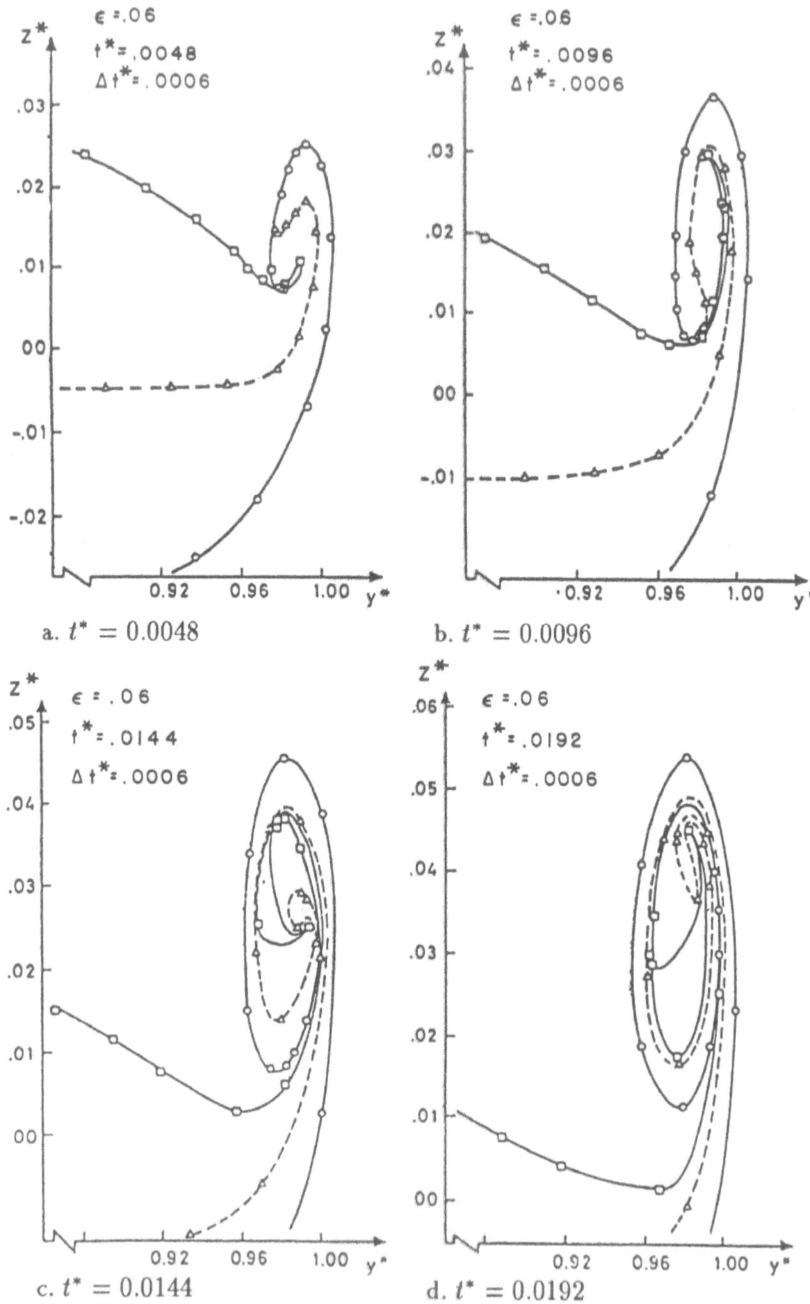

a. $t^* = 0.0048$ b. $t^* = 0.0096$

c. $t^* = 0.0144$ d. $t^* = 0.0192$

Fig. 5.10. The initial roll-up of a 6% thick elliptic wake (Portnoy 1975)

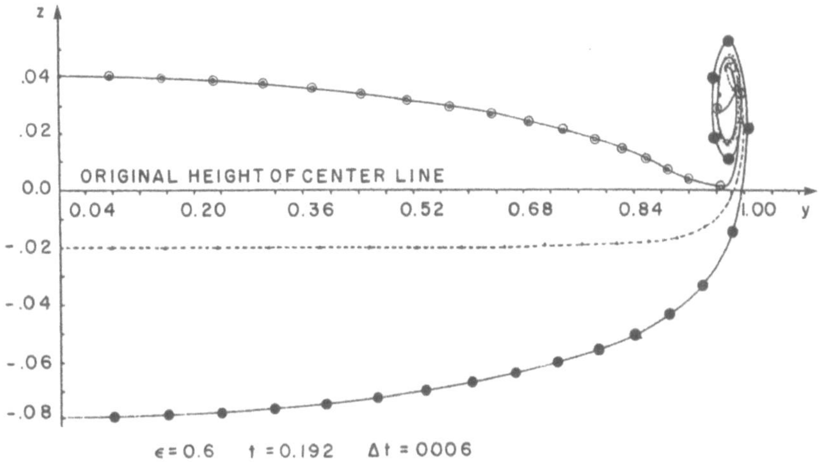

Fig. 5.11. The shape of the 6% thick elliptic wake at $t^* = 0.0192$ (Portnoy 1975)

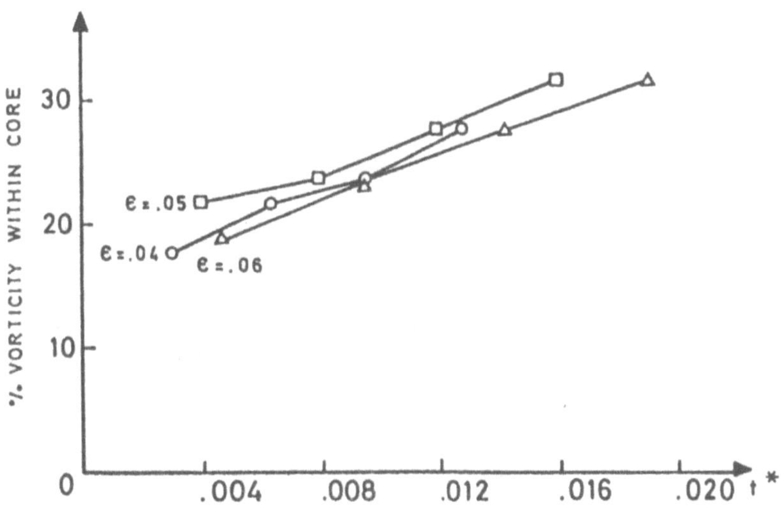

tion of close to two turns of the spiral of the tip vortex. The introduction of vorticity distributed continuously throughout the cross-section of the thick wake allows us to obtain a reasonable number of turns during the development of initial rolling-up of the tip of the finite-thickness vortex sheet, in contrast to the infinite spiral predicted by Kaden's analysis.

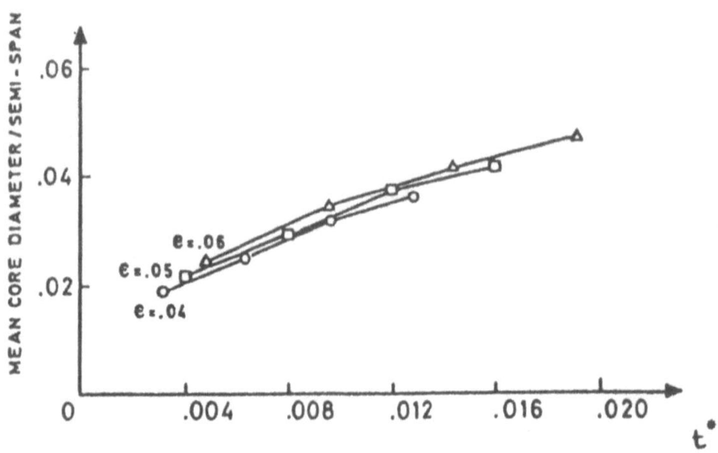

Fig. 5.13. Mean core diameter as a function of t^* (Portnoy 1975)

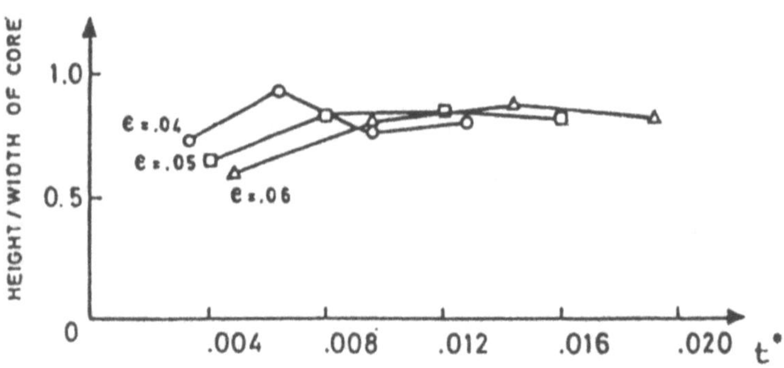

Fig. 5.14. The height-to-width ratio of the core as a function of t^* (Portnoy 1975)

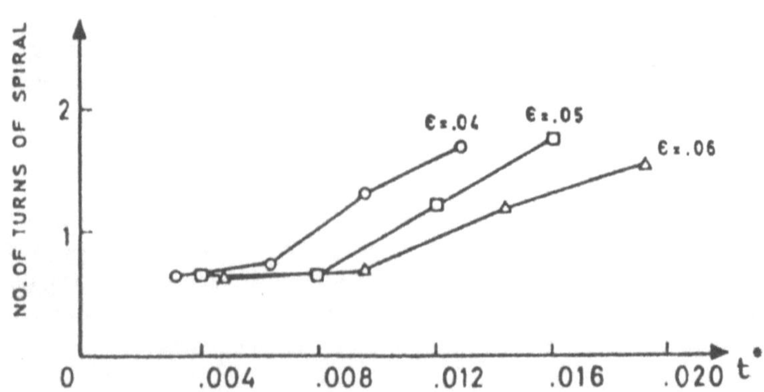

Fig. 5.15. Number of turns of the tip spiral as a function of t^* (Portnoy 1975)

5.4 The breakdown of Rolled-Up Vortices

In order to complete the description of of the phenomena associated with the rolling-up of vortices, it should be noted that vortex flows over slender bodies, as well as those over highly swept wings at high angles of attack, are known to become unstable to the point of causing a complete breakdown in the orderly flow of the rolled-up vortex. The onset of instabilities and the process of breakdown leading to the mechanisms of vortex "bursting" are the subject of extensive research by Sarpkaya (1971), Faler and Leibovich (1978), Wedemeyer (1982), Nakamura et al. (1986) and Hafez et al. (1987). Experimental measurements show that two types of vortex breakdown can be distinguished: the bubble and the spiral. Some of the experimental findings are discussed in Part 2.6. Various effects of the phenomena are demonstrated in Figs. 2.41 to 2.46. The bubble type breakdown is shown in Fig. 2.45 and the spiral type breakdown in Fig. 2.46. Analyses of the instability of vortices are presented by Benjamin (1962) and by Leibovich and Stewartson (1983). The growth of three-dimensional disturbances is analyzed, based on concepts of control theory, by Rusak (1989). Although much progress has been achieved in research on the mechanism of vortex breakdown, it still remains one of the intriguing unresolved problems in fluid mechanics. More insight to the "breakdown" mechanism can be seen from the numerical solutions of the Euler and the Navier-Stokes equations discussed in Chapters 8 and 9.

References

1. Almosnino, D. (1985), "High Angle of Attack Calculations of the Subsonic Vortex Flow on Slender Bodies", AIAA Journal, Vol. 23, No. 8, pp. 1150–1156.

2. Almosnino, D., Zorea, C. and Rom, J. (1978), "A Method for Calculating Longitudinal Characteristics of Wings and Multiple Lifting Surfaces in Subsonic Flow, and at High Angles of Attack", Israel Journal of Technology, Vol. 16, pp. 132–141.

3. Benjamin, (1962), "Theory of the Vortex Breakdown Phenomenon", J.F.M. vol. 4, Part 4, pp. 593–629.

4. Betz, A. (1933), "Behaviour of Vortex Systems", NACA TM 713.

5. Butter, D.J. and Hancock, G.J. (1971), "A Numerical Method for Calculating the Trailing Vortex System Behind a Swept Wing at Low Speed", The Aero. Journal of the Royal Aero. Soc., Vol. 75, pp. 564–568.

6. Clements, R.R. and Maul, D.J. (1973), "The Rolling up of a Trailing Vortex Sheet", The Aeronautical Journal, Vol. 77, No. 745, pp. 46–51.

7. Crow, S.C. (1970), "Stability Theory for a Pair of Trailing Vortices", AIAA Journal, Vol. 8, pp. 46–51.

8. Faler, J.H. and Leibovich, S. (1978), "An Experimental Map of the Internal Structure of a Vortex Breakdown", Journal of Fluid Mechanics, Vol. 86, pp. 313–335.

9. Fink, P.T. and Soh, W.K. (1978), "A New Approach to Roll-Up Calculations of Vortex Sheets", Proc. of the Royal Soc. A, Vol. 362, pp. 195–209.

10. Greene, G.C. (1985), "An Approximate Model of Vortex Decay in the Atmosphere", AIAA Paper 85-1835 CP.

11. Hackett, J.E. and Evans, M.R. (1971), "Vortex Wakes Behind High Lift Wings", Journal of Aircraft, Vol. 8, No. 5, pp. 334–340.

12. Hafez, A.M., Kuruvila, J.G. and Salas, M.D. (1987), "Vortex Breakdown Simulation", AIAA Paper 87-1343.

13. Hall, (1967), "A New Approach to Vortex Breakdown", Proceedings of the 1967 Heat and Fluid Mechanics Institute, Stanford University Press.

14. Hoeijmakers, H.W.M. (1989), "Computational Aerodynamics of Ordered Vortex Flows", NRL TR 88088U, Amsterdam, The Netherlands.

15. Kaden, H. (1931), "Aufwicklung einer unstabilen unstetigkeitsfläche", Ing. Arch., Vol. 2, pp. 140–168.

16. Labrujere, Th. E. (1972), "A Numerical Method for the Determination of the Vortex sheet Location Behind a Wing in Incompressible Flow", NRL TR 7209U.

17. Leibovich, S. and Stewartson, K. (1983), "A Sufficient Condition for the Instability of Columnar Vortices", Journal of Fluid Mechanics, Vol. 126, pp. 335–356.

18. Ludwieg, H. (1970), "Vortex Breakdown", DFVLR 70-40.

19. Maskew, B. (1973), "The Calculation of Potential Flow Aerodynamics Characteristics of Combined Lifting Surfaces with Relaxed Wakes", Hawker-Siddeley Aviation Ltd., (Brough) Aerodynamic Design Dept., Note YAD 3192.

20. Moore, D.W. (1974), "A Numerical Study of the Rollup of a Finite Vortex Sheet", Journal of Fluid Mechanics, Vol. 63, pp. 225–235.

21. Moore, D.W. (1975), "The Rolling Up of a Semi-Infinite Vortex Sheet", Proc. of the Royal Soc. A, Vol. 345, pp. 417–430.

22. Moore, D.W. (1979), "The Spontaneous Appearance of a Singularity in the Shape of an Evolving Vortex Sheet", Proc. of the Royal Soc. A, Vol. 365, pp. 105–119.

23. Moore, D.W. (1981), "On the Point Vortex Method", SIAM Journal of Science & Stat. Comp., Vol. 2, No. 1, pp. 65–84.

24. Nakamura, Y., Leonard, A. and Spalart, P.R. (1986), "Internal Structure of Vortex Breakdown", AIAA Paper 86-1074.

25. Portnoy, H. and Rom, J. (1971), "The Flow Near the Tip and Wake Edge of a Lifting Wing With Trailing-Edge Separation", TAE Report 132, Department of Aeronautical Engineering, Technion, I.I.T., Haifa, Israel.

26. Portnoy, H. (1975), "The Initial Roll-Up of a Thick, Two-Dimensional Wake Behind a Wing of Finite Span", TAE Report 262, Department of Aeronautical Engineering, Technion, I.I.T., Haifa, Israel.

27. Portnoy, H. (1977), "Thick, Two-Dimensional Wake Roll-Up Behind a Wing of Finite Span – Extended Calculations", Aeronautical Journal, pp. 460-463.

28. Rodgers, A. (1954), "Application of Two Dimensional Vortex Theory to the Prediction of Flow Fields Behind Wings of Wing-Body Combinations at Subsonic and Supersonic Speeds", NACA TN 3227.

29. Rom, J. and Zorea, Z. (1973), "The Calculation of the Lift Distribution and the Near Vortex Wake Behind High and Low Aspect Ratio Wing in Subsonic Flow", TAE Report 168, Department of Aeronautical Engineering, Technion, I.I.T., Haifa, Israel.

30. Rom. J., Zorea. C. and Gordon, R. (1974),Ön the Calculation of Non-Linear Aerodynamic Characteristics and the Near Vortex Wake", The 9th Congress of ICAS, ICAS paper 74-27.

31. Rom, J., Portnoy, H. and Zorea, C. (1976), "Investigations of the Rolling-Up of the Vortex Wake and Calculation of Non-Linear Aerodynamic Characteristics of Wings", TAE Report 277, Department of Aeronautical Engineering, Technion, I.I.T., Haifa, Israel.

32. Rosenhead, L. (1931), "The Formation of Vortices from a Surface Discontinuity", Proc. Roy. Soc. London, Ser. A. Vol. 134, pp. 170–192.

33. Rossow, V.J. (1973), "On the Inviscid Rolled-Up Structure of Lift-Generated Vortices", Journal of Aircraft, Vol. 10, No. 11, pp. 647–650.

34. Rossow, V.J. (1975), "Theoretical Study of Lift-Generated Vortex Wakes Designed to Avoid Rollup", AIAA Journal, Vol. 13, No. 4, pp. 476–484.

35. Rusak, Z., (1989), "Three-Dimensional Analysis of Equilibrium and Stability of Discrete Thin Vortices", D.Sc. Thesis, Technion, I.I.T., Faculty of Aerospace Engineering, Haifa, Israel.

36. Sarpkaya, T. (1971), "Vortex Breakdown in Swirling Conical Flows", AIAA Journal, Vol. 8, pp. 1792–1799.

37. Siddiqi, S. (1987), "Trailing Vortex Rollup Computations using the Point Vortex Method", AIAA Paper 87-2479.

38. Smith, J.H.B. (1984), "Theoretical Modelling of Three-Dimensional Vortex Flows in Aerodynamics", Aeronautical Journal, pp. 101–116.

39. Spreiter, J.R. and Sacks, A.H. (1951), "The Rolling Up of the Trailing Vortex Sheet and its Effects on the Downwash Behind Wings", Journal Aeronautical Sciences, Vol. 18, No. 1, pp. 21–32.

40. Wedemeyer, E. (1982), "Vortex Breakdown" in "High Angle of Attack Aerodynamics", AGARD-LS-121, 9-1 & 9-17.

41. Wentz, W.H. Jr. and Kohlman, D.L. (1972), "Vortex Breakdown on Slender Sharp-Edged Wings", Journal of Aircraft, Vol. 8, No. 3, pp. 156–161.

42. Westwater, P.L. (1935), "The Rolling Up of the Surface Discontinuity Behind an Airfoil of Finite Span", British A.R.C. R&M 1692.

Chapter 6

Nonlinear Aerodynamics of Wings and Bodies at High Angles of Attack

In the development of the linear lifting surface theory, it is assumed that the trailing vortices remain in the plane of the lifting surface and leave the surface only at the trailing edge as a plane vortex sheet (Fig. 6.1a). As the angle of attack is increased, the flow tends to separate from the side and trailing edges of the lifting surface and roll-up into concentrated vortices. For relatively sharp edges, the separation manifests itself in the generation of a viscous shear layer. As this shear layer leaves the surface, it behaves as a viscous sheet that rolls-up above the lifting surface into concentrated vortices. In this case, the nonlinear lifting surface model must allow the separation of the trailing vortices from the surface or, at least, from the side and trailing edges (Fig. 6.1b). The induced velocities due to these vortices affect the pressure distribution on the wing, so as to induce additional lift force. The lift force, due to the leading and side edge vortices, increases as a nonlinear function of the increasing angle of attack.

Although the initiation of the leading edge vortices results from the separation of viscous shear layers, it is important to note that once these shear layers leave the wing surfaces and roll-up into concentrated vortices, they induce a potential vortex velocity field as if these vortices are inviscid potential vortices. Thus, various methods based on modeling the rolled-up vortices by potential concepts for the evaluation of the nonlinear aerodynamic characteristics have been developed. The methods, which are based on analytical models and on semi-empirical approaches, are presented in this chapter, while those which involve numerical calculation methods based on the nonlinear panel methods will be presented in Chapter 7.

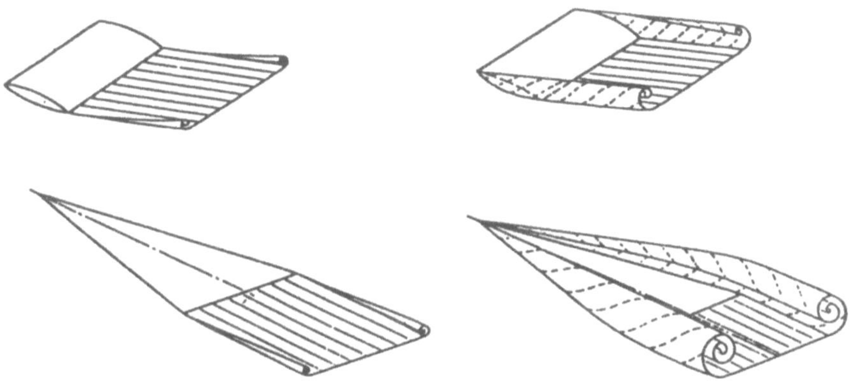

a. Linear theory b. Nonlinear theory

Fig. 6.1. The trailing vortex sheet models in the linear and nonlinear lifting surface theories

6.1 Analytical and Semi-Empirical Methods for Calculations of the Nonlinear Aerodynamic Characteristics

6.1.1 Cross-flow Model

Considering the question of the aerodynamic properties of wings of large chord relative to their span (small AR), Betz (1935) proposed that the aerodynamic force on such a wing is due mainly to the drag force which is caused by the normal component of the velocity in the cross-plane, as indicated in Fig. 6.2. In this case, the normal component of the velocity in the cross plane is $V' = V \sin\alpha$ and the normal force is

$$P = (1/2)\rho V'^2 S C_{D_c} = (1/2)\rho V^2 S C_{D_c} \sin^2\alpha \tag{6.1}$$

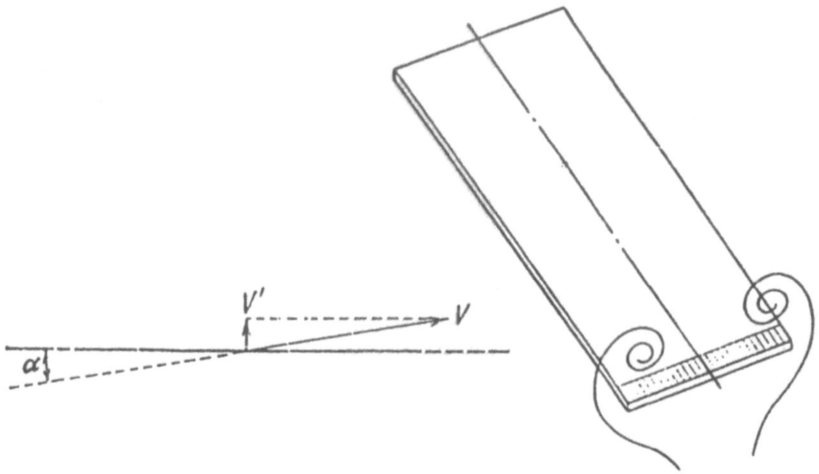

Fig. 6.2. Normal components of velocity and force in the cross plane (Betz 1935)

The lift force is $L = P\cos\alpha$. The drag coefficient for a flat rectangular wing positioned normal to the flow is $C_{D_c} \cong 2$. Using this value for the drag coefficient in Eq. 6.1, the lift coefficient for the low AR wing is

$$C_L = 2\sin^2\alpha\cos\alpha \tag{6.2}$$

It is interesting to note that this extremely simplified analysis gives a reasonable first order estimation for the nonlinear lift of small AR wings.

6.1.2 Nonlinear Lifting Surface Theory for Low AR Wings

The "classical" application of the lifting surface theory, based on the developments of Prandtl and Glauert, is in the calculation of the *linear* aerodynamic characteristics at small angles of attack of medium and high AR wings. Concepts of the lifting surface theory can also be used as a basis for the calculations of the *nonlinear* aerodynamic characteristics of low AR wings. An analytical evaluation of the nonlinear characteristics of rectangular wings of low AR was developed by Bollay (1937) and extended to wings of other low AR planform geometries by Gersten (1963).

Let us first consider a rectangular wing of small span and large chord, i.e. low AR wing. Following Bollay (1937), let us assume constant spanwise vorticity distribution so that γ is a function of x only, $\gamma(x)$, and in this case the downwash is constant in the spanwise direction. Then, let us assume that the trailing vortices leave the surface from the side edges of the wing with an angle θ measured from the wing surface, which is assumed to be constant, as seen in Fig. 6.3.

The downwash along the chord can be calculated for this system of vortices, where the value of γ at every chord position is equated to the value of the vorticity calculated along the chord of the zero AR wing. Applying the boundary condition on the wing, that the normal component of the flow velocity, $V_\infty \sin\alpha$, must be canceled at every point on the surface, results in an integral equation relating $\gamma(x)$, α and θ,

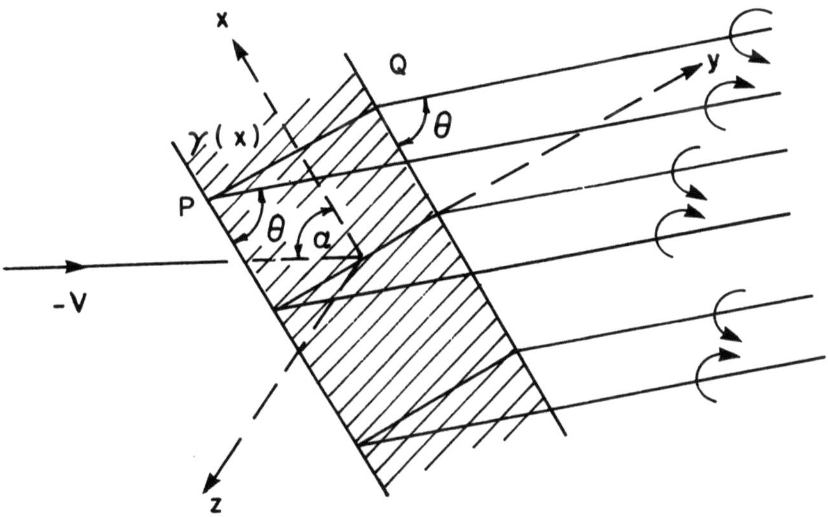

Fig. 6.3. Nonlinear lifting surface vortex model for low AR rectangular wing

$$V_\infty \sin\alpha = \frac{\mathrm{AR}}{2\pi}\left[\int_{-1}^{1} \frac{\gamma(\xi)}{(x-\xi)F(x,\xi)}\,d\xi + \int_{-1}^{1} \frac{\gamma(\xi)\cos\theta}{G(x,\xi)}\,d\xi - \int_{-1}^{1} \frac{(x-\xi)\gamma(\xi)\cos^2\theta}{F(x,\xi)G(x,\xi)}\,d\xi\right] \tag{6.3}$$

where

$$F(x,\xi) = [(\mathrm{AR})^2 + (x-\xi)^2]^{1/2}, \qquad G(x,\xi) = (\mathrm{AR})^2 + (x-\xi)^2 \sin^2\theta,$$

and x, ξ are normalized by $c/2$. In general, θ may vary along the chord due to the variation of the vorticity $\gamma(x)$. Because of the effects of the addition of their induced velocities, these free trailing vortices, shed at the angle θ (in general $0 < \theta < \alpha$), cause the nonlinear variation of the loading on the wing at increasing angles of attack. Bollay (1937) showed that the integral equation 6.3 can be solved when $\gamma(\xi)$ has the form given by Eq. 6.4, and the total lift of the low AR rectangular wing can be then calculated, using $\theta = \alpha/2$. Comparisons with experimental data show very high correlation for the range of validity of this analysis for AR$< 2/3$, as shown in Fig. 6.4. In the limiting case of AR=0, the lift force calculated by Bollay's analysis is equal to $C_L = 2\sin^2\alpha$, which is in agreement with the cross-flow calculation of Betz, as discussed in Section 6.1.1. In the linear lifting surface theory, the trailing vortices are forced to remain in the plane of the wing, $\theta = 0$; they leave the lifting surface only at the trailing edge and are required to stay as a vortex sheet wake in the plane of the wing. This flow model gives reasonable results for high AR wings at small angles of attack. The additional nonlinear lift observed on low AR rectangular wings can be attributed to the free trailing vortices shed from the side edges, as shown by Bollay (1937).

$$\gamma(\xi) = \gamma_o \left[\frac{1+\xi}{1-\xi}\right]^{1/2} \tag{6.4}$$

The application of the nonlinear lifting surface theory to wings with various planforms, of low or medium AR, was presented by Gersten (1961, 1963). In Gersten's analysis, the wing planform is substituted by a number of elementary wings of infinitesimal span, (dy), where the trailing vortices are allowed to leave the surface at an angle of $\alpha/2$, as indicated in Fig. 6.5.

The induced velocity due to this vortex model is calculated by Gersten (1961) to be

$$\begin{aligned} w(x,y) = &-\frac{1}{4\pi}\iint\limits_S \frac{\frac{d}{d\eta}\left[\gamma(\xi,\eta)\left(1+\frac{x-\xi}{R}\right)\right]}{y-\eta}\, d\eta\, d\xi - \\ &-\frac{1}{8}\frac{\alpha}{\mid\alpha\mid}\alpha\int_{x_{TE}}^{x_{LE}}\left(1+\frac{x-\xi}{\mid x-\xi\mid}\right)(x-\xi)\frac{\partial^2}{\partial y^2}[\gamma(\xi,y)]\, d\xi \end{aligned} \tag{6.5}$$

where $R = [(x-\xi)^2 + (y-\eta)^2]^{1/2}$, the first integral in Eq. 6.5 being identical to the linear value of the induced velocity, represented by $L_1(\gamma)$, and the second term in Eq. 6.5, indicating the effect of the free trailing vortices which cause the nonlinear variation at increasing angles of attack, represented by $L_2(\gamma)$.

It is now assumed that the induced angle of attack $\alpha_w = w/V_\infty$ can be expressed in the form

$$\alpha_w = L_1(\gamma) + \frac{\alpha}{\mid\alpha\mid}\alpha L_2(\gamma) + \ldots \tag{6.6}$$

In this case, γ can be expressed by the relation

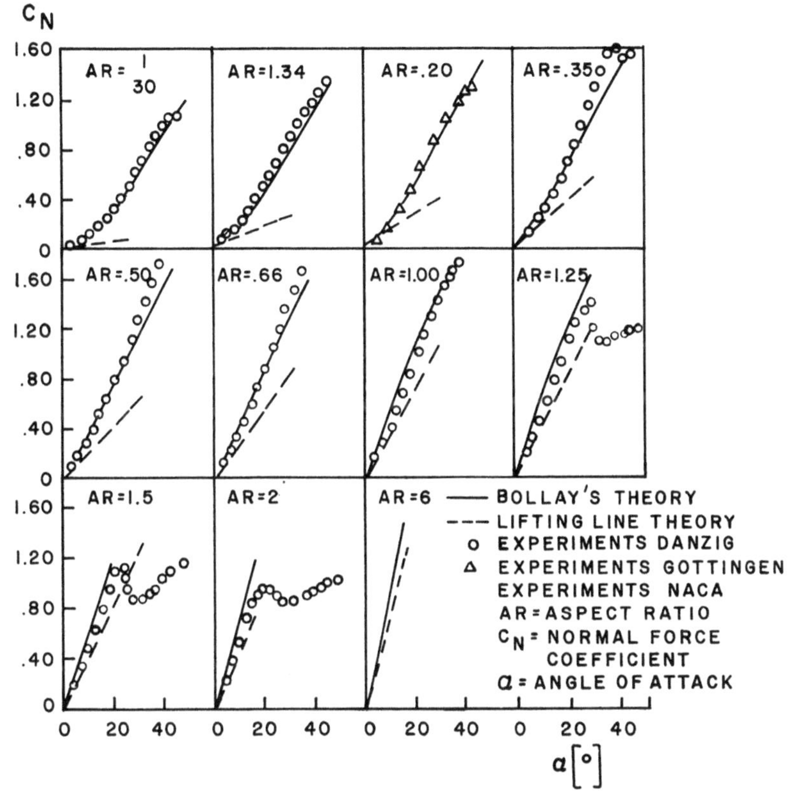

Fig. 6.4. Comparison of experimental results with the lift coefficients calculated by the nonlinear lifting surface theory for low AR rectangular wings (Bollay 1937)

(6.7)
$$\gamma = \gamma_1 \alpha + \gamma_2 \frac{\alpha}{|\alpha|} \alpha^2$$

The values of γ_1 and γ_2 are evaluated by the relations

(6.8)
$$L_1(\gamma_1) = \alpha_w / \alpha$$

(6.9)
$$L_2(\gamma_2) = -(\alpha_w' / \alpha) L_2(\gamma_1)$$

Eq. 6.8 is identical to the linear lifting-surface equation which can be solved for the value of γ_1. Using this value of γ_1, Eq. 6.9 can be solved for γ_2, while $\gamma(x, y)$ is evaluated by Eq. 6.7. Gersten (1961) used the linear lifting-surface theory of Truckenbrodt (1953) for the evaluation of the linear

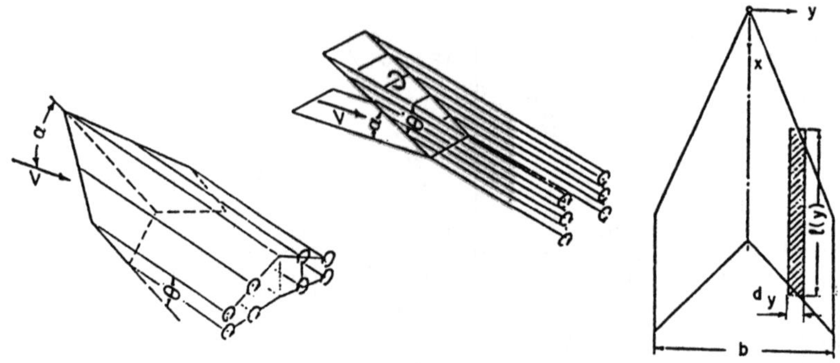

Fig. 6.5. The vortex model for the nonlinear lifting surface theory of Gersten (1961, 1963)

terms in his method. A different approach is investigated by Garner and Lehrian (1963) where the linear lifting-surface theory of Multhopp (1950) is introduced into the nonlinear method of Gersten. The lift and pitching moment coefficients are then calculated by

$$C_L = \frac{2}{V_\infty A_W} \iint \gamma(x, y)\, dx\, dy \tag{6.10}$$

$$C_m = -\frac{2}{V_\infty \bar{c} A_W} \iint x\gamma(x, y)\, dx\, dy \tag{6.11}$$

$$C_{Di} = C_L \alpha \tag{6.12}$$

The nonlinear aerodynamic coefficients can be expressed by the relations

$$C_L = C_{L_\alpha}|_{\text{linear}}\, \alpha + C_1 \frac{\alpha}{|\alpha|} \alpha^2 \tag{6.13}$$

$$C_m = C_{m_\alpha}|_{\text{linear}}\, \alpha + C_2 \frac{\alpha}{|\alpha|} \alpha^2 \tag{6.14}$$

where

$$C_{L_\alpha}|_{\text{linear}} \text{ and } C_{m_\alpha}|_{\text{linear}}$$

are calculated by the linear theory and the functions C_1 and C_2, calculated from the nonlinear terms in Eqs. 6.7 to 6.11 for rectangular, delta and swept back wings, are shown in Fig. 6.6.

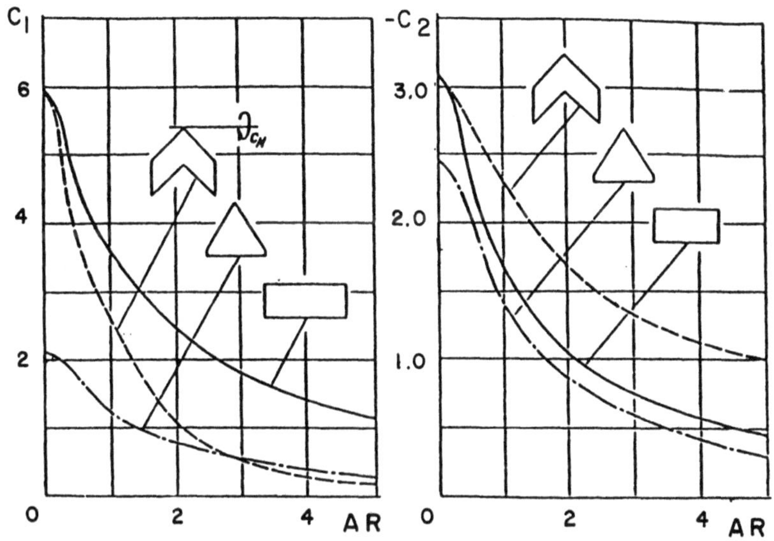

Fig. 6.6. The functions C_1 and C_2 for wings of various planform shapes (Gersten 1961)

The results of the calculations of the aerodynamic coefficients using this nonlinear lifting-surface method are compared with experimental data for a cropped delta and a swept back wing, both with low AR, and show very good agreement, as seen in Fig. 6.7. The above results are very important for the understanding of the nonlinear behavior of slender configurations at high angles of attack. It is found that allowing the trailing vortices to separate from the side edges (Bollay 1937 model) or leave the surface continuously (Gersten 1961 model) causes additional induced velocities on the wing, which result in additional lifting force as the angle of attack increases – i.e. the nonlinear effect. The additional nonlinear lift force results from the fact that the free vortices' trajectories flow over the wing, and their induced velocities on the surface are relatively large. Therefore, the magnitude of the additional lift depends on the shape and position of these free-trailing vortices. The models of Bollay and Gersten assume a predetermined vortex trajectories. Bollay (1937) assumes straight lines leaving the side edges at an angle of $\alpha/2$ relative to the wing; Gersten (1961) extended this model to the uniform separation of straight line vortices from the entire surface at a local angle of $\alpha/2$. The effect of the different relative angles of the straight line vortex trajectories was studied by Ermolenko (1966) and by Ermolenko and Barinov (1966), who found that the value of $\alpha/2$ for the inclination of the vortices relative to the wing is about the average of the local inclination of the shed vortices along the chord of the wing. It is important to note that these analyses clearly show that the nonlinear variation of the lift force with increasing angles of attack is due to the velocities induced by the free vortices that are released from the surface and side edges and flow over the wing.

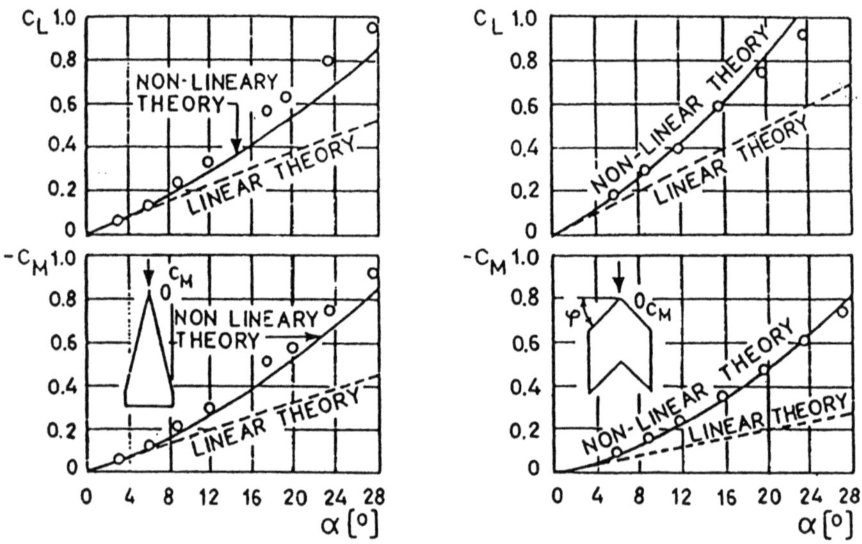

a. Cropped delta wing, AR=0.78, taper ratio=0.125

b. Swept back wing, AR=1.0, sweep angle = 45°

Fig. 6.7. Comparisons of calculations of the lift and pitching moment coefficients using the nonlinear lifting surface theory with experimental data (Gersten 1961)

6.1.3 Leading-Edge Vortex Models for Slender Delta Wings

The experimental observation of the rolled-up vortices above the leading edges of slender delta wings (for example, Fig. 2.8) led to investigations of simplified models of such flows. These models utilize the linearized conical flow model for slender delta wings of R.T. Jones (1946) and the concept of Legendre (1952) to add concentrated line vortices into the flow above the wing.

6.1.3.1 Concentrated line vortices with plane feeding sheet. The model proposed by Brown and Michael (1955) is illustrated in Fig. 6.8. The compressible linearized three-dimensional perturbation potential equation $(1 - M^2)\phi_{xx} + \phi_{yy} + \phi_{zz} = 0$, is reduced, in the case of slender configurations, to the two-dimensional Laplace equation in the cross-flow plane, $\phi_{yy} + \phi_{zz} = 0$, since $(1 - M^2)\phi_{xx} \ll \phi_{yy}$ or ϕ_{zz}. In the case of the flow over a slender delta wing with leading-edge separation, it can be expected that the separated layer at the sharp leading-edges will roll up into a spiral vortex above the wing, as illustrated in Fig. 6.8a. For such a flow, the boundary conditions should be no-flow through the wing surface, with the pressure across the free rolled-up vortex sheet being continuous. A solution of this problem is difficult since it requires providing both the shape and the strength of the free vortex sheet. A possible simplifying assumption is used by Brown and Michael (1955) where the rolled-up spiral sheet is replaced by a concentrated vortex near the spiral center. For conical flow, the net vorticity in the free vortex linearly increases in the flow direction; therefore, the strength of the concentrated vortex must also increase linearly in the chord direction. This increase in strength can be accomplished by assuming a plane feeding vortex sheet, an assumption satisfying Kelvin's theorem, between the leading edge and the concentrated line vortex, as shown in Fig. 6.8b. This flow model is amenable to simplified analysis for the calculations of the aerodynamic forces and pressure distributions on the slender delta wing.

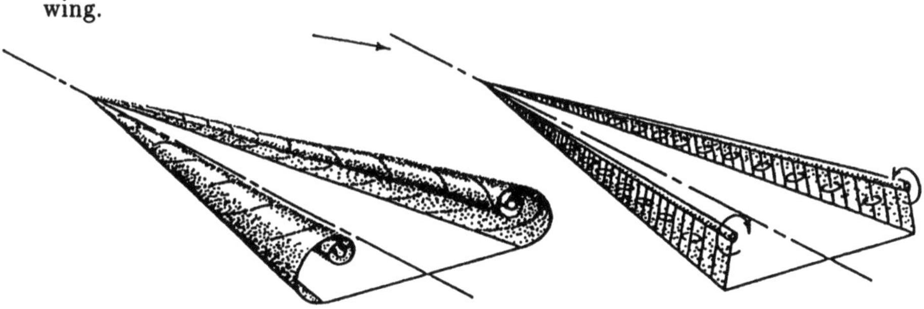

a. Assumed flow field b. Approximated flow field

Fig. 6.8. Flow model for the concentrated vortices over a slender delta wing according to Brown and Michael (1955)

The solution of the Laplace Equation in the cross-flow plane (the y-z plane), presented by the model of Fig. 6.8b, can be obtained for the following boundary conditions: on the wing, the normal velocity must be zero due to the no-flow condition, and the flow must separate tangentially at the wing's leading-edges. In the far field, the disturbances must vanish at infinity, and the fluid pressure must be continuous everywhere in the field. It is, therefore, assumed that the forces on the feeding vortex sheet be canceled by the equal but opposite forces on the concentrated vortex. That is to say, the net force on the combination of the feeding vortex sheet and the concentrated vortex is equal to zero. The solution is obtained by using conformal mapping of the flow over a plate with two symmetrically placed vortices of equal but opposite strength, as shown in Fig. 6.9.

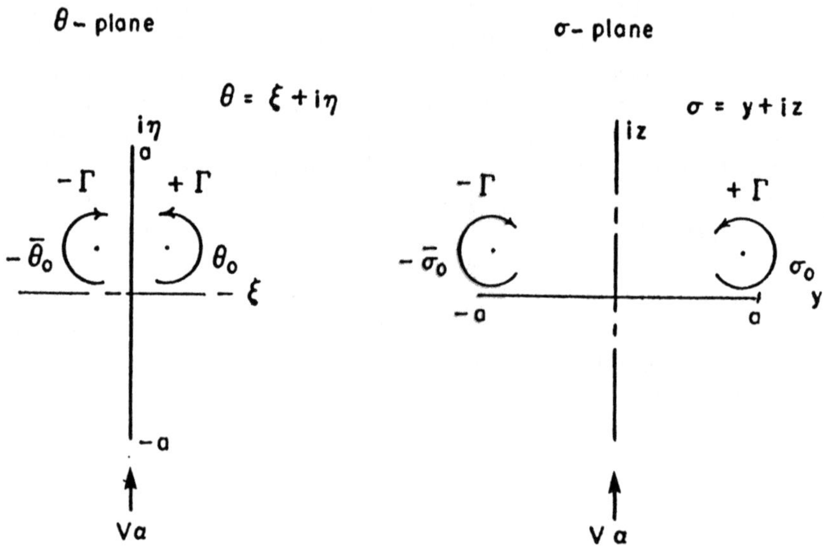

Fig. 6.9. The conformal transformation coordinates and notations

The complex velocity $W(\theta)$ is then

(6.15)
$$W(\theta) = -\frac{i\Gamma}{2\pi}\log_e\frac{\theta-\theta_o}{\theta+\bar{\theta}_o} - iV\alpha\theta$$

In the σ plane

(6.16)
$$W(\sigma) = -\frac{i\Gamma}{2\pi}\log_e\frac{(\sigma^2-a^2)^{1/2}-(\sigma_o^2-a^2)^{1/2}}{(\sigma^2-a^2)^{1/2}+(\bar{\sigma}^2-a_o^2)^{1/2}} - iV\alpha(\sigma^2-a^2)^{1/2}$$

Applying the boundary conditions, the equation for the complex position of the vortex σ_o can be written

(6.17)
$$V\varepsilon\left(\frac{2\bar{\sigma}_o}{a}-1\right) = \frac{i\Gamma}{2\pi}\left[\frac{\sigma_o}{\sigma_o^2-a^2+[(\sigma_o^2-a^2)(\bar{\sigma}_o^2-a^2)]^{1/2}} - \right.$$
$$\left. - \frac{\sigma_o}{[(\sigma_o^2-a^2)(\bar{\sigma}_o^2-a^2)]^{1/2}} - \frac{\sigma_o}{\sigma_o^2-a^2} + \frac{a^2}{2\sigma_o(\sigma_o^2-a^2)}\right]$$

The position of the vortex, $\sigma_o = y_o + iz_o$ can be calculated by the numerical solution of Eq. 6.17. An approximate solution for the vortex position is estimated assuming that $(\sigma_o - a)/a \ll 1$, then

$$\frac{y_o - a}{a} = \frac{1}{2}\left(\frac{z_o}{a}\right)^{2/3}\left[1 - \frac{3}{4}\left(\frac{z_o}{a}\right)^{2/3}\right] \tag{6.18}$$

The calculated positions of the vortex by the numerical solution of Eq. 6.17 and by the linearized approximation of Eq. 6.18 are presented in Fig. 6.10a, and the variation of the vertical position z_o with increasing angles of attack is presented in Fig. 6.10b. The approximate relation between the vertical position of the vortex and the angle of attack is

$$\frac{\alpha}{\varepsilon} = 4\frac{z_o}{a}\left[1 + \frac{1}{2}\left(\frac{z_o}{a}\right)^{2/3}\right] \tag{6.19}$$

The lift coefficient is calculated to second order in $(\sigma - a)/a$ to be

$$\frac{C_L}{\varepsilon^2} = \frac{2\pi\alpha}{\varepsilon} + 16\pi\left(\frac{\alpha}{4\varepsilon}\right)^{5/3}\left[1 + \frac{2}{3}\left(\frac{\alpha}{4\varepsilon}\right)^{2/3}\right] \tag{6.20}$$

Comparison of the calculated lift coefficients with experimental measurements on various slender delta wings is presented in Fig. 6.11. The method is capable of calculating the pressure distributions and the moments acting on the wing as well. It is seen that this method of calculation overestimates the nonlinear effect and that the overestimation increases as the AR increases.

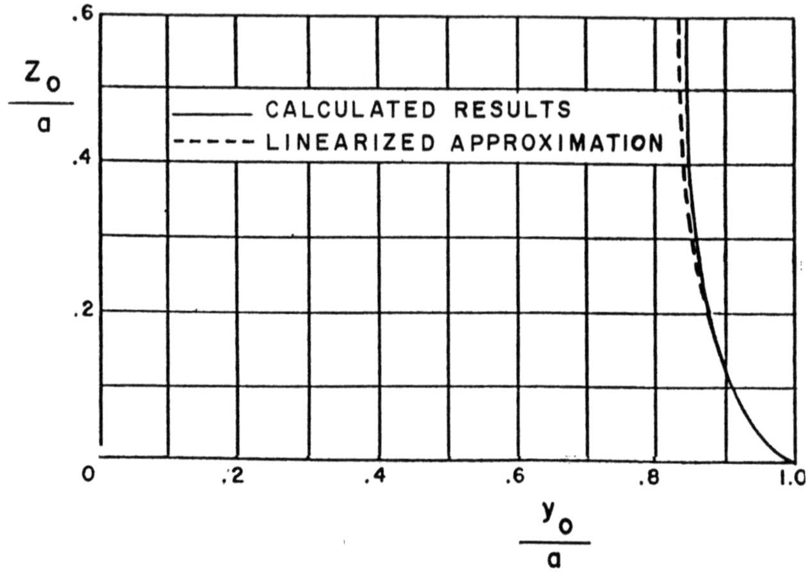

Fig. 6.10a. Calculations of vortex positions – lateral vortex positions (Brown and Michael 1955)

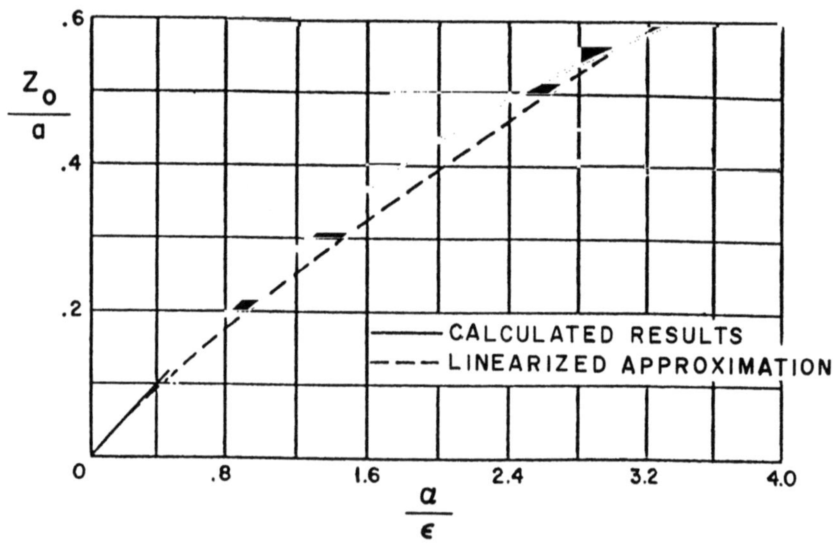

Fig. 6.10b. Variation of vortex position with increasing angles of attack (Brown and Michael 1955)

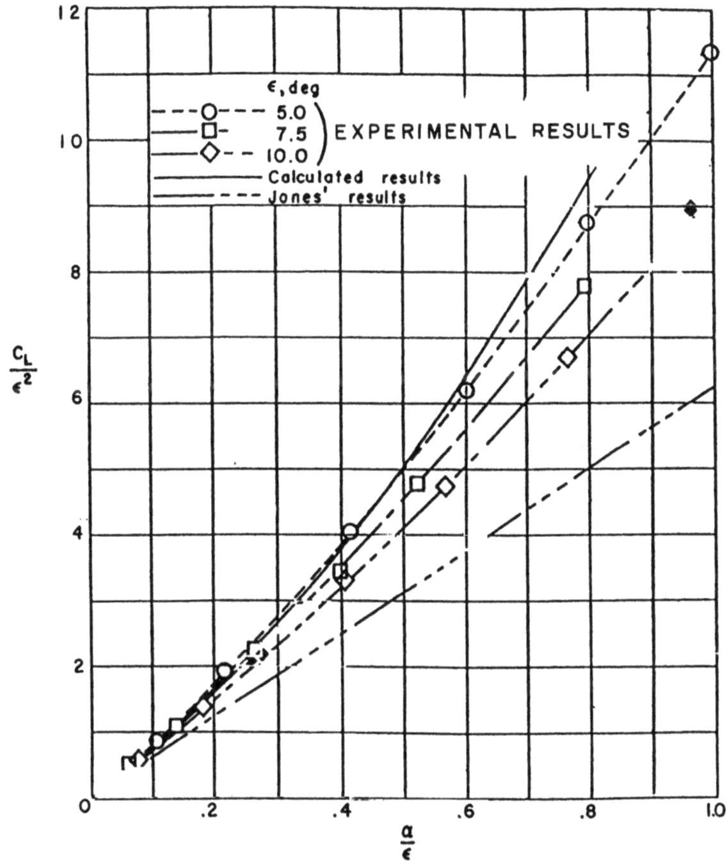

Fig. 6.11. Comparison of calculated and experimental results for the lift coefficients on delta wings (Brown and Michael 1955)

6.1.3.2 The Spiral Vortex Sheet. It may be expected that the simplified method of Brown and Michael (1955) can be improved by allowing the vortex sheet, which is separating from the leading-edge of the delta wing, to form the shape of a spiral vortex sheet. The calculation method for a slender delta wing with a spiral vortex sheet was presented by Mangler and Smith (1959). The flow over the slender delta wing is assumed to be conical, so that the velocity potential function satisfies the Laplace equation, $\nabla^2\Phi = 0$, in the y-z cross-flow plane. A conical stream surface, such as the wing and/or the vortex sheet, is defined in polar coordinates, Fig. 6.12, as

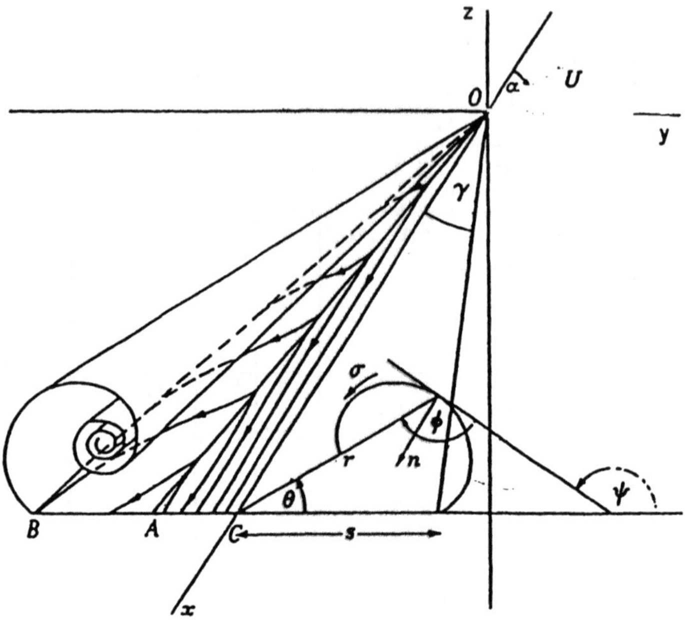

Fig. 6.12. The coordinates of the delta wing with the spiral vortex sheet (Mangler and Smith 1959)

$$S(x, r, \theta) = r - xf(\theta) = 0 \tag{6.21}$$

On this surface, the normal velocity component is then,

$$\frac{\partial \Phi}{\partial n} = -V_\infty \frac{r}{x} \sin\phi = -\varepsilon V_\infty \frac{r}{s} \sin\phi \tag{6.22}$$

On the wing, $\phi = 0$ or π, therefore $\frac{\partial \Phi}{\partial n} = 0$, fulfills the boundary condition there. On the vortex sheet, the pressure must be continuous, therefore, $\Delta C_p = 0$, and the condition on the potential function is then

$$\Delta\Phi = \Delta\left(\frac{\partial \Phi}{\partial \sigma}\right)\left[r\cos\phi - \frac{s}{\varepsilon V_\infty}\left(\frac{\partial \Phi}{\partial \sigma}\right)_m\right] \tag{6.23}$$

In the crossflow plane, this problem can be looked on as corresponding to the two-dimensional, time-dependent problem; thus Eqs. 6.22 and 6.23 can be considered as the boundary conditions which must be satisfied on the vortex sheet which expands from its origin steadily with time in two-dimensional flow.

It is possible to solve this problem for the shape of the spiral vortex sheet and for the pressure distribution on the wing and to calculate the nonlinear lift coefficients by the method presented by Mangler and Smith (1959). Comparison of the calculated normal force coefficients by the method of Mangler and Smith (1959) with experimental data and also with the results of the calculations of Brown and Michael (1955) are presented in Fig. 6.13. The conical flow assumption requires that the normal force coefficient be proportional to ε^{-2} as indicated by the calculated curves in Fig. 6.13. The experimental measurements of the lift coefficients of delta wings of various AR are scattered over this figure showing that the conical parameter does not correlate well the experimental data. In this figure the results of the calculations of Mangler and Smith (1959) fall within the scatter of the data, while the values of Brown and Michael (1955) are too high.

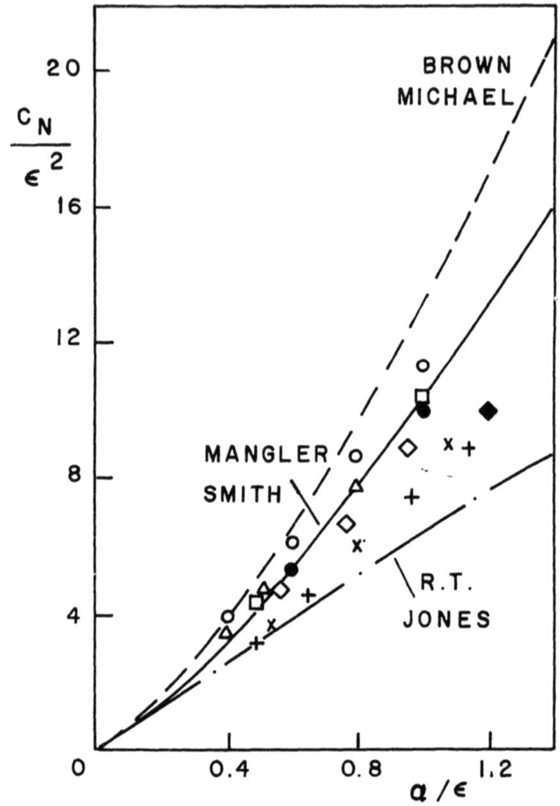

Fig. 6.13. Comparison of the calculated normal force coefficients by the Mangler-Smith method with experimental data and with results of the Brown-Michael method (Mangler and Smith 1959)

6.1.4 Leading-Edge Suction Analogy

This analysis, developed by Polhamus (1966), is based on the assumption that the total lift can be calculated as a sum of a potential-flow lift together with the lift associated with the existence of the separated leading-edge spiral vortices. The potential flow lift is modified to include the effects of high angles of attack and leading-edge conditions. As a result, the vortex lift is determined by a method in which the vortex flow is assumed to be related to the potential flow about the leading-edge.

The potential lift calculation on a wing having sharp leading-edges must be modified for the effect of this sharp edge at high angles of attack. In order to account for the leading-edge separation, a Kutta type flow condition is assumed to exist at the sharp leading-edge so that no leading-edge suction can be developed. Since the flow reattaches downstream of the separation vortex, the potential flow lift is diminished only by the loss of the lift component of the leading-edge suction. Therefore, for this condition of zero leading-edge suction, the resultant force (neglecting friction drag) for the planar wings is perpendicular to the wing chord and is equal to the potential flow normal force. Thus, the potential flow lift coefficient is

$$C_{L,p} = C_{N,p} \cos\alpha \tag{6.24}$$

The normal force, evaluated by using the Kutta-Joukowski theorem with the velocity component parallel to the wing chord, is

$$N = \rho \Gamma b (V_\infty \cos\alpha) \tag{6.25}$$

where Γ is the total effective circulation, which is

$$\Gamma = K_p \frac{S V_\infty}{2b} \sin\alpha \tag{6.26}$$

It is assumed that K_p depends only on the planform shape, thus

$$C_{N,p} = K_p \sin\alpha \cos\alpha \tag{6.27}$$

and using equation 6.24,

$$C_{L,p} = K_p \sin\alpha \cos^2\alpha \tag{6.28}$$

For small angles of attack, Eq. 6.28 reduces to $C_{L,p} = K_p \alpha$ so that K_p can be identified with the lift curve slope at small angles of attack. Therefore, the values of K_p for the study of the aerodynamics of delta wings can be evaluated from any suitable linear lifting-surface theory. The variation of the values of K_p as a function of the Aspect Ratio of delta wings, calculated by the modified Multhopp lifting-surface theory, is presented in Fig. 6.14.

The vortex lift is evaluated utilizing the leading-edge flow model proposed by Polhamus (1966), where the force required to maintain the equilibrium of the flow over the separated spiral leading edge vortex (provided that the flow reattaches on the upper surface) is equated with the force associated with the theoretical leading-edge singularity for thin wings in potential flow. The assumed leading edge flow model and the associated forces are shown in Fig. 6.15.

This flow model indicates that for attached flow conditions, flow accelerates around the leading-edge and develops the leading-edge suction force C_s. As the leading-edge radius is increased, the potential flow suction force remains essentially constant as suction pressures vary inversely to the leading-edge radius. It is further assumed that for the sharp leading-edge condition, the force acting on the wing will act primarily over the upper surface of the wing, rather than on the leading-edge itself. Therefore, the assumed vortex lift force is equal to the theoretical suction force of the

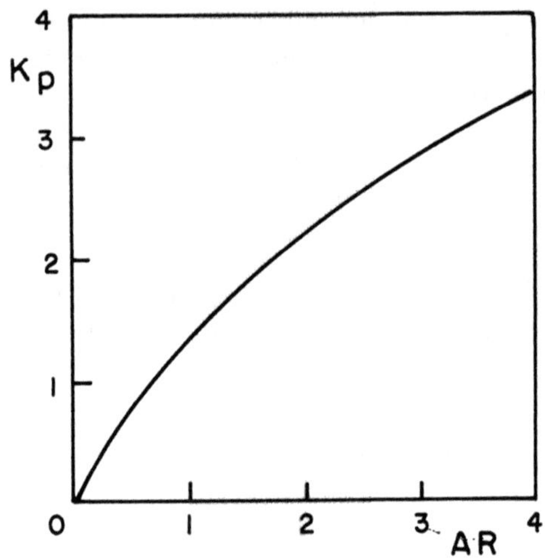

Fig. 6.14. Variation of K_P as a function of AR for delta wings (Polhamus 1966)

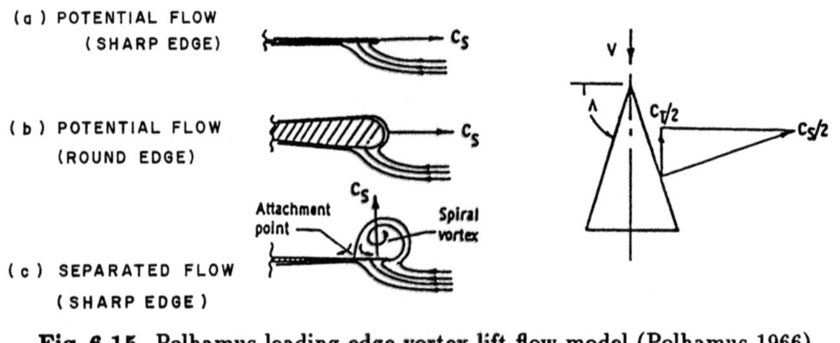

Fig. 6.15. Polhamus leading-edge vortex lift flow model (Polhamus 1966)

leading-edge. As the theoretical leading-edge suction force is essentially independent of the leading-edge radius, the normal force associated with the separated spiral vortex sheet should be equivalent to the leading-edge suction force, as calculated by an appropriate thin wing linear lifting-surface theory. This suction force is perpendicular to the leading-edge and only the component parallel to the direction of flight contributes to the vortex lift. This component of the leading-edge suction in the flight direction is known as the leading-edge thrust coefficient C_T, as indicated in Fig. 6.15. Accordingly, the normal force coefficient associated with the leading-edge vortex is given by

$$C_{L,v} = C_{N,v} \cos\alpha = C_s \cos\alpha = \frac{C_T \cos\alpha}{\cos\Lambda} \tag{6.29}$$

The value of C_T, which is associated with the potential flow around the leading-edge, is calculated by Polhamus (1966), and it is expressed in the terms of the functions $K_p = \partial C_L/\partial\alpha$ and $K_i = \partial C_{Di}/\partial C_L^2$ as follows

$$C_{L,v} = (K_p - K_p^2 K_i)\frac{\cos\alpha \sin^2\alpha}{\cos\Lambda} = K_v \cos\alpha \sin^2\alpha \tag{6.30}$$

where $K_v = (K_p - K_p^2 K_i)/\cos\Lambda$, and the calculated variation of K_v for sharp edge delta wings of varying AR is shown in Fig. 6.16.

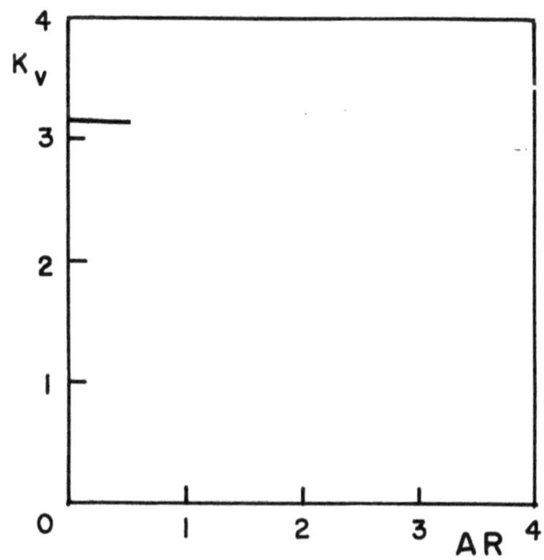

Fig. 6.16. Variation of K_v as a function of the aspect ratio for sharp edge delta wings (Polhamus 1966)

The total lift on the sharp leading-edge delta wing is obtained by the addition of potential lift (Eq. 28) and the vortex lift (Eq. 30). Therefore, the nonlinear lift coefficient for the delta wing with sharp leading-edges can be given as

$$C_L = K_p \sin\alpha \cos^2\alpha + K_v \cos\alpha \sin^2\alpha \tag{6.31}$$

The values of K_p and of K_v are obtained from Figs. 6.14 and 6.16, respectively. The lift coefficients determined by the leading-edge suction analogy due to Polhamus, are found to compare very well with experimental data for sharp edge delta wings of Aspect Ratios from 0.5 to 2.0, as shown in Fig. 6.17. These results clearly reveal the nonlinear contribution due to the vortex lift. The portion of the vortex lift increases as the Aspect Ratio decreases. The Polhamus theory predicts the lift coefficient quite well for low Aspect Ratio wings; but this agreement deteriorates as the Aspect Ratio increases beyond 2.0.

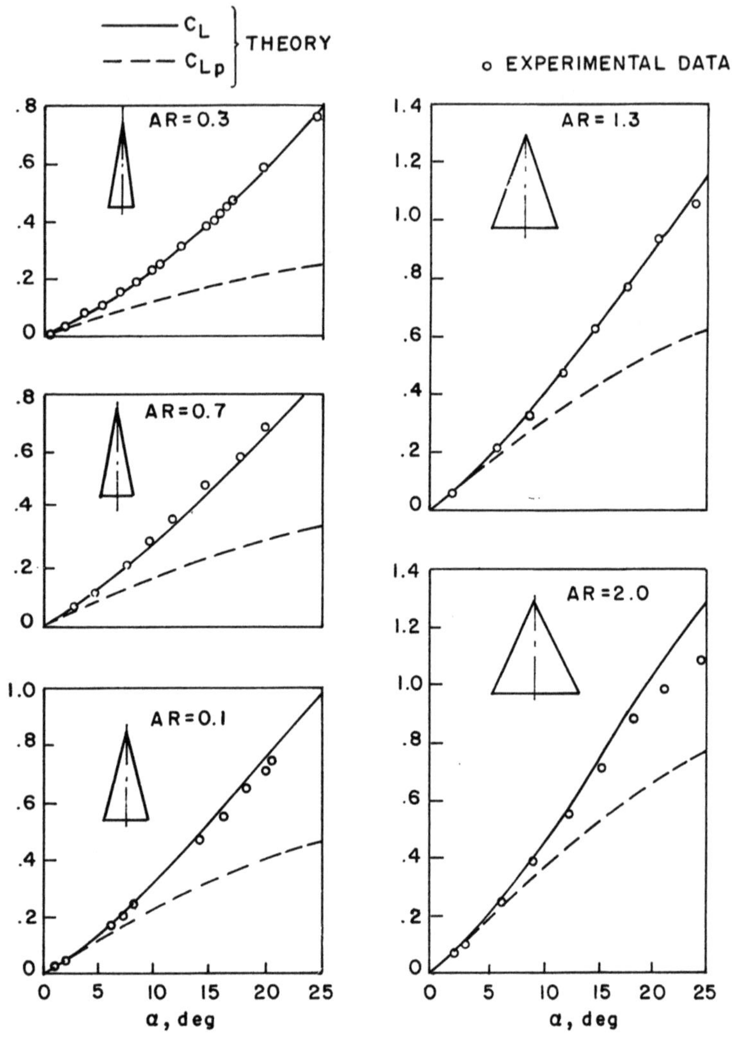

Fig. 6.17. Comparison of Polhamus theory with experimental data for sharp edge delta wings of AR=0.5 to 2.0

The induced drag is given by the relation $C_{D_i} = C_L \tan\alpha$. Therefore, the induced drag is

$$C_{D_i} = K_p \sin^2\alpha\cos\alpha + K_v \sin^3\alpha \tag{6.32}$$

These equations for calculating induced drag compare well with experimental results measured on low AR sharp edge delta wings.

6.1.5 Semi-Empirical Correlation for Nonlinear Lift

Various empirical and semi-empirical relations correlating experimental measurements of the lift coefficient on low Aspect Ratio wings of various planform geometries have been developed. Most are based on formulas obtained by adding an exponent n variation of the lift coefficient with angle of attack to the linear term, in the form given by

$$C_L = A_1\alpha + A_2\alpha^n \tag{6.33}$$

A_1 can be evaluated to equal $(\pi/2)$ AR from the theory for low AR wings, while A_2 and n must be evaluated from correlation with experimental results. Edwards (1954) proposed the relation

$$C_L = (\pi/2)\mathrm{AR}\ \alpha + \pi(\mathrm{AR})^{1/3}\alpha^{5/3} \tag{6.34}$$

The values of $A_2 = 4$ and $n = 2$ were obtained using the analysis of Mangler and Smith (1959). The calculated lift coefficient using these values in Eq. 6.35, is much too high when compared with experimental results, and is

$$C_L = (\pi/2)\ \mathrm{AR}\ \alpha + 4\alpha^2 \tag{6.35}$$

Better agreement with experimental data is obtained by using the formula first developed by Lawrence and Flax in 1951 and later adopted by Kuchemann (1962)

$$C_L = (\pi/2)\ \mathrm{AR}\ \alpha + 2\alpha^2 \tag{6.36}$$

An empirical relationship was developed by Alexander (1971) using the measurements of Peckham (1961), Kirby (1969), Hummel and Srinvasan (1967) and Bartlett and Vidal (1955) to correlate the constants in the equation given by

$$C_L = 1.23(\mathrm{AR})^{0.77}\alpha + 1.59(\mathrm{AR})^{0.60}\alpha^{1.74} \tag{6.37}$$

Poisson-Quinton and Erlich (1965) proposed an equation which provides a good correlation for wings with $AR < 0.5$

$$C_L = 0.915\frac{4.9}{4.9 + AR}\left(\frac{\pi}{2}\ \mathrm{AR}\ \alpha + \pi(\mathrm{AR})^{1/3}\alpha^{5/3}\right) \tag{6.38}$$

Bergessen and Porter (1960) developed a correlating equation for wings of AR>1, including the effect of wing thickness ratio, t/c, which gives reasonable results

$$C_L = \frac{2\pi\ \mathrm{AR}}{\frac{P^* AR}{b} + 2}\alpha + 0.0995\bar{\alpha}^2 - 0.0146\bar{\alpha} - (0.529\alpha - 0.034)(t/c)^{1/2} \tag{6.39}$$

where P^* is half of the wing circumference and $\bar{\alpha}$ is α/ε.

Figures 6.18a to 6.18d present the lift coefficients evaluated by these equations as well as those evaluated by the various nonlinear methods discussed in this chapter with experimental data on sharp edge delta wings of AR=0.5, 1.0, 1.5 and 2.0 for angles of attack up to 30°.

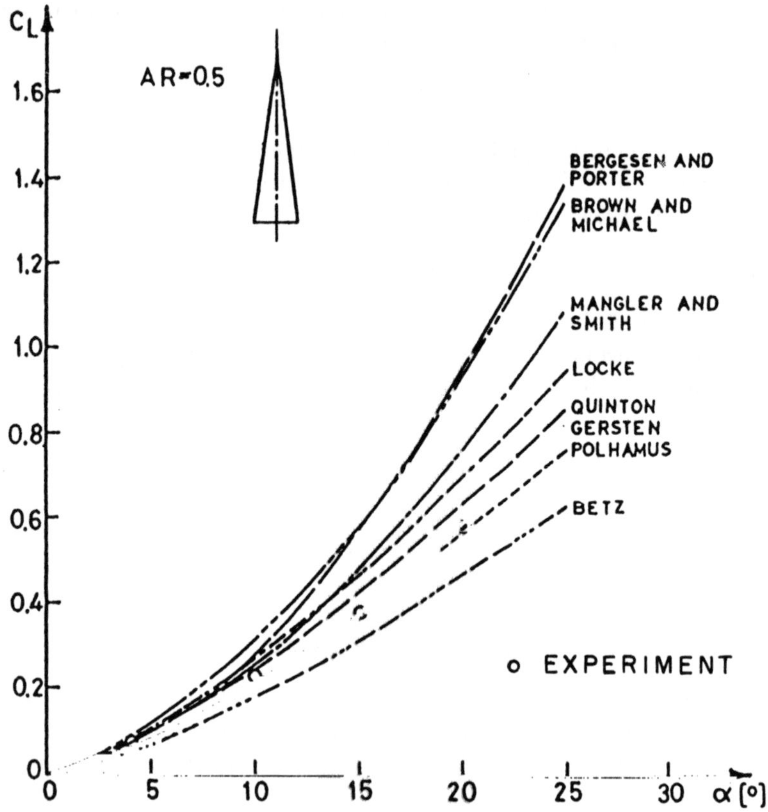

Fig. 6.18a. Comparison of calculated lift coefficient with experimental data for sharp edge delta wings of AR=0.5

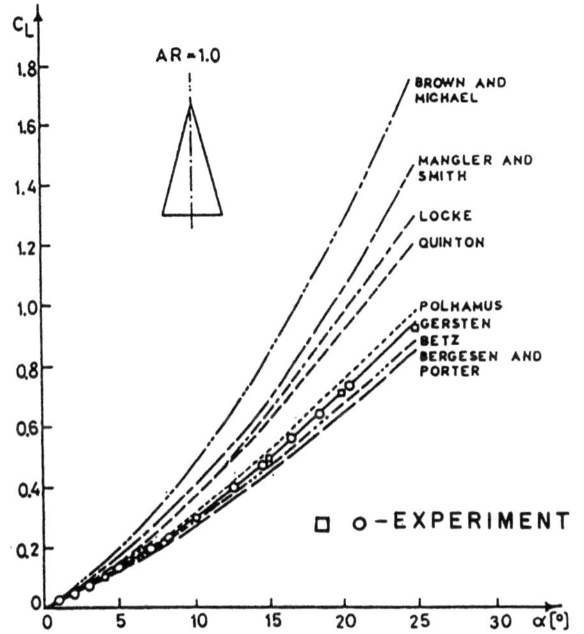

Fig. 6.18b. Comparison of calculated lift coefficient with experimental data for sharp edge delta wings of AR=1.0

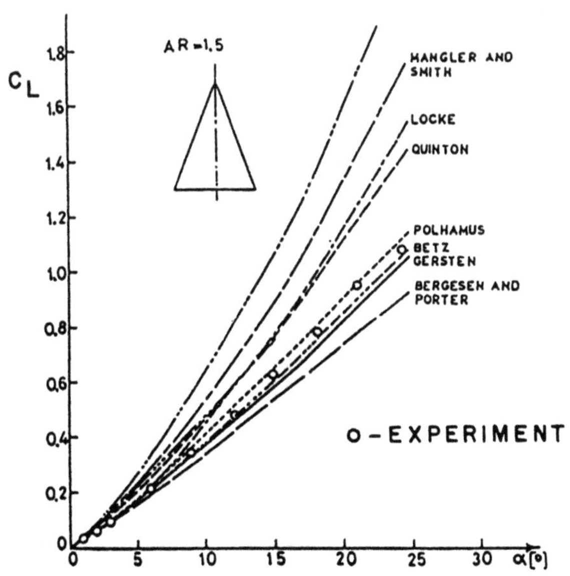

Fig. 6.18c. Comparison of calculated lift coefficient with experimental data for sharp edge delta wings of AR=1.5

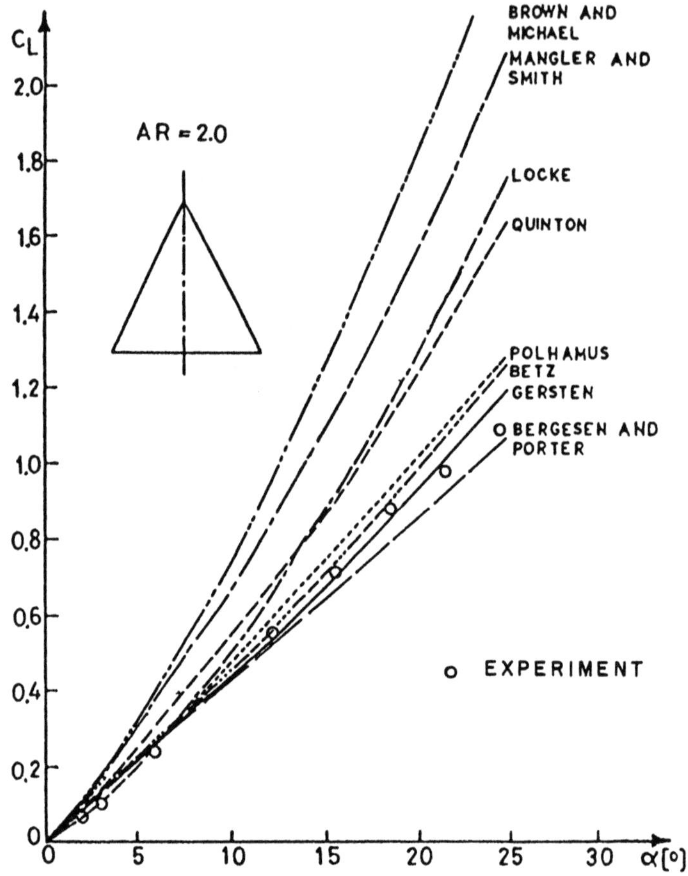

Fig. 6.18d. Comparison of calculated lift coefficients with experimental data for sharp edge delta wings of AR=2.0

References

1. Alexander, A.J. (1971), "Correlation of Total Lift Data for Thin, Sharp-Edged, Low Aspect Ratio Delta Wings at Low Speeds", The Aero. Journal, Royal Aeronautical Society, pp. 569–570.

2. Barlett, G.E. and Vidal, R.J. (1955), "Experimental Investigation of Influence of Edge Shape on the Aerodynamic Characteristics of Low Aspect Ratio Wings at Low Speeds", Journal of the Aeronautical Sciences, Vol. 22, No. 8, pp. 517–533.

3. Bergesen, A.J. and Porter, J.D. (1960), "An Investigation of the Flow Round Slender Delta Wings with Leading Edge Separation", O.N.R. Report No. 510, Princeton University.

4. Betz, A. (1935), "Applied Airfoil Theory", in *Aerodynamic Theory*, Vol. IV, Durand, W.F. (editor), pp. 69–72.

5. Bollay, W. (1937), "A Theory for Rectangular Wings of Small Aspect Ratio", Journal of Aeronautical Sciences, Vol. 4, pp. 294–296.

6. Bollay, W. (1939), "A Non-Linear Wing Theory and its Application to Rectangular Wings of Small Aspect Ratio", Z. Angew. Math. Mech., Vol. 19, pp. 21–35.

7. Brown, C. E. and Michael, W.H. Jr. (1955), "On Slender Delta Wings with Leading-Edge Separation", NACA TN 3430.

8. Edwards, R.H. (1954), "Leading Edge Separation from Delta Wings", Journal of Aeronautical Sciences, Vol. 21, No. 2, pp. 134–135.

9. Ermolenko, S.D. (1966), "Non-Linear Theory of Small Aspect Ratio Wings", Reports of Higher Educational Institutes, Technic of Aviation, No. 2, pp. 9–18 (in Russian).

10. Ermolenko, S.D. and Barinov, M.T. (1966), "Calculation of the Stream Deflection Angles Behind Rectangular Wings of Small Aspect Ratio at Subsonic Speeds Using Non-Linear Theory", Reports of Higher Educational Institutes, Technic of Aviation, No. 3, pp. 3–10 (in Russian).

11. Flax, A.H. and Lawrence, H.R. (1951), "The Aerodynamics of Low Aspect Ratio Wings and Wing-Body Combinations", Third Anglo American Aeronautical Conference.

12. Garner, H.C. and Lehrian, D.E. (1963), "Non-Linear Theory of Steady Forces on Wings with Leading-Edge Flow Separation", A.R.C. R&M No. 3375.

13. Gersten, K. (1961), "Nichtlineare Tragflächentheorie insbesondere für Tragflügel mit kleinem Seitenverhältnis", Ingenieur Archiv, Vol. 30, pp. 431–452.

14. Gersten, K. (1963), "A Non-Linear Lifting Surface Theory Specially for Low Aspect Ratio Wings", AIAA Journal, Vol. 1, No. 4, pp. 924–925.

15. Hummel, D. and Srinivasan, P.S. (1967), "Vortex Breakdown Effects on the Low Speed Aerodynamic Characteristics of Slender Delta Wings In Symmetrical Flows", Journal of the Royal Aeronautical Society, Vol. 71, No. 676, pp. 319–322.

16. Kirby, D.A. (1969), "An Experimental Investigation of the Effect of Planform Shape on the Subsonic Longitudinal Stability Characteristics of Slender Wings", A.R.C. R&M 3568.

17. Kuchemann, D. (1962), "Aircraft Shapes and their Aerodynamics for Flight at Supersonic Speeds", *Advances in Aeronautical Sciences*, Vol. 3, Pergamon Press.

18. Legendre, R. (1952), "Ecoulement au Voisinage de la Pointe avant d'une ail à forte flèche aux incidences moyennes", ONERA Rech. Aeron., No. 30, Nov.-Dèc., pp. 3–8, and No. 31, Jan.-Fèv., 1953, pp. 3–6.

19. Locke, F.W.S. Jr. (1949), "An Empirical Study of Low Aspect Ratio Lifting Surfaces with Particular Regard to Planing Craft", Journal of the Aeronautical Sciences, Vol. 16, No. 3, pp. 184–188.

20. Mangler, K.W. and Smith, J.H.B. (1959), "A Theory of Slender Wings with Leading Edge Separation", Proc. Royal Soc. London, Series A, Vol. 251, pp. 200–217.

21. Multhopp, H. (1950), "Methods for Calculating the Lift Distribution of Wings (Subsonic Lifting-Surface Theory)", A.R.C. R&M No. 2884.

22. Peckham, D.H. (1961), "Low Speed Wind Tunnel Tests on a Series of Uncambered Slender Pointed Wings with Sharp Edges", British A.R.C. R&M No. 3186.

23. Polhamus, E.G. (1966), "A Concept of the Vortex Lift of Sharp Edge Delta Wings Based on a Leading Edge Suction Analogy", NASA TN-D-3767.

24. Polhamus, E.G. (1968), "Application of the Leading Edge Suction Analogy of Vortex Lift to the Drag Due to Lift of Sharp Edge Delta Wing", NASA TN-D-4739.

25. Polhamus, E.G. (1971), "Prediction of Vortex Lift Characteristics by a Leading Edge Suction Analogy", Journal of Aircraft, Vol. 8, No. 4, pp. 193–199.

26. Poisson-Quinton, Ph. and Erlich, E. (1965), "Hyper-Lift and Balancing of Slender Wings", NASA TT F-9523.

27. Smith, J.H.B. (1968), "Improved Calculations of Leading-Edge Separation from Slender, Thin, Delta Wings", Proc. Roy. Soc. A, Vol. 306, pp. 67–90.

28. Tosti, L.P. (1947), "Low Speed Static Stability and Damping in Roll Characteristics of some Swept and Unswept Low Aspect Ratio Wings", NACA TN 1468.

29. Truckenbrodt, E. (1953), "Tragflächentheorie bei inkompressibler Strömung", Jb. Wiss. Ges. Flugtech. (Luftf.), pp. 40–65.

Chapter 7

The Nonlinear Panel Methods for Aircraft and Missile Configurations at High Angles of Attack

It was demonstrated in Chapter 6 that the nonlinear variation of the aerodynamic coefficients as a function of the angle of attack, is caused by the effects of the rolled-up free vortices generated by the vortex sheets which are separated from the leading- and side edges of the lifting surfaces and from the surfaces of the bodies. This result is obtained from observation of experimental measurements and is further supported by the results of theoretical analysis. One analytical approach is that of Bollay (1937), Gersten (1961) and Ermolenko (1966) who that releasing the free trailing line vortices from the demonstrated surfaces and/or from the leading- and side edges generated nonlinear lift forces. Another approach for the evaluation of the nonlinear lift on slender delta wings by the introduction of concentrated line vortices into the conical flow field was presented by Legendre (1952), Brown and Michael (1955) and Mangler and Smith (1959). In these analyses, it is required to assume the position and the shape of the separated vortex sheet from the leading-edges of slender delta wings which roll-up into the concentrated line vortices.

In all of these flow models, the induced flow, arising from the free vortices whose trajectories cover the lifting surface, causes the additional lift force which results in the nonlinear aerodynamic characteristics of these configurations. The methods of analysis, presented in Chapter 6, all assume the a priori determination of the shape and position of the separated vortex flow over the configuration and in its wake, as well as the position of the center of the rolled-up vortex.

Therefore, for the methods based on vortex line separation, Bollay (1937) assumed free straight line vortices released from the side edges of a slender rectangular wing at a fixed angle relative to the wing surface equal to $\alpha/2$. Gersten (1955) extended this analysis to arbitrarily shaped wings by allowing the free trailing vortices to be released from all over the upper surface of the wing at a fixed angle of $\alpha/2$ relative to the wing surface. Ermolenko (1966) also used straight free vortices released from the wing surface but calculated the angle of angle varied along the wing chord and that the value of $\alpha/2$ (used by Bollay and Gersten) is a reasonable approximate average value. The vortex wakes presented in these investigations of Bollay, Gersten and Ermolenko are all modeled as wakes composed of straight line vortices issuing from the lifting surface at predetermined angles into the flow above and behind the lifting surface. However, this model does not represent the flow interactions which reshape the vortex layer into the rolled-up concentrated vortices. Therefore, these methods are not able to simulate this important feature of the high angle of attack flow over slender lifting surfaces. Consequently, although these analyses did derive the nonlinear variation of the aerodynamic coefficients with angles of attack, they cannot be expected to be in good agreement with experimental results without proper representation of the interactions leading to the roll-up process.

The other analytical approach accentuates the model of the separating vortex sheets which roll-up into the leading-edge vortices. Legendre (1952) assumed that this flow can be represented by the straight conical line vortices which leave the delta wing apex and are positioned above the wing. Brown and Michael (1955) added the planar feeding vortex sheet from the leading-edge to the concentrated line vortex. Mangler and Smith (1959) and Smith (1966), assumed a spiral vortex sheet leaving the leading-edge of the delta wing and rolling-up into a concentrated vortex with a core whose diameter increase in the chord direction. In these analyses, the positions of the concentrated vortices are determined by the solution of the conformal transformation of the cross-flow potential equations assuming conical flow geometry and proper boundary conditions on the wing, on the free vortex sheet and on the concentrated rolled-up vortices. Therefore, for configurations where the flow cannot be analyzed by assuming conical flow, this calculation of the vortex shape and position is inadequate and, therefore, the aerodynamic coefficients which are determined by this method cannot be expected to accord well with experimental results.

It can be, therefore, stated that in order to simulate correctly the flow at high angles of attack, the shapes and positions of the free vortex sheets and the rolled-up concentrated vortices, which are not known a priori, must be determined by including the results of the interaction of the flow induced by the aerodynamic configuration and the velocities induced by the trailing vortices which are released as free vortices from various parts of the lifting surfaces. The aerodynamic forces and moments caused by these free vortices become the dominating factors, particularly for slender elongated bodies and wings of low Aspect Ratios.

It is, therefore, necessary to use models of the flow with the appropriate methods of calculations that simulate the separating vortex sheet, the roll-up process, and describe the proper shape of the vortical flow field over the lifting surface. These requirements become more demanding when configurations of new high performance aircraft, such as wing-canard and/or wings with leading-edge extensions (LEX) etc., are studied. The nonlinear methods, presented in Chapter 6, are limited in their application and cannot compute the flows over complex configurations.

It can be anticipated that panel methods, which involve dividing the aerodynamic configuration into many small panels and distributing the singularities on these panels, can be designed to conform to complex aircraft or missile configurations. Panel methods for the calculations of the linear aerodynamic characteristics are presented in Chapter 4. It is important to note that in the linear panel methods, discussed there, the trailing vortices from all the panels are required to remain attached to the lifting surface; it is further assumed that these trailing vortices are separated only from the trailing edge and constitute a wake whose shape is a flat sheet in the plane of the wing. In order to simulate the flow field in the nonlinear case, the panel methods must be modified. The free vortices must be allowed to separate from the lifting surfaces. In addition to the vortex separation from the trailing edge, it is usual to allow for vortex separation from leading- and side edges and from separation lines on body surfaces. These methods can be also modified to include in their calculations the determination of the shape and position of the free rolled-up vortex lines or vortex sheets. Generally, such calculations must involve certain iteration procedures.

One group of panel methods is based on the vortex lattice concept. Here, the lifting surface is divided into many panels, where a horseshoe vortex is imbedded into each panel, the bound vortex being attached to the panel, and where the trailing vortices are allowed to leave the panels as free vortices into the flow. In most methods, only the trailing vortices from the panels along the leading, side and trailing edges are released into the flow. In deference to the assumptions of Bollay (1937), Gersten (1961) and Ermolenko (1966), the trailing vortices are not kept as straight lines issuing at predetermined fixed angles, but each free vortex line is divided into a finite number of finite segments followed by one semi-infinite segment. Then, the trajectories of these free vortices are allowed to follow the flow streamlines, so that the rolling-up of leading-edge or tip vortices is observed, as presented by Belotserkovskiy (1969). This is accomplished by a calculation method which requires certain iterative procedures until the convergent rolled-up vortex shape and the corresponding nonlinear aerodynamic coefficients are determined, as presented by Rom, Zorea and Gordon (1974). This version of the Nonlinear Vortex Lattice Method (NLVLM) is applicable in the subsonic speed range; however, the supersonic version of the vortex lattice method can be modified to enable also a supersonic version of the NLVLM, as discussed in Section 7.3 of this chapter.

A second group of methods is based on the free vortex sheet (FVS) concept. These methods require, first, the introduction of the rolled-up vortex sheet of given shape and the position; then in principle, the free vortex sheet shape and position can be modified in accordance with the flow field solutions by an iterative computational process. The initial shape of the vortex sheet can be assumed to conform to the conical solution of the spiral vortex sheet of Mangler and Smith (1959) and Smith (1966), separating from the sharp edges. Sources and doublet singularities are distributed on panels on the lifting surfaces and doublet singularities alone on the separating free vortex sheets. The new shapes of these free vortex sheets can be determined by utilizing a calculation method which involves some iterative procedures which are required in order to account for the interactions of the singularities on all the panels, those on the surface and those on the free rolled-up vortex sheets.

These two groups of nonlinear panel methods, developed for the calculations of the aerodynamic characteristics of complex configurations, will be presented in the following sections.

7.1 The Nonlinear Vortex Lattice Method (NLVLM) for Subsonic Flows

Using the vortex lattice concept, the aerodynamic configuration, such as an aircraft or a missile, is divided by a suitable surface grid structure into a large number of small panels. A horseshoe vortex and a source distribution are introduced into each of these panels. The nonlinear variation of the aerodynamic coefficients as the angle of attack increases is obtained when a method of calculation is devised by which both the strength of the bound vortices and the source distribution in each panel (for wings of finite thickness) are determined in a unified procedure. In this method, the trajectories of the free trailing vortices shed from the lifting surface into the flow over and behind the surface are also calculated. Since the strengths of the vortices in each panel depend on the interactions caused by the velocities induced by the free trailing vortices released from the leading, side and trailing edges (there is also an option to release free trailing vortices from all panels) and since the trajectories of these free vortices follow the streamlines over the lifting surface which, in turn, depend on the strength distribution of the vortices bound to the surface, the calculation method must, therefore, include some iterative procedures that relate these interrelations.

A representative flow chart for such a calculation developed by Rom, Zorea and Gordon (1974) is shown in Fig. 7.1. Computer programs were developed for wings of any planform as well as for combinations of multi-surfaces, such as wing-canard configurations (Almosnino, Zorea and Rom, 1978), for sharp leading-edge wings with thickness and camber (Rom and Gordon, 1981) and for bodies of various shapes (Rom and Almosnino 1980).

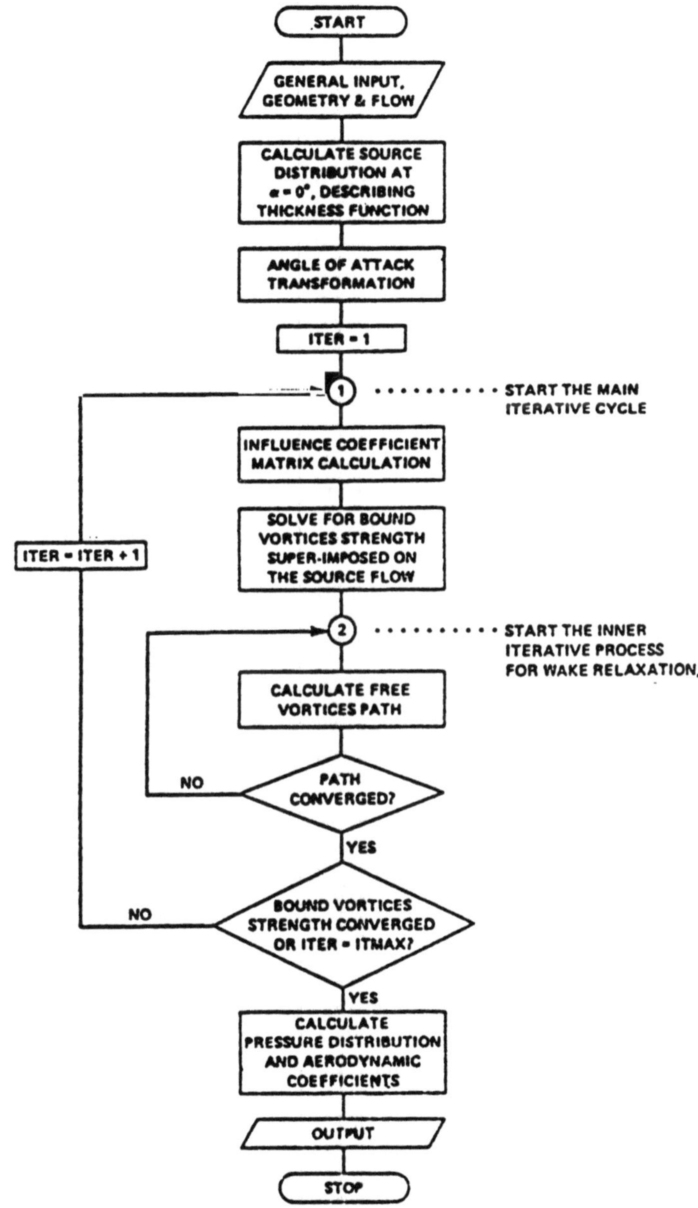

Fig. 7.1. Flow Diagram of the NLVLM calculation (Rom, Zorea and Gordon 1974)

7.1.1 Mathematical Formulation of the NLVLM

The mathematical formulation of the Nonlinear Vortex Lattice Method (NLVLM) is formally similar to the linear VLM; however, the nonlinear behavior is introduced through the boundary conditions. Assuming the flow to be steady, incompressible, [1] inviscid and irrotational except for isolated vortices, the basic equation for the perturbation velocity potential is the Laplace equation

$$\nabla^2 \phi = 0$$

and the boundary conditions are

on solid surfaces, the velocity must satisfy the tangency condition

$$\frac{\partial \phi}{\partial n_s} + V_\infty \cdot \mathbf{n}_s = 0$$

while at infinity, the disturbances must vanish, $\nabla\phi \Rightarrow 0$ as $\mathbf{r} \Rightarrow \infty$;

the free vortex wake shed from the edges of the lifting configuration follow the free stream surfaces, so that on the wake

$$\frac{\partial \phi}{\partial n_w} = V_\infty \cdot \mathbf{n}_w = 0$$

and the free vortex wake cannot support any pressure jump, $\Delta Cp_w = 0$.

The Kutta condition must be satisfied on all sharp edges – the leading, side and trailing edges of the lifting surfaces that shed the free vortices. This condition requires that there cannot exist a pressure jump across these sharp edges of the lifting surface.

The solution for the Laplace Equation with the above specified boundary conditions constitutes a well-defined boundary value problem. Even though the Laplace Equation is linear, the problem of evaluating the potential ϕ becomes nonlinear because the spatial position and shape of the free vortex wake is dependent on the induced velocities determined from the perturbation velocity potential.

The perturbation velocity potential function ϕ can be evaluated utilizing Green's theorem, which states that the function ϕ that satisfies the integral equation is a solution of the Laplace Equation

$$\phi(P) = \frac{1}{4\pi} \iint\limits_B \left[\phi \frac{\partial}{\partial n}\left(\frac{1}{R}\right) - \left(\frac{1}{R}\right)\frac{\partial \phi}{\partial n} \right] dS \tag{7.1}$$

where P is any point in the flow field, $P(x, y, z)$, R is the distance from this point to a point on the integration surface B that encloses all the singularities in the flow field, and $\partial/\partial n$ is the derivative in the direction of the surface vector normal $\mathbf{n}$ pointing into the fluid. Thus, the perturbation

[1] compressibility effects in subsonic flow can be included by the use of the Prandtl-Glauert-Goethert corrections.

velocity potential can be evaluated for any lifting configuration (consisting of bodies and wings in any combinations) by superposition of a source distribution σ of strength $\partial\phi/\partial n$ and doublet distribution μ of strength ϕ on the boundary of the fluid. The doublet distribution can be replaced by an equivalent vortex distribution which is the basis for the vortex lattice formulation.

The surface B consists of the solid surfaces of the wings and bodies, as applicable for the specified configuration, and the free vortex sheets shed from the lifting surfaces. The thickness effects of the bodies and the wings with finite thickness is simulated by the source distribution; the doublets/vortices can be simulated in the vortex lattice formulation by bound vortices attached to the surface, while the free trailing vortices simulate the separated vortex sheets. Since the free vortex lines follow streamlines, their trajectories can be calculated from the relations

$$\frac{dy}{dx} = \frac{v}{V_\infty + u} \quad \text{and} \quad \frac{dz}{dx} = \frac{w}{V_\infty + u} \tag{7.2}$$

The strengths of the source and vortex distributions and the trajectories of the free trailing vortices are determined by the solution of a matrix equation which equates the condition of tangency of the velocity vector at the control points on the surface and simultaneously solves for the vortex trajectories from Eq. 7.2. This is accomplished by an iterative procedure as indicated in Fig. 7.1. The calculation methods for lifting surfaces and for bodies will be presented in the following sections.

7.1.2 The Nonlinear Calculation Method for Lifting Surfaces

7.1.2.1 Calculation of the Source and Vortex Strength Distributions and the Trajectories of the Free Vortices. The source distribution required to simulate the effect of thickness of the wing is evaluated while setting the wing to zero angle of attack. In the NLVLM, the surface of the wing is divided into a large number of panels. Sources of constant strength are distributed over each of the panels, while a control point is placed at the 3/4 point of the center chordof each panel, to coincide with the control point of the horseshoe vortex element. The strength of the source distribution in the panels determined by the solution of a system of linear equations set up to satisfy the boundary conditions of tangential flow at the control point of each of the panels. Therefore, the wing's upper (+) and lower (-) surfaces are defined by the function $f(x, y) = c(x, y) \pm t(x, y)$, where $c(x, y)$ is the mean camber surface and $t(x, y)$ is the thickness function for the wing. The source strengths in each panel are evaluated at zero angle of attack, while the thick wing is placed with its chord plane at the desired angle of attack α. The perturbation velocities induced by these source distributions are calculated and superimposed on the induced velocities generated by the vortices.

The strength of the vortices required to simulate the lift force of the configuration is calculated by placing a horseshoe vortex in each panel, imbedding the bound vortex in the 1/4 chord line of each panel, placing the control point at the 3/4 point at the center chord of the panel, and allowing the trailing vortices to leave the surface from the leading, side and trailing edges or by any other option to be determined in the program.

In the case of zero thickness wing ($t = 0$), the source distribution is identically zero. In this case, the mean camber surface is set at the desired angle of attack α, and the chord plane of the wing is divided into a large number of panels. A horseshoe line vortex is imbedded into each panel with the bound vortex set at the 1/4 chord line of the panel and a control point located at the 3/4 chord of the middle chord of each panel. The trailing vortices are allowed to leave the surface either at the leading, side and trailing edges or by any other option determined in the program.

The calculation program starts by specifying the input for the flow conditions and the configuration geometry, executing the division of the surfaces into panels and specifying the options of calculating the source distribution for the thick wing or setting the source distribution to zero for zero thickness wing. Then the configuration is set to the desired angle of attack. The next step is to set the free trailing vortices to the selected initial trajectories - generally, these are set as straight lines separating from the surface in the flow direction but inclined at a local angle of $\alpha/2$ to the wing chord plane.

Afterwards, the strength of the vortices in each panel is calculated by the solution of the system of linear algebraic equations set up to satisfy the boundary conditions of tangential flow over the panel surface at its control point. The resulting matrix equation has the general form (similar to that in the linear VLM)

$$[H] \cdot \Gamma_\alpha = -\mathbf{V}_{\infty,n} \tag{7.3}$$

where $\mathbf{V}_{\infty,n}$ is the vector of the components of the free stream velocity normal to the upper or lower wing surface for the thick wing case and is the component normal to the mean chord plane for zero thickness wing, as evaluated at the control point of each panel. Γ_α is the vector of the strengths of the bound vortices in each panel. The influence coefficients matrix $[H]$ is evaluated by applying the Biot-Savart law for the calculation of the velocities induced by the line vortices from all panels at each of the control points in the direction normal to the wing surface. The influence coefficient matrix is calculated by using the trajectories of the free vortices as determined in the previous iteration, starting in the first iteration with the assumed initial trajectories.

After each iteration, the trajectories of the free vortices are updated by using the calculated new trajectories. These new trajectories are the streamline shapes evaluated in the previous iteration. These streamline trajectories are determined by the interactions of the induced flow of the bound vortices and the free vortices, and, for thick wings, also by the effect of the induced flow of the source distribution. Each free vortex line is divided into a number of straight line segments followed by a semi-infinite straight line segment. The new position of each segment is calculated from the latest positions of all the line vortices segments, in accordance with the following relations

$$
\begin{aligned}
y_j^{(i)} &= y_{j-1}^{(i)} + \frac{v_{j-1}^{(i)}}{V_\infty + u_{j-1}^{(i)}}(x_j - x_{j-1}) \\
z_j^{(i)} &= z_{j-1}^{(i)} + \frac{w_{j-1}^{(i)}}{V_\infty + u_{j-1}^{(i)}}(x_j - x_{j-1}^{(i)})
\end{aligned}
\tag{7.4}
$$

where $u_{j-1}^{(i)}$, $v_{j-1}^{(i)}$, $w_{j-1}^{(i)}$ are the Cartesian components of the velocity induced by the line vortices at $x_{j-1}^{(i)}, y_{j-1}^{(i)}, z_{j-1}^{(i)}$ in the i-th iteration. In the case of a thick wing, these induced velocities also include the velocity components induced by the source distribution calculated in the previous iteration. This calculation of the trajectory is repeated for each free vortex line that is separated from the lifting surface.

Now, the trajectories that were calculated in this i-th iteration are used as input for the $i+1$ iteration in which both the new vortex strength distribution and new trajectories are computed. The process is repeated until both the vortex strengths in each panel and the free vortex trajectories converge to within a predetermined criterion for deviation between subsequent iterations. This new convergence criterion, where both the vortex strength in each cell *and* the free vortex trajectories must converge simultaneously, is an improvement on previous calculation methods where convergence is determined by a single convergence criterion.

7.1.2.2 Calculations of the Pressure Distribution and Aerodynamic Coefficients. The previously described iterative procedure yields the local values of the vortex strengths (and the source strengths for thick configurations) in each panel on the lifting surface. It is then possible to evaluate the local induced velocities at the control points of each panel (u_i, v_i, w_i at the control point of panel i). Thus, the local pressure coefficient can be determined by the Bernoulli Equation

$$
C_{P_i} = 1 - [(V_\infty + u_i)^2 + v_i^2 + w_i^2)]/V_\infty^2 \tag{7.5}
$$

The total induced velocity vector at the control point of panel i is composed of the sum of the velocity vector induced by all line vortices, $\mathbf{V}_{\mathrm{L.V.},i}$, and half of the velocity vector induced by the local distributed vorticity vector, $\mathbf{V}_{\mathrm{D.V.},i}$ – all evaluated at the control point of panel i. For thick lifting surfaces, the induced velocities due to the sources distribution, $\mathbf{V}_s$, which is calculated at the control point, is added. The details of this calculation program are presented by Rom, Almosnino and Gordon (1981).

The aerodynamic coefficients are calculated by the summation of the forces acting on each panel

$$
(7.6)\quad
\begin{aligned}
C_L &= \frac{1}{S_{\text{ref}}}\sum_{i=1}^{2N} S_i C_{P_i}\cos\gamma_{z_i}, \qquad C_D = \frac{1}{S_{\text{ref}}}\sum_{i=1}^{2N} S_i C_{p_i}\cos\gamma_{x_i},\\
C_m &= \frac{1}{S_{\text{ref}}c_{\text{ref}}}\sum_{i=1}^{2N} S_i C_{P_i}[\cos\gamma_{z_i}(x_i - x_{\text{ref}}) - \cos\gamma_{x_i}(z_i - z_{\text{ref}})]
\end{aligned}
$$

where S_{ref} and c_{ref} are the reference area and cord, respectively, of the lifting surface, N is the number of panels per semi-span of the lifting surface, S_i is the area of panel i, γ_{z_i} and γ_{x_i} are the inclination angles of the normal to the panel surface with respect to the z and x axis, respectively, at the control point of panel i.

This calculation program can be used for single and for multi-lifting surface configurations, such as wing-canard (Almosnino, Zorea and Rom 1978) or other combinations of lifting surfaces. Since the computer program includes criteria for minimum "cut-off" distances for the joining of line vortices or finite core vortices, this program can handle complex multi-surface configurations without some of the numerical difficulties experienced in other panel methods.

7.1.2.3 Schemes for Division of Delta Wings Surfaces to Panels. The division of the lifting surface into panels and the distribution of the singularities in each of the panels may greatly affect the results of the calculations of the aerodynamic characteristics. The division of a rectangular or trapezoidal planform can follow the patterns used in the linear panel methods, by setting panels of rectangular or trapezoidal shape. In this case, the bound vortex is set at the 1/4 chord line of the panel. However, contrary to the linear case, the trailing vortices from the panels of the side edges and of the trailing edge are set free at an arbitrary initial angle. Each of these line vortices is divided into a finite number of finite segments followed by a semi-infinite segment. Through an iterative calculation procedure, the vortex line segments are allowed to follow the roll-up of the streamlines over the lifting surface and its wake. There is also an option to release the trailing vortices from every one of the panels. It will be shown that in this case, some of the vortices from the inner panels tend to rotate in a direction counter to the vortices from the leading edge panels, so that this structure can be taken as a prediction of the secondary vortices observed on wings at large angles of attack.

The option for the division of delta wings is more problematical. The delta wing surface can be divided by equally spaced spanwise divisions and by an equal number of equally spaced chordwise divisions, as seen in Fig. 7.2. This division scheme results in a definition of rectangular panels in the inner parts of the planform and triangular ones along the leading-edges. In the rectangular panels, the vortex is placed in the usual form: a horseshoe line vortex is placed in the 1/4 chord line of each panel and a control point at the 3/4 point on the panel's middle chord. The trailing vortices from all of these inner panels can be either kept on the wing surface and allowed to separate from the trailing edge or allowed to separate from the surface from each panel. The problematical arrangement is that required for the triangular panels. Here, there are various possibilities for the positioning of the bound vortex line and the position of the control point. The various arrangements of the horseshoe vortex in a triangular panel proposed by Zorea and Rom (1978), Kandil et al. (1977) and Gordon and Rom (1985) are indicated in Figs. 7.2 a, b and c, respectively. Fig. 7.2d presents an application of the trapezoidal panelling to delta planform.

The drawings in Fig. 7.2 describe fully the geometrical arrangements for each of these panelling schemes, both for the placing of the bound vortex as well as the positioning of the free trailing vortex from the leading-edge in the Zorea-Rom panelling (Fig. 7.2a), from a point ahead of the actual surface in the Kandil-Mook-Nayfeh panelling (Fig. 7.2b) and Gordon-Rom panelling (Fig. 7.2c), or from points on the surface as in the trapezoidal panelling (Fig. 7.2d). In the cases of shedding the trailing vortices from points ahead of the surface, the pressure loading on the panel is calculated only on the actual surface of the wing.

That part of the "panel" off the surface is envisaged only for the purpose of determining a point to fix the coordinates of the free trailing vortices. Since the geometry of the panelling scheme and the positions of the bound vortex and the control points have an important effect on the results of these calculations, it is of interest to examine the relative merit of the various panelling schemes.

7.1.2.4 The Relative Merit of the Various Panelling Schemes. In the linear panel methods, the geometry of the panels does not effect the results of the calculations, since the trailing vortices are attached to the surface of the wing and/or body and are shed only at the trailing edge of the surface. Therefore, the strength of the imbedded singularities depends only on the geometry of the configuration.

It was shown that the shedding of the trailing vortices from the surface into the free stream and the induced flow due to these free vortices, with their trajectories over the lifting surface, cause the nonlinear behavior of the aerodynamic coefficients at increased angles of attack. Therefore, the nonlinear component of the aerodynamic coefficient is dependent on the strength of the free trailing vortices and on their trajectories over the lifting surface. Examining the case of the delta wing, the strength and positions of the line vortices shed at the leading-edges of the delta wing depend on the panelling geometry and on the positioning of the bound vortex and the points of release of the free trailing vortex, as well as on the position of the control point in the panel. The four panelling schemes indicated in Figs. 7.2a to 7.2d have different characteristics for these parameters. Three of the panelling schemes, those of Zorea-Rom (7.2a), Kandil-Mook-Nayfeh (7.2b) and Gordon-Rom (7.2c) all release vortices from the leading-edge panels which are equal to the total strength of their corresponding bound vortices. The differences between these panelling schemes is in the positioning of the bound vortex and, therefore, in the position of the release of the free vortices. In the Zorea-Rom case, the bound vortex stays in the triangular panel and the free vortex leaves the surface at the leading-edge of the wing. In the Kandil-Mook-Nayfeh case, the bound vortex is placed at 90° to the leading-edge at the 1/4 chord point, so that most of the bound vortex is projected into the flow and the points of release of the free trailing vortices are fixed to points ahead of the leading-edge. In the case of the Gordon-Rom panelling, the bound vortex is placed at the 1/4 chord of the fictitious rectangular cell composed of a triangle on the wing and a triangle in the free stream, with the points of release of the free trailing vortices at the edges of the bound vortices at points ahead of the leading-edge, as seen in Fig. 7.2c. It can be shown that the strength of the bound vortex in the leading-edge panel is proportional to the distance between the line of the bound vortex to the control point of the corresponding panel, $\Gamma_i \propto \Delta r_i$. In the Zorea-Rom case, the distance between the control point and the bound vortex is the smallest; this means that the bound and free vortices from these leading-edge panels can be expected to be smaller than those of the two other panelling schemes. Since the largest distance between the control points to the bound vortices is in the Kandil-Mook-Nayfeh case, we can expect that, for a given wing planform, the strongest leading edge vortices will be obtained from the Kandil-Mook-Nayfeh panelling scheme than the somewhat weaker vortices from the Gordon-Rom scheme and the further weaker vortices from the Zorea-Rom scheme. In the trapezoidal division, shown in Fig. 7.2d, the strength of the trailing vortices to be released from the leading-edge panels are obtained from the difference of the strengths of the bound vortices in the adjacent panels, so these vortices are expected to be much weaker.

The relative merit of these panelling schemes can be examined by computing the aerodynamic characteristics and comparing their results with experimental data. For slender configurations with relatively small Aspect Ratios, the results for both the Kandil-Mook-Nayfeh and the Gordon-Rom panels are very close and tend to overestimate the nonlinear component of the aerodynamic coefficient, while the Zorea-Rom panelling seems to accord well for the lift and induced drag coefficients.

7.1.3 Results of the NLVLM Calculations for Subsonic Flows

7.1.3.1 Rectangular and Trapezoidal Wings With or Without Sweep. The rectangular or trapezoidal planform can be divided into equal chords and equal span panels by selecting N_c and N_s divisions in the chordwise and spanwise directions, respectively. For high Aspect Ratio wings, N_c values of 2 to 4 and N_s values of 6 to 20 have been used. The horseshoe vortex is introduced into each panel with the bound vortex being placed on the 1/4 chord line of the panel and the control point at the 3/4 point of the mid-chord line of the panel. It is possible to account for the thickness effect by adding a distribution of sources over the upper and lower surfaces of the thick wing. The nonlinear program can include the following three options: 1. keeping the trailing vortices attached to the wing surface with separation of these vortices into the wake at the trailing edge of the wing (linear case); 2. allowing the trailing vortices to separate from the side edges and trailing edge, or from the leading, side and trailing edges; and 3. allowing the trailing vortices to leave the wing surface from all panels.

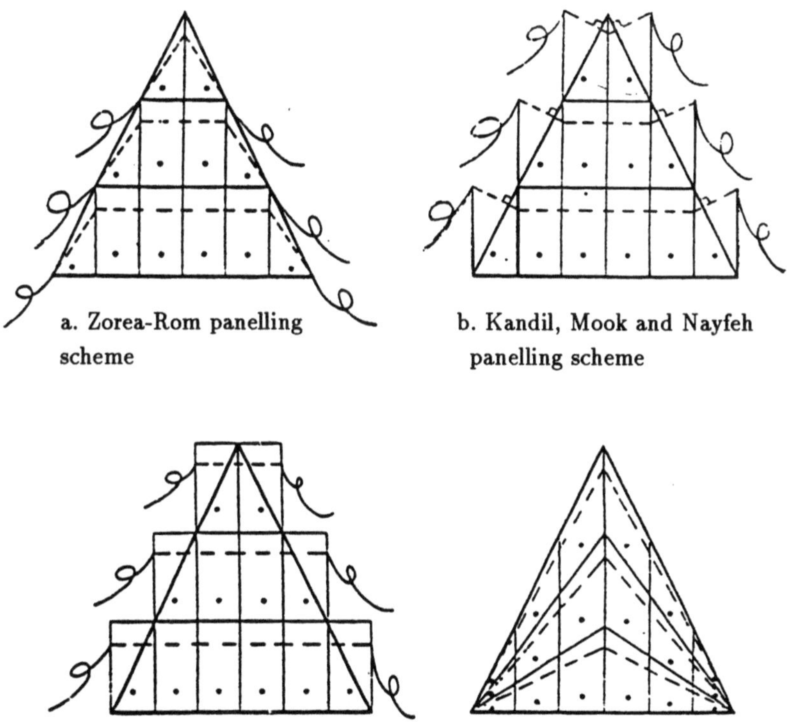

a. Zorea-Rom panelling scheme

b. Kandil, Mook and Nayfeh panelling scheme

c. Gordon-Rom panelling scheme

d. Trapezoidal panelling

Fig. 7.2. Panelling schemes for delta wings

7.1.3.1.1 High Aspect Ratio Wings. The roll-up of the tip vortices has a relatively small effect on the aerodynamic coefficients of the high Aspect Ratio wing. Therefore, calculating the aerodynamic coefficients, including the various available options for the release of the vortices from the wing surface and all its edges, results in linear variation of these coefficients with

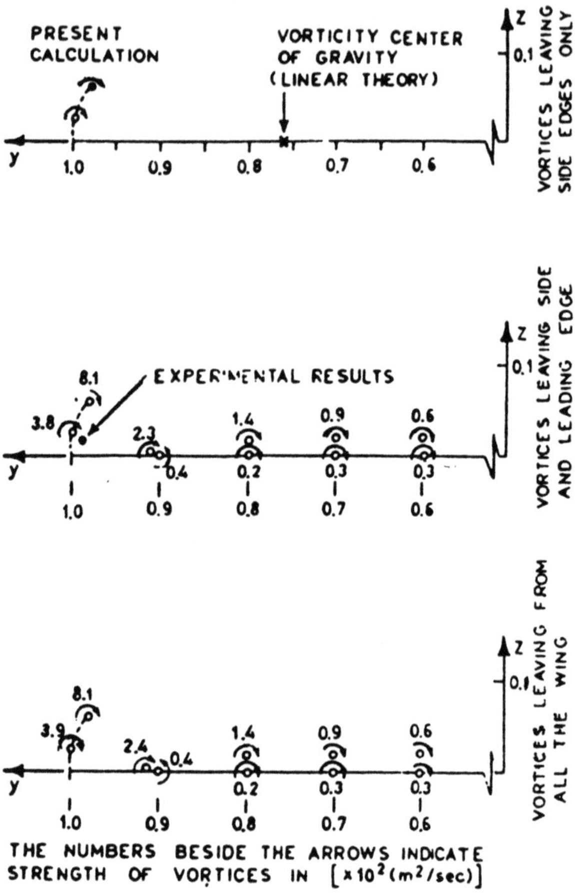

Fig. 7.3. The calculated strengths and positions of vortices shed from a rectangular wing of AR=5.3 at $\alpha = 12^\circ$ in comparison with experimental results of Chigier and Corsiglia (1972) (Zorea and Rom 1978)

angle of attack. However, the significant new results in these "nonlinear" calculations is the formation of the rolled-up wake with the distinctive tip vortices, as shown in Figs. 5.6 and 5.7. It is of interest to compare the calculated positions and strength of the vortices at the trailing edge of the wing with the experimental data. This was done for a rectangular wing with AR of 5.33 and the results are presented in Fig. 7.3. In the case of vortices leaving the side edges only, the results already show the roll-up of the side edge vortices. The cases of vortices leaving the leading- and side edges and that of vortices leaving from all panels are shown, and the strengths and direction of the vortices are indicated in this figure. Comparison of the vortices shed in these last two cases indicate that the strength and direction of these vortices are almost equal in both cases. Therefore, the determining factor is the vortices shed from the leading- and side edges, with most of the vorticity being concentrated near the tips so that the vortices in the mid-wing are much weaker. The core size of the concentrated tip vortex is evaluated to be 2.7% of the semi-span and its position is very near the tip of the wing, both results according well with the measurements of Chigier and Corsiglia (1972). The measured position of the center of the tip vortex is indicated in Fig. 7.3 at $y/b = 0.98$, while the estimation of the center of vorticity of the linear theory is inboard, at $y/b = 0.76$. Another interesting result obtained in this calculation is the appearance of a counter-rotating vortex line, near $y/b = 0.9$, which may be identified with the secondary vortex found in the experimental investigation.

7.1.3.1.2 Low Aspect Ratio Rectangular Wings. The nonlinear variation of the aerodynamic coefficients with angle of attack is very well demonstrated in the cases of slender wings. In particular, we will present both experimental and computed results for rectangular wings. The calculations were performed with chordwise divisions of $N_c = 2$ and 4, and spanwise divisions of $N_s = 10$, 20 and 40. The comparison of the calculated and experimental results for the lift and pitching moment coefficients is presented in Fig. 7.4, and the induced drag coefficient is presented in Fig. 7.5, where the computed values are shifted by the value of the drag at zero lift.

The positions of the vortices near the wing tip at various angles of attack showing the roll-up process is shown in Fig. 7.6, where the position of the center of the tip vortex is shown by the flagged symbol. The development of the wake downstream of the trailing edge, with the interaction of the vortices shed from the side edges with those shed from the trailing edge, is shown in Fig. 7.7.

The comparison between the measured and calculated pressure distributions is an indicative test of the calculation method. The measured pressure distribution made by Wickens (1967) on a rectangular wing of AR=0.25 is compared with the calculated values, using $N_c = 8$ and $N_s = 6$, in Figs. 7.8a and 7.8b, respectively. The measured pressure distribution on an AR=1 rectangular wing of Scholz (1950) is compared with the calculated values, using $N_c = 4$ and $N_s = 10$ in Figs. 7.9a and 7.9b, respectively. For both wings, that of the AR=1 and the AR=0.25, the calculated pressure distributions are generally in good qualitative and even quantitative agreement with the measured data, except for the differences in the peaks of the pressure near the leading edge.

The capability of the NLVLM to evaluate the aerodynamic coefficients of very slender aerodynamic configurations is illustrated in Fig. 7.11. The calculation of the lift coefficient of a rectangular wing of AR=1/30 for angles of attack up to 40° accord well with the experimental results of Winter (1937). It is found that the results using NLVLM are in better agreement with the experimental data at the high angles of attack than those using the calculations of Bollay (1937). These results indicate that

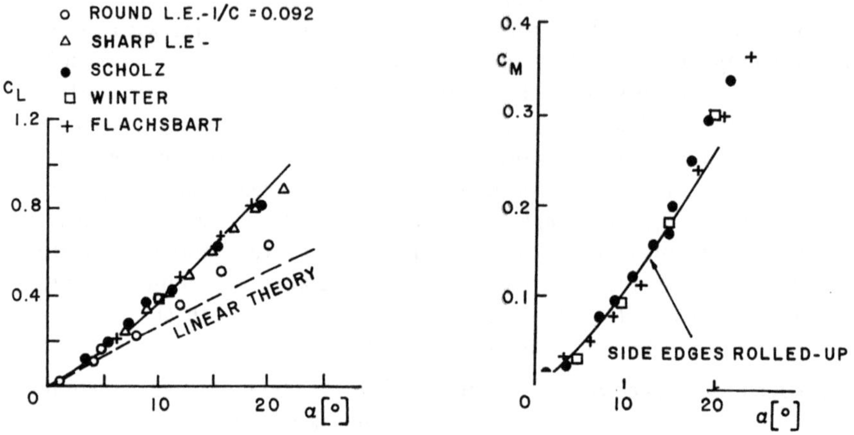

Fig. 7.4. The lift and pitching moment coefficients for a rectangular wing of AR=1 (Zorea and Rom, 1978)

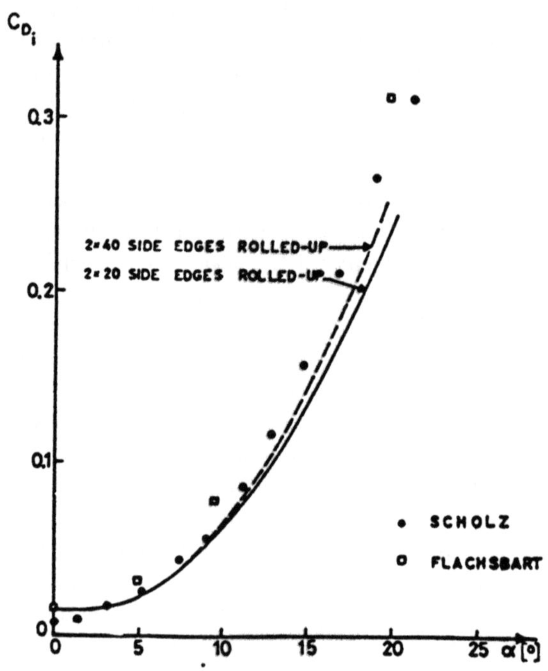

Fig. 7.5. The induced drag coefficient for a rectangular wing of AR=1 (Zorea and Rom, 1978)

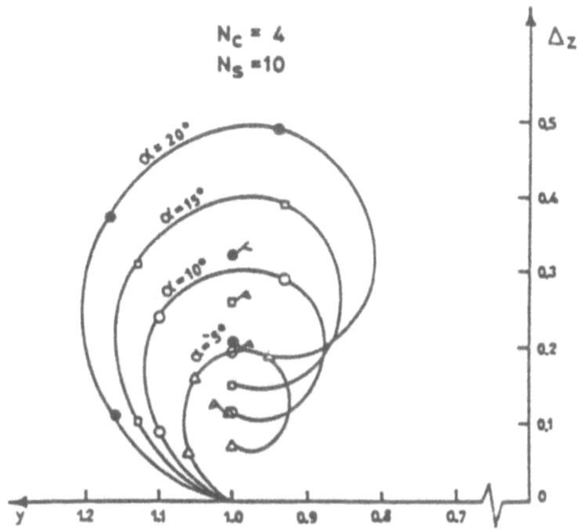

Fig. 7.6. The rolling-up of the side vortices near the tip of the rectangular wing of AR=1 at various angles of attack (Zorea and Rom, 1978)

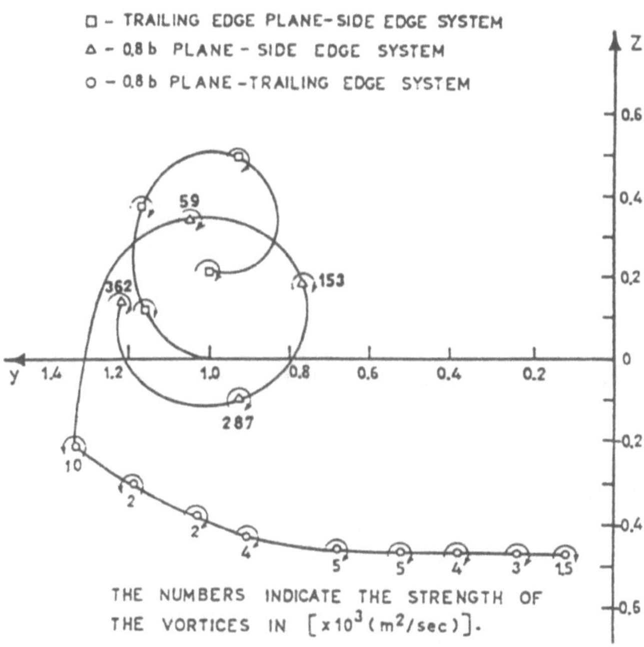

Fig. 7.7. The development of the vortex wake behind the trailing edge of the rectangular wing of AR=1 at $\alpha = 20^o$ (Zorea and Rom, 1978)

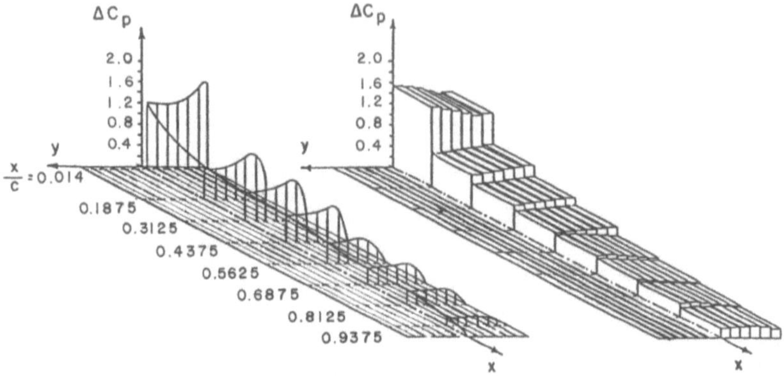

a. Experimental (Wickens 1967) b. Calculated (Zorea and Rom 1978)

Fig. 7.8. Pressure distribution for AR=0.25 rectangular wing at $\alpha = 20^o$

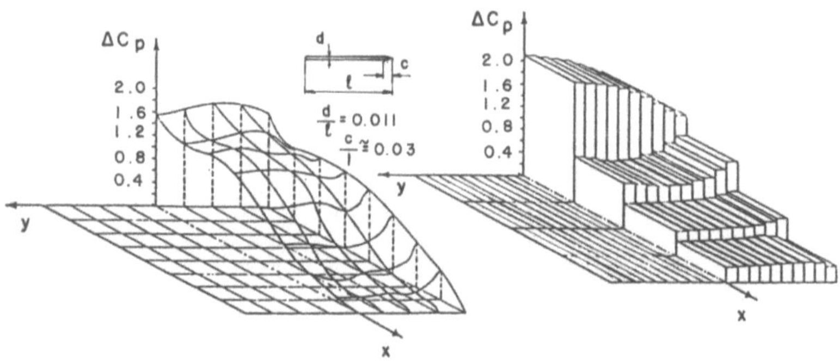

a. Experimental (Scholz 1950) b. Calculated (Zorea and Rom 1968)

Fig. 7.9. Pressure distribution for AR=1 rectangular wing at $\alpha = 20^o$

the accounting for the roll-up of the side edges vortices, which is done in the NLVLM calculation, does improve the evaluation of the aerodynamic coefficients over the straight line trailing vortices from the side edges as assumed in Bollay's nonlinear model.

The capability of the nonlinear calculation method to evaluate the aerodynamic coefficients of a swept back rectangular wing is presented in Fig. 7.10. The values of the normal force and pitching moment coefficients variation with angles of attack calculated by Belotserkovskiy (1969) and Kandil et al. (1976) are found to accord well with the experimental data of Gersten (1961).

7.1.3.2 Delta Wings in Subsonic Flows. The results of the NLVLM calculations for rectangular and trapezoidal wings encouraged the application of this method to triangular geometries – delta-type wings. Nevertheless,

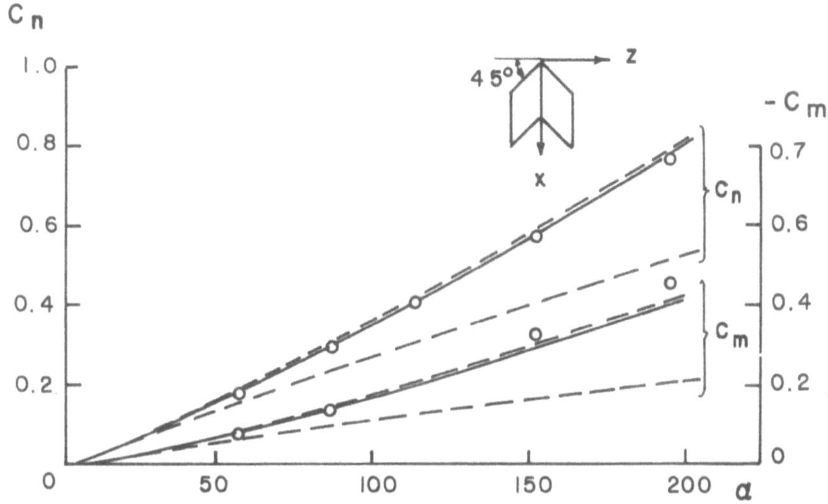

Fig. 7.10. Variation of the normal force and pitching moment coefficients for a 45° swept back AR=1 rectangular wing

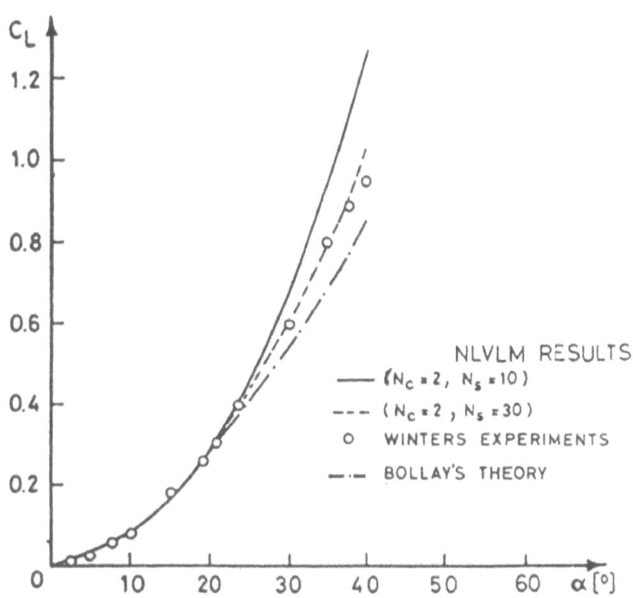

Fig. 7.11. Comparison of the calculated and experimental lift coefficients for a rectangular wing of AR=1/30 at angles of attack up to 40° (Zorea and Rom 1978)

imbedding the bound line of the horseshoe vortex and deciding on the position of the control point in a triangular panel presents a difficult problem. An interesting approach was presented by Rehbach (1973) where the flow over the delta wing is obtained by calculating the flow over a series of wings starting with a rectangular geometry and progressively deforming the shape by narrowing the leading-edge to a pointed apex, until a triangular geometry is obtained, as shown in Fig. 7.12. The triangular geometry obtained by this process determines the positions of the imbedded horseshoe vortices and the control points in each deformed panel, as seen in Fig. 7.12, which in the last stage shows a panelling pattern which corresponds to the trapezoidal panelling method of Fig. 7.2d.

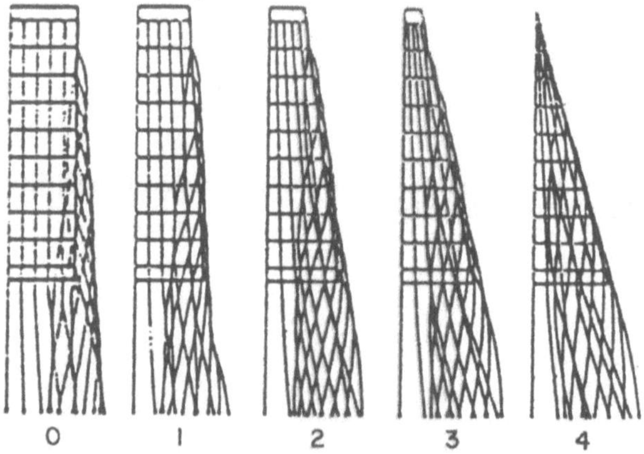

Fig. 7.12. The flow over a delta wing obtained by the progressive deformation of a rectangular wing as proposed by Rehbach (1973)

The application of the NLVLM for the calculations of the aerodynamic characteristics, following the computation program represented in Fig. 7.1 and using any one of the panelling schemes for delta wings as in Fig. 7.2a to 7.2d, enables the calculations of the flow over delta wings directly.

The velocity vector at all points on and about the wing are calculated, so that the shape of the streamlines – vortex-lines of the roll-up of the separating leading-edge vortices is obtained, as illustrated for the case of AR=1 wing in Fig. 7.13. The free vortex lines are seen leaving the leading-edges and rolling-up over the wing, and their trajectories are in their general shape similar to those seen in the flow visualization photograph. The rolling-up is well demonstrated in the visualization experiment on a 75°/60° double-delta wing, presented in Fig. 7.14. However, a closer examination indicates that the rolling-up of the calculated trajectories occurs further downstream, and the vortex center is inboard in comparison with the visualization experiment photograph.

The results of the calculations of the aerodynamic coefficients of flat,

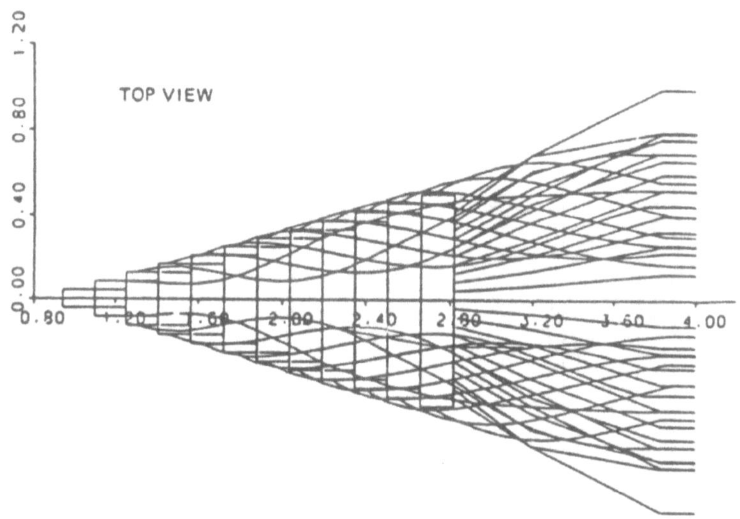

Fig. 7.13. Calculated vortex trajectories over a delta wing of AR=1 (Gordon and Rom 1985)

Fig. 7.14. Flow visualization of the rolled-up vortices over a delta wing of AR=2 (Rehbach 1973)

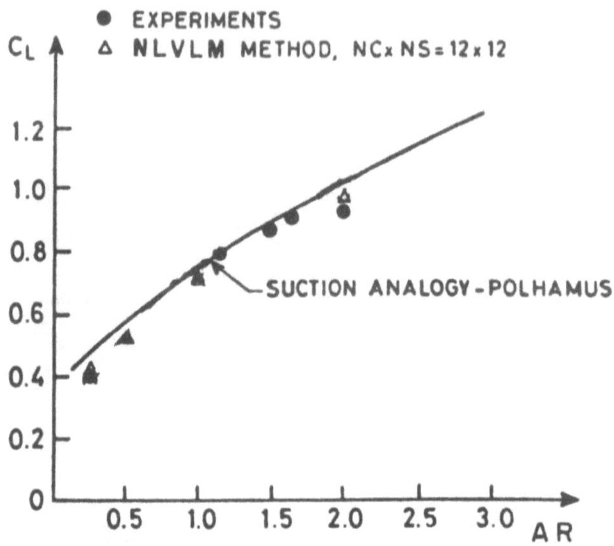

Fig. 7.15a. Variation of the lift coefficient with Aspect Ratio for flat delta wings at $\alpha = 20^o$ (Rom et al. 1976)

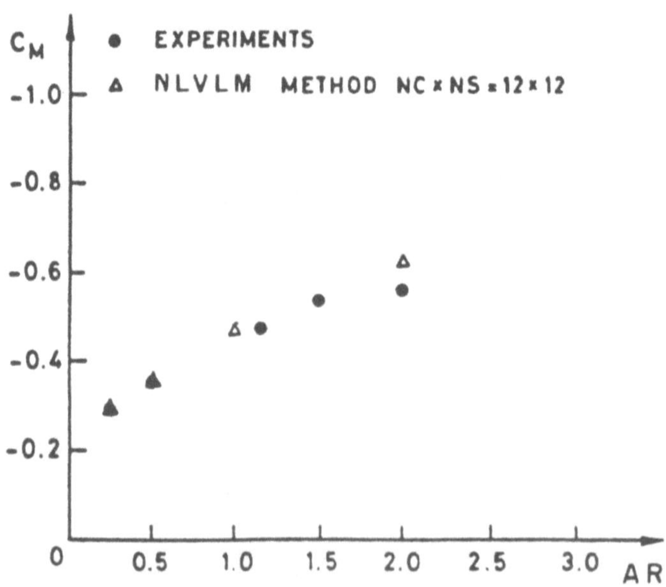

Fig. 7.15b. Variation of the pitching moment coefficient with Aspect Ratio for flat delta wings at $\alpha = 20^o$ (Rom et al. 1976)

zero thickness delta wings of Aspect Ratios in the range of 0.25 up to 2.0 are compared with experimental data in Figs. 7.15a and 7.15b for the lift and pitching moment, respectively.

The calculated values by the NLVLM and the experimental data of Peckham (1961), Bartlett and Vidal (1955), Fox and Lamar (1974), Wentz (1972), Wentz and Kohlman (1972) and Kirkpatrick (1970), which are presented in these figures, seem to accord well for all the low AR delta wings, except that there is a fall in the experimental lift coefficient of the AR=2 wing. This is probably due to the loss of lift caused by the effect of the movement of the vortex breakdown into the wing planform which is observed by Wentz and Kohlman (1972) to begin at $\alpha = 18^o$ in the case of the delta wing of AR=2.

The effect of thickness of the delta wing can be included in the NLVLM calculation. Such calculations were performed for both flat zero thickness and 12% thickness delta wings of AR=1, and the results were compared with the experimental data of Peckham (1961), shown in Fig. 7.16. The calculated and experimental lift, drag and pitching moment coefficients are presented in Figs. 7.17a, 7.17b and 7.17c respectively.

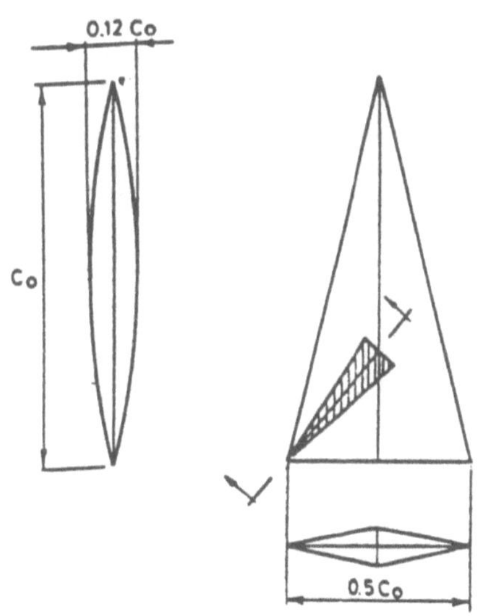

Fig. 7.16. The 12% thickness delta wing geometry (Gordon and Rom 1985)

The lift coefficient for the thick wing is found to be about 15% to 20% below that for the zero thickness delta wing, and the pitching moment is about 10% more positive in the thick wing case. Examination of these results indicate that the NLVLM calculations accord well with the experimental data. Very good agreement is obtained for the lift and the induced drag coefficients for both the zero thickness and for the thick delta wings, while the experimental pitching moments are a few percent more positive than the calculated values. However, these differences in the pitching moment are rather small and even these results can be accepted to be in reasonable agreement.

When the option of releasing the trailing vortices from inbound panels, in addition to those released from the leading-edge panels, is used, it is found that some of the vortices in the second row are found to rotate in the opposite direction to that of the leading-edge vortices in a way that can be interpreted as showing the formation of a secondary vortex separation as indicated in Fig. 7.18, where the direction and positions of the vortices are shown.

7.1.3.3 Multi-Lifting Surface Configurations in Subsonic Flows. The integration of the calculations of the vortex wake shapes with the aerodynamic loading on the lifting surfaces and the definition of the "cut-off" distance for the joining of vortex lines make it possible to extend the NLVLM calculations to complex configurations. The application of the NLVLM to calculate the aerodynamic characteristics of lifting surfaces of simple shapes, as discussed previously, can be extended to practically any reasonable geometrical shape. It is of interest to examine the application of the NLVLM to configurations with any reasonable number of lifting surfaces, since one of the useful methods for enhancement of the delta wing performance is the use of a combination of multi-lifting surfaces. A useful combination is the wing-canard configuration which is used in a number of modern fighter aircraft such as the Swedish Draken J-35 and the new Gripen JAS-39, the proposed European Fighter Aircraft (EFA), the Israeli Kfir C-2 and the Lavi, the French Rafale as well in some versions proposed for the USAF Advanced Tactical Fighter (ATF). Another method of improving the aerodynamic performance of slender wings is the addition of the LEX (leading edge extension) in front of the main delta wing. Such a configuration is used in the F-18 aircraft. The evaluation of the aerodynamic characteristics of these configurations by the NLVLM and comparison with experimental data will be presented.

7.1.3.3.1 The Wing-Canard Configuration at Subsonic Speed. Improved aerodynamic performance of slender configurations at high angles of attack can be achieved by the enhancement of the strength of the rolled-up vortices in a way which will delay vortex breakdown. At increasing angles of attack, the induced lift forces due to these vortices increases as long as the strength of the rolled-up vortices continues to increase. This increase of the nonlinear lift is finally diminished when the bursting of the concentrated vortices moves over the wing surface, so that further increase of the angle of attack will accelerate the effect of "vortex breakdown" until the wing stalls. The enhancement of the strength of the concentrated vortex from the leading-edges of the slender wing can be achieved by generating a strong concentrated vortex, which will interact with the rolled-up wing vortex. The devices which were proposed for the generation of a strong vortex

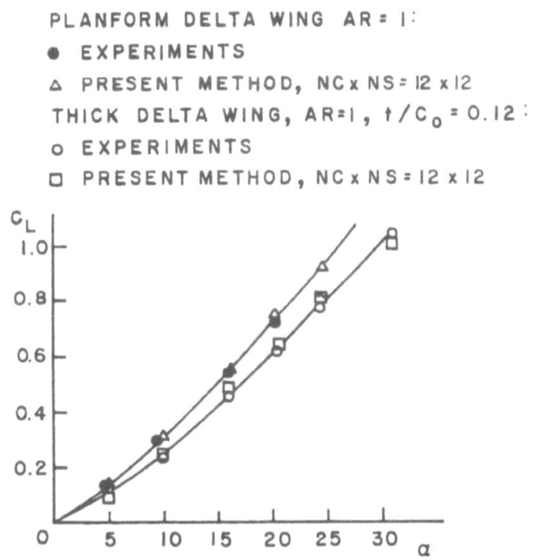

Fig. 7.17a. Lift coefficient variation with angle of attack for the 12% thick delta wing (Gordon and Rom 1985)

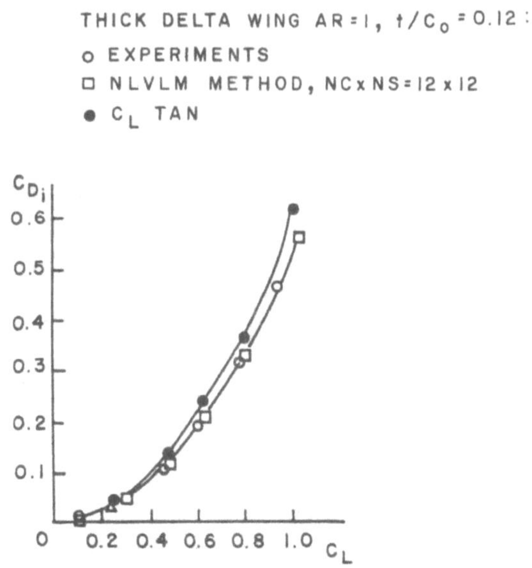

Fig. 7.17b. Drag coefficient variation with lift for the 12% thick delta wing (Gordon and Rom 1985)

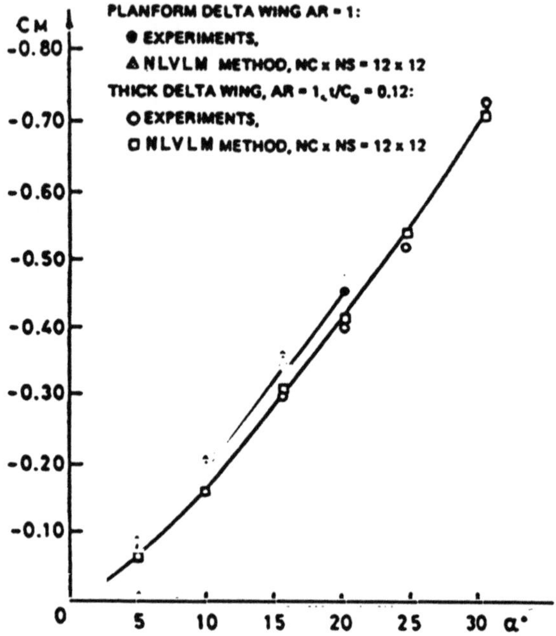

Fig. 7.17c. Pitching moment coefficient variation with angle of attack for the 12% thick delta wing (Gordon and Rom 1985)

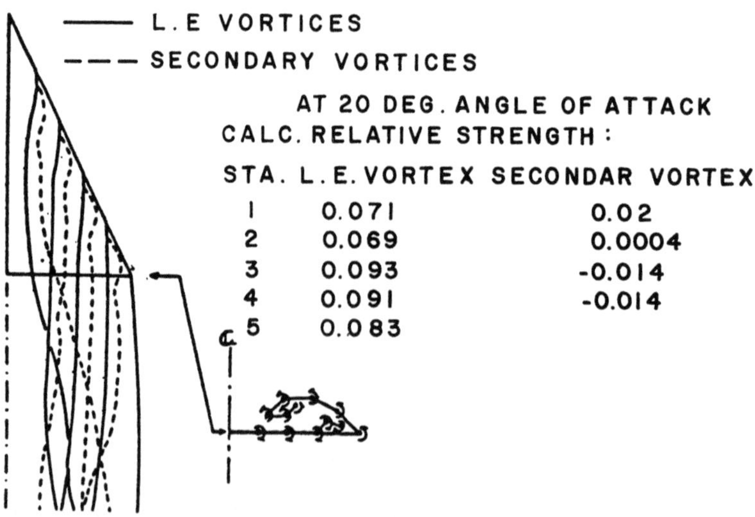

Fig. 7.18. NLVLM calculation of the trajectories and strengths of the leading edge panel and the second row panel vortices indicating secondary vortex separation (Almosnino et al. 1978)

include the leading edge extensions (LEX), strakes and "saw tooth" modifications to the wing leading-edges and winglets of various shapes placed as vortex generators on the wing. One of the most effective methods for such aerodynamic enhancement is by using the vortex generated by the slender canard placed just ahead of and above, the main wing – the closed coupled wing-canard configuration.

The closed coupled wing-canard configuration has the additional advantage of utilizing the canard as a movable aerodynamic control surface. In this case, the strength of the vortices generated by the movable canard can be increased by higher canard deflections. The close coupling between the canard and the wing will cause an upwash on the canard because of the wing and a downwash on the wing because of the canard. These induced flows will affect the lift forces and moments on each one of these surfaces and, therefore, on the total lift force and the trim control effectiveness of the wing-canard configuration. We will examine the capabilities of the NLVLM to evaluate these aerodynamic characteristics of the wing-canard configuration at relatively high angles of attack until vortex breakdown dominates the flow.

The effects of the interaction between the flows over the canard and the wing are clearly seen in flow visualization photographs, an example being presented in Figs. 7.19a and 7.19b. The trajectories of the cores of the free vortices from the canard and the wing are obtained by the schlieren photographs of the top view (obtained using a transparent wing model), Fig. 7.19a, and of the side view, Fig. 7.19b. These figures include a comparison with the calculated trajectories by the NLVLM which are superimposed on the experimental schlieren photographs. These show that the vortex from the canard is pulled inside and pushed below the wing leading-edge vortex. The trajectories calculated by the NLVLM are found to be in good qualitative agreement with the visualization photographs.

However, the calculated shape of the rolling-up of the leading-edge vortex indicates that over the wing surface, this vortex center is inboard from the measured vortex center and completes the rolling-up only further in the wake. It can then be expected that the calculated pressure distributions will show a lower suction peak positioned inboard in comparison with the experimental results, as seen in Fig. 7.21.

The aerodynamic coefficients and the pressure distributions measured on a wing-canard model, designated as European Model B (shown in Fig. 7.20), are presented in Figs. 7.20 and 7.21, respectively. The NLVLM calculation results are included for comparison with the experimental data. The lift and the induced drag are in very good agreement; the pitching moment is evaluated reasonably well by the NLVLM at low angles of attack, but the deviation from the experimental values becomes significant above 15°–20°.

The calculations using the NLVLM evaluate the three-dimensional flow field, so it is possible to map the velocity vector and the dynamic pressure values at each point around the configuration. A graphic program for the

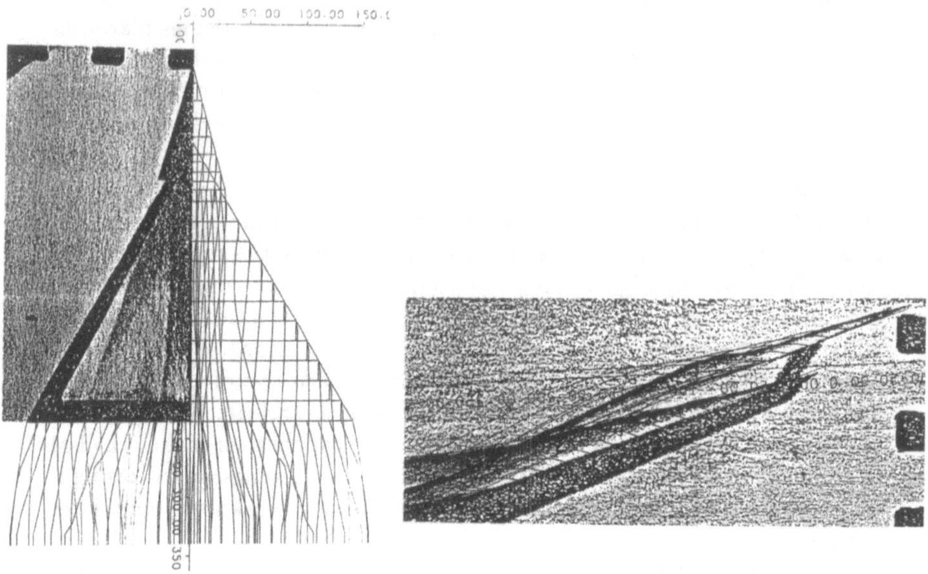

a. Top View b. Side View

Fig. 7.19. Schlieren photograph visualization on a wing-canard model at $\alpha = 17^o$ (Rom et al. 1990)

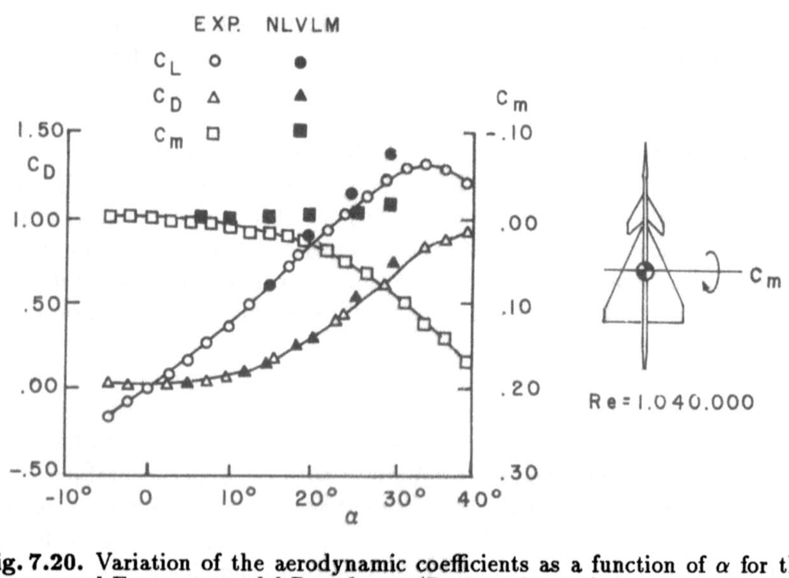

Fig. 7.20. Variation of the aerodynamic coefficients as a function of α for the wing-canard European model B at $\delta_c = 0$ (Rom et al. 1990)

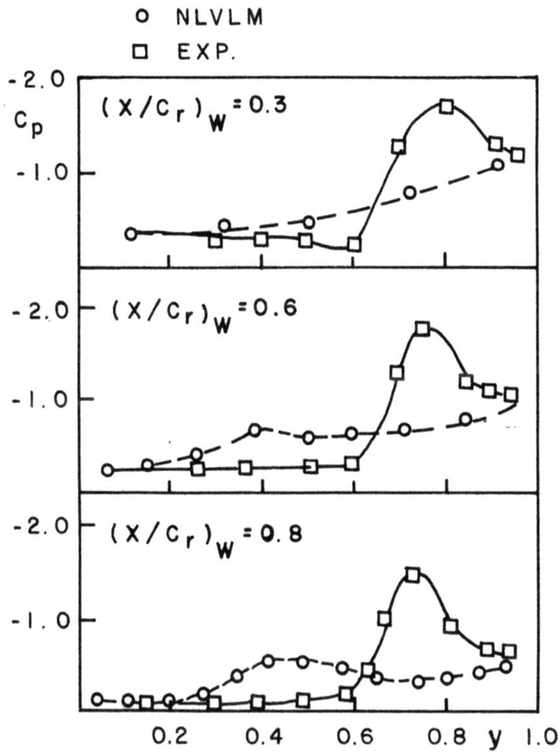

Fig. 7.21. Pressure distribution on the wing of the European model B configuration at $\alpha = 15^o$ (Rom et al. 1990)

presentation of these parameters was developed by Gordon (1990). The results of such calculations performed on a wing-canard configuration which has a 60° delta wing and a 60° delta canard having a span ratio of 0.4 are presented in Fig. 7.22 and 7.23, respectively. This model has been designated as European Model A and a top view of an oil flow visualization on this model is included in Figs. 7.22 and 7.23. The dynamic pressure isobars are calculated in the cross plan at $x/c = 0.8$ of the wing and are compared with the experimental measurements of Oelker and Hummel (1988) in Fig. 7.22.

The comparison of the calculated velocity vector in the cross plan at $x/c = 0.8$ of the wing with the experimental data is presented in Fig. 7.23. There is a very good qualitative agreement between the calculations using the NLVLM and the experimental data. The positions of the dynamic pressure peaks at the cores of the wing leading-edge vortex and at the canard leading-edge vortex are clearly seen to be very similar in both the calculated and measured results, seen in Fig. 7.22. Similarity, good qualitative correspondence is seen in Fig. 7.23 for the results of the calculated and measured velocity vectors in the cross plan on the wing.

The deflection of the canard results in variation of the lift on the canard and, therefore, this has a strong effect on the pitching moment which trims the configuration. An additional effect which was observed in the wind-tunnel measurements on the wing-canard configurations was the variation of the lift to drag ratio as the canard is deflected at low and moderate angles of attack as reported by Rom, Er-El and Gordon (1988). In this case, the canard deflections can lower the L/D ratio by 25% to 30% for angles of attack of 2° to 6°.

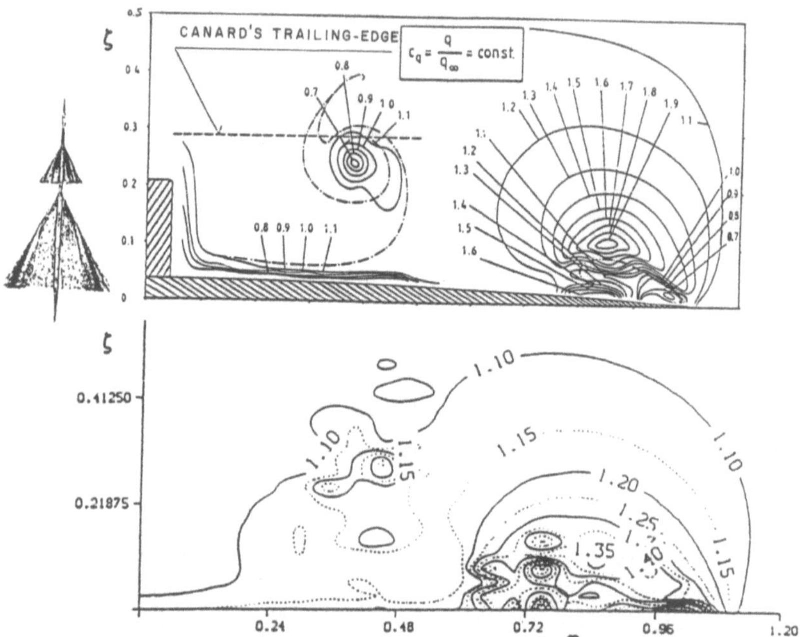

Top – Oelker and Hummel 1988 Bottom – Gordon 1990

Fig. 7.22. Comparison between the NLVLM calculated dynamic pressure and experimental results at the cross plan $X/C = 0.8$ and $\alpha = 8.7°$

7.1.3.3.2 The Strake-Delta Wing Configuration at Subsonic Speed. The vortex generated by adding a slender delta strake as an extension on the leading-edge of the delta wing is found to enhance the strength of the main wing leading edge vortex, so as to delay vortex breakdown and increase the high angle of attack performance of the configuration. The aerody-

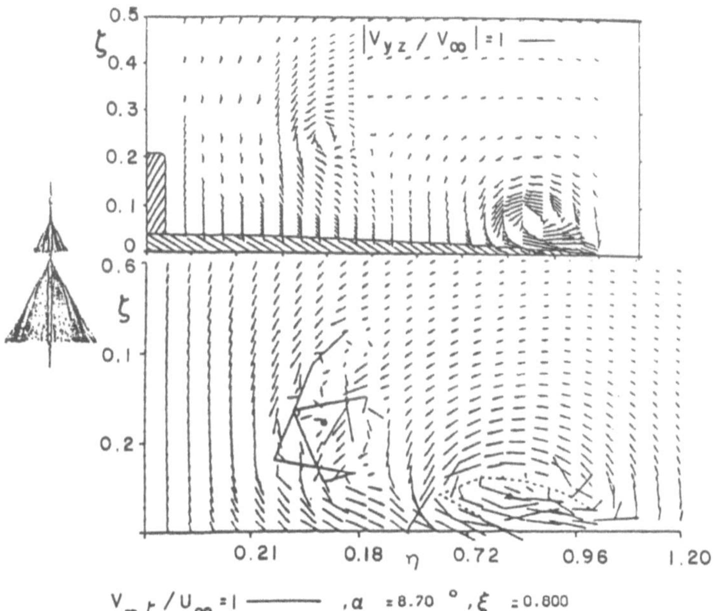

Fig. 7.23. Comparison of the NLVLM calculated local velocity vectors and the experimental results at the cross plan $X/C = 0.8$ and $\alpha = 8.7^o$ (Top - Oelker and Hummel 1988, Bottom - Gordon 1990)

namic characteristics of such configurations, specified as double-delta or strake-delta wings, were measured by Brennenstuhl and Hummel (1982) and were calculated by Gordon (1990) using the NLVLM and by Fuji and Schiff (1989) solving the thin layer Navier-Stokes equations. The appearance of the concentrated vortex from the strake leading-edge and from the wing leading-edge and their interactions is well demonstrated in the oil flow visualization shown in Fig. 7.24.

Applying the NLVLM, Gordon (1990) calculated the lift and pitching moment coefficients for this strake-delta wing, and these results are compared with the experimental data in Fig. 7.25.

The NLVLM program was used also for the calculation of the velocity vector field at various cross-sections over the strake-delta model (Gordon 1990), and these results are compared with the measured data of Brennenstuhl and Hummel (1982) in Fig. 7.26, indicating reasonable qualitative agreement.

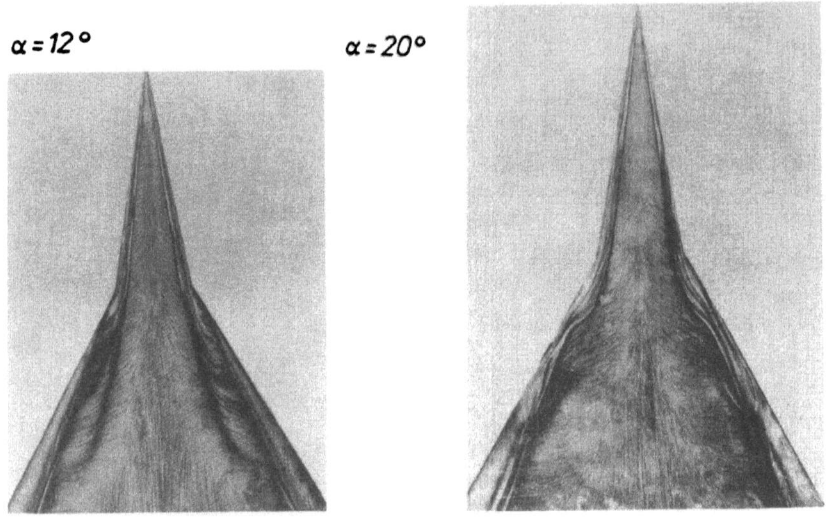

Fig. 7.24. Oil flow visualization on the surface of a 80°/60° strake-delta wing at low speed (Brennenstuhl and Hummel 1982)

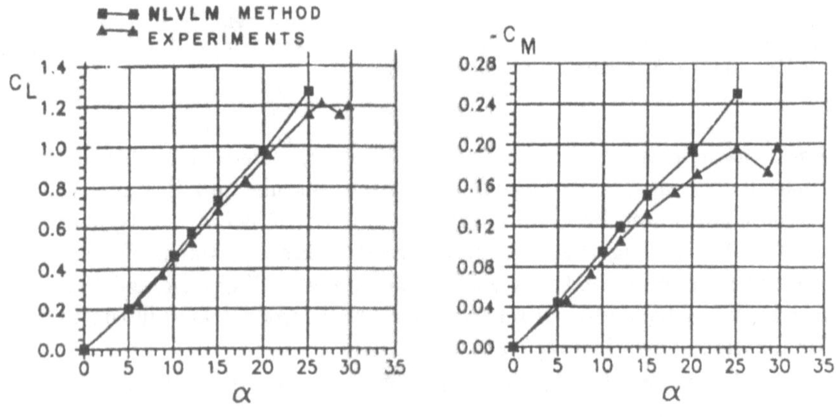

Fig. 7.25. Variation of the lift and pitching moment coefficients with angle of attack for strake-delta wing (Gordon 1990, Brennenstuhl and Hummel 1982)

The comparison between the experimental and calculated pressure distributions on the upper surface of the strake-delta wing is presented in Fig. 7.27. Here again, the NLVLM results follow the general trend of the experimental data, indicating the two peaks due to the strake and the wing leading edge vortices; however, the positions of the negative peaks are inboard for the NLVLM results, and the level of these peaks are lower for the NLVLM calculations in comparison with the experimental data.

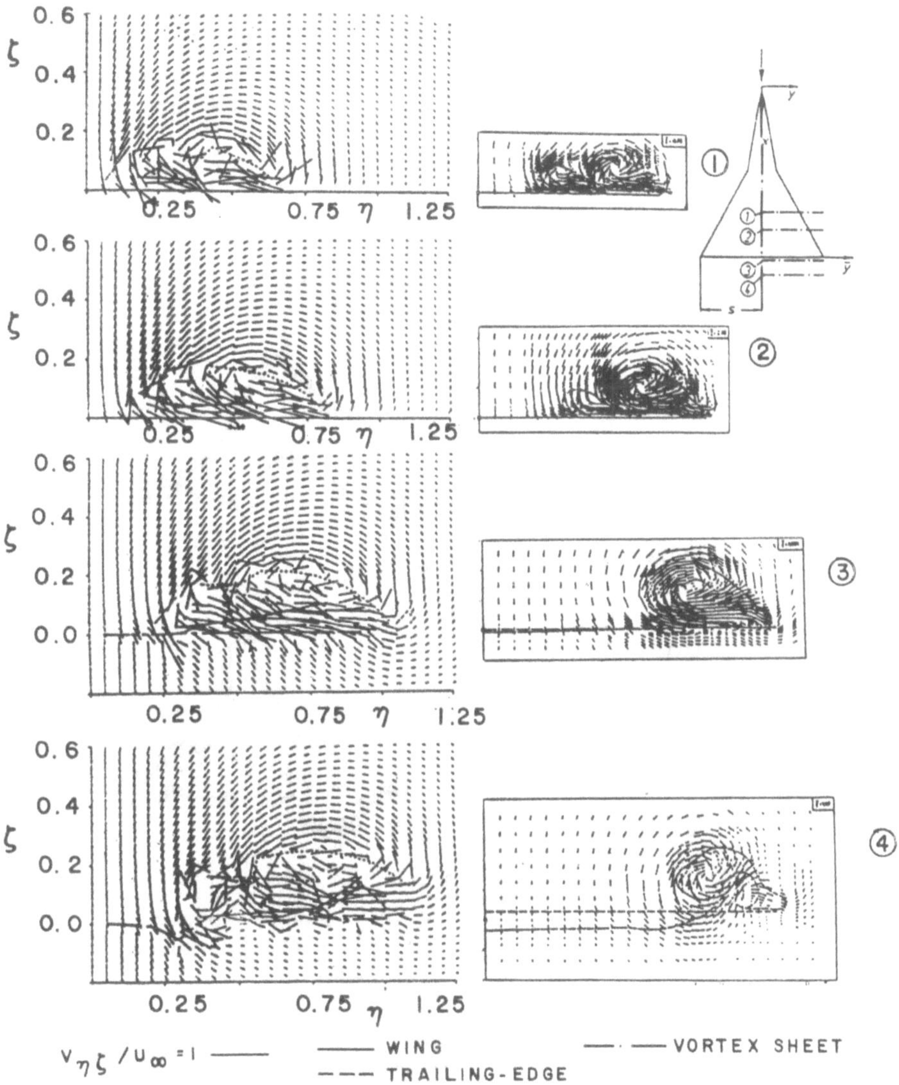

NLVLM – (Gordon 1990); Experiment – (Brennenstuhl and Hummel 1982)

Fig. 7.26. The velocity vector field at various cross-sections over the strake-delta wing

Calculations for this strake-delta wing by the use of the Euler and the thin layer Navier-Stokes equations are also available and the results will be discussed in Chapters 8 and 9, respectively. It should be noted that the NLVLM computation requires only 15 to 20 minutes on an IBM 3081D computer and probably less then a minute on a Cray 2 class supercomputer.

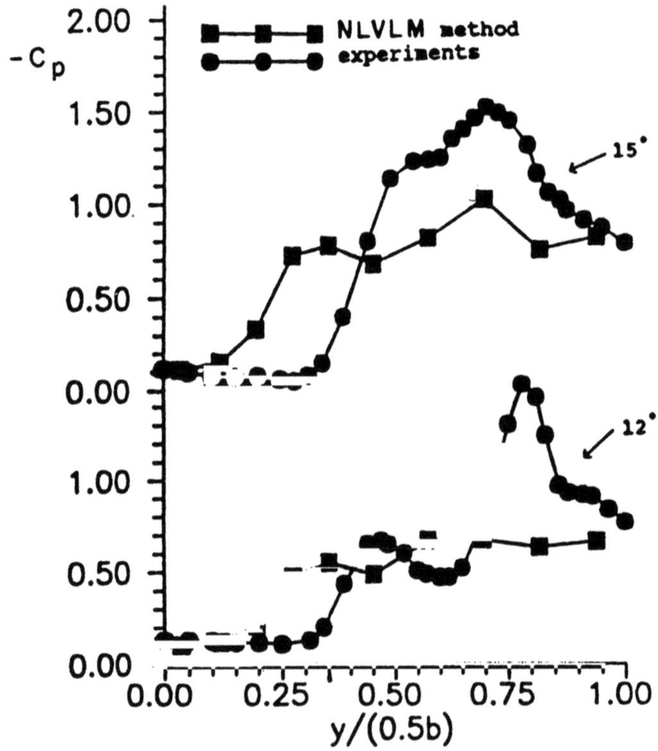

Fig. 7.27. The pressure distributions on the upper surface of the strake-delta wing at $X/C = 0.75$ (Experiment – Brennenstuhl and Hummel 1982, NLVLM – Gordon 1990)

7.1.4 Calculations of Subsonic Flows over Bodies at High Angles of Attack

In essence, the application of the NLVLM for the calculation of the aerodynamic characteristics of bodies is very similar to the procedure used for the thick wing case. As in the thick wing case, the body thickness is simulated by source distribution over the panels which are spread on the body surface, while the lifting forces are simulated by imbedding the bound vortex of each elemental horseshoe vortex in every panel on the body surface. The computer program is then constructed following the flow diagram presented in Fig. 7.1. The main difference between the wing calculation and that for the body is in the conditions for the release of the trailing vortices from the body surface. In the case of the wing with sharp leading-edges, it is easily justifiable to separate the trailing vortices from the sharp edges for all angles of attack below vortex breakdown and wing stall. However, in the case of the slender body, the lines of separation are determined by the flow conditions, and their shape and position on the body vary as the angle of attack increases. Therefore, a major effort must be devoted to the determination of the position and shape of the separation line on the body before the calculation can be completed. In principle, once the separation line is determined, it is possible to release the trailing vortices from the panels positioned along the separation line and, in this way, simulate the free vortex which is shed into the free stream from the separation line.

The NLVLM calculation begins by dividing the body surface into a large number of panels and imbedding into them the source distribution and the vortices, with all trailing vortices, initially, being required to remain attached to the body surface. Examples of the separation line on an axially symmetric body, and of panel divisions on a flattened ellipsoid and a Shuttle type configuration are shown in Figs. 7.28a, 7.28b and 7.28c, respectively.

The low speed aerodynamic characteristics of slender bodies, such as in the lifting surface case, can be obtained from the solution of the Laplace Equation, which requires the superposition of singularities fulfilling Green's theorem expressed in Eq. 7.1. The appropriate boundary conditions are that of tangency of the perturbation velocity vector on the body and the decay of the perturbation velocities far away from the body.

In the NLVLM, these conditions are fulfilled approximately by discretization of the surface, i.e. dividing it into many panels, distributing the singularities (sources and vortices) onto each panel and requiring the tangency condition at the control point of each panel. This procedure results in a set of linear algebraic equations from which the strengths of the vortices are calculated by the solution of the matrix equation (Eq. 7.3). The problem becomes nonlinear when the trajectories of the separated free vortices are included since their positions and strengths, which are related to the strengths of the bound vortices in the panels, are inter-related and can both be evaluated by an iterative calculation, as indicated in Fig. 7.1. This procedure is similar to that used in the case of the lifting surface with thickness.

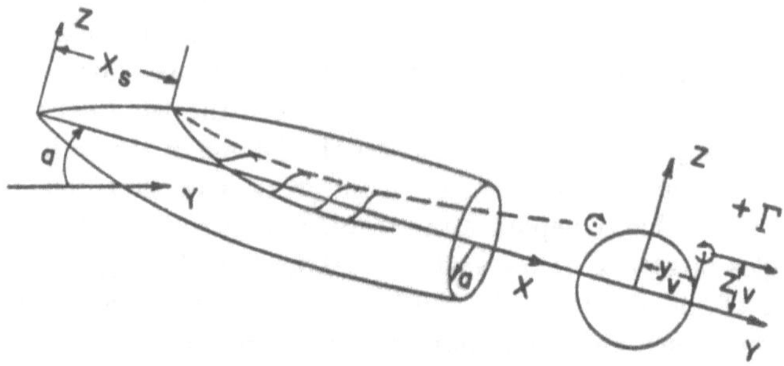

a. Elongated axially symmetric body

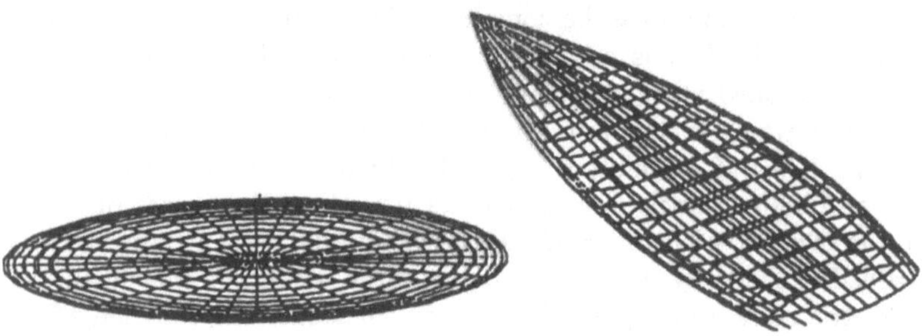

b. Ellipsoid with a fineness ratio 12 and cross section axis ratio of 4 (Rom, Almosnino and Gordon 1987)

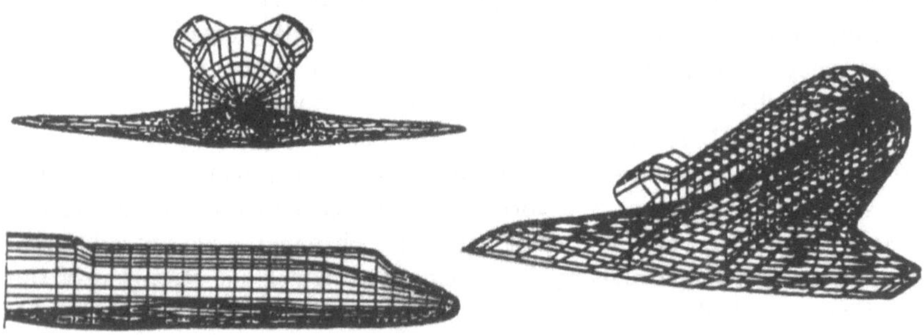

c. Shuttle type configuration (Almosnino and Rom 1988)

Fig. 7.28. Panelling of Various Bodies for NLVLM Calculations

The three-dimensional body geometry results in curved nonrectangular panel surfaces, as indicated in Fig. 7.29. It is then necessary to calculate the position and the curved shape of the bound vortex for each cell. The bound vortex is placed (by convention) at the 1/4 chord of the panel, and the paths of the trailing vortices, those remaining attached to the body surface, are directed along the boundaries of the panels up to the trailing edge of the body. The control point is positioned at the middle of the 3/4 chord line (again by convention, as in the wing panel case), and both its coordinates, and more importantly, the direction of the unit vector normal to the surface must be evaluated in accordance with the local surface curvature, as indicated in Fig. 7.29. In cases where there is separation from the body surface, the position and shape of the separation line must be evaluated independently. Since the NLVLM is not capable of determining the occurrence of separation, the position and shape of the separation line must be established from experimental or other independent results, as indicated in Fig. 7.28a. Given a separation line, it is then possible to release into the flow the trailing vortices from the panels bordering on this separation line, as indicated in Fig. 7.29. The trajectories of these free vortices are then calculated, as in the case of the lifting surface, by determining the streamline trajectories by Eq. 7.2.

Based on the new path of the free vortices, the calculations proceed to evaluate the new influence coefficient matrix in accordance with the NLVLM iterative procedure until the calculations converge and the local velocities in each panel are evaluated. It is then possible to calculate the pressure coefficient in each panel by the relation

$$C_p = \frac{2u}{V_\infty} - \frac{u^2 + v^2 + w^2}{V_\infty^2}$$

7.1.4.1 The Position of Separation on Elongated Bodies. At present, we do not have the capability to predict accurately the characteristics of the flow over slender bodies at increasing angles of attack, such as the onset of symmetric vortices, the position of the separation and the shape, for example, of the three-dimensional lines of separation. Therefore, it is of importance to present the available experimental measurements of flows over slender bodies of various shapes with their three-dimensional separation so that they may be used as guidelines for some reasonable evaluation of such flows which are dominated by the effects of separation.

Data on the onset of symmetric concentrated vortices over Ogive-Cylinder bodies is presented in Fig. 7.30, It is seen that such vortices appear on a moderately slender body of x/d of 8 at an angle of attack of less than 3°–5°, which accentuate the dominance of these vortices on this flow at higher angles of attack.

On various axially-symmetric bodies at increasing angles of attack, the measurements of the position of separation indicate a very large scatter in the evaluation of the separation point. Results for sharp-nosed Ogive-Cylinder, cone-cylinder and parabolic bodies are presented in Fig. 7.31a and in Fig. 7.31b for similar bodies with blunt noses.

The measured vertical and horizontal positions of the concentrated vortices over these bodies are presented in Figs. 7.32a and 7.32b, respec-

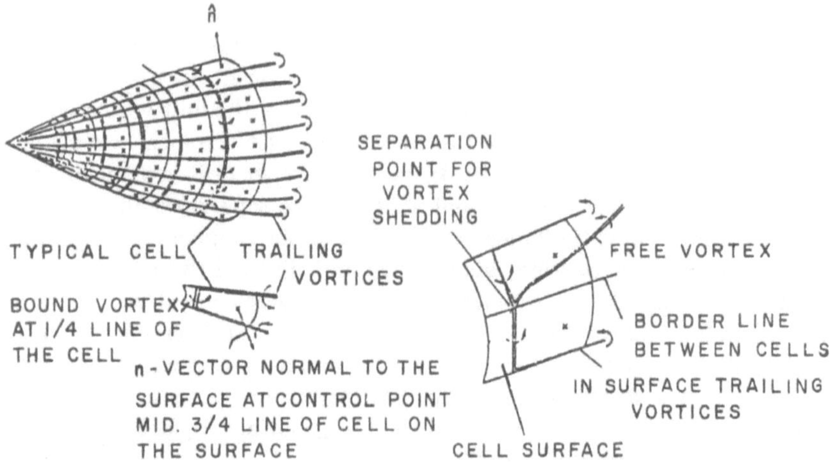

Fig. 7.29. Vortex arrangements and control point positions in the panels on the surface of a sharp nose body (Rom and Almosnino 1980)

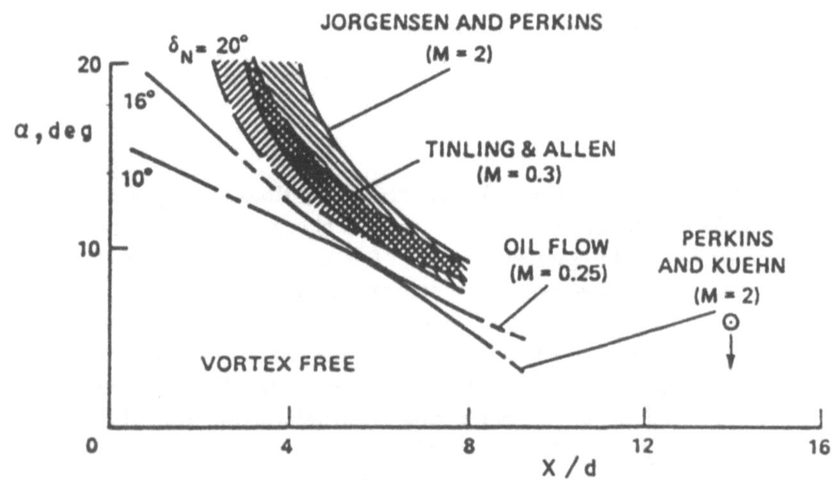

Fig. 7.30. Onset of symmetric vortices on Ogive-Cylinder bodies

tively. The measured strength of these vortices is presented in Fig. 7.33. In spite of the relatively large scatter in the experimental measurements, conclusions as to general trends in the variation of the physical parameters can be drawn from these results. Furthermore, these results can also be used for comparison with the results of the numerical methods.

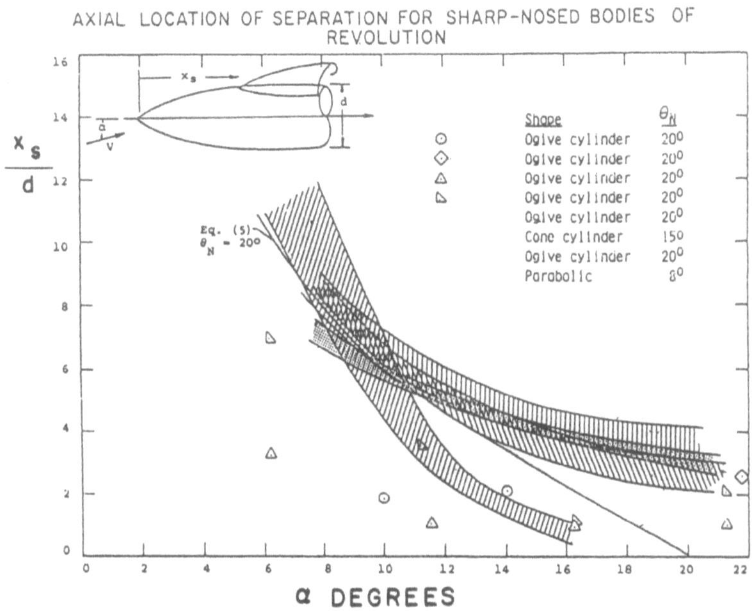

Fig. 7.31a. Axial location of separation for sharp-nosed bodies

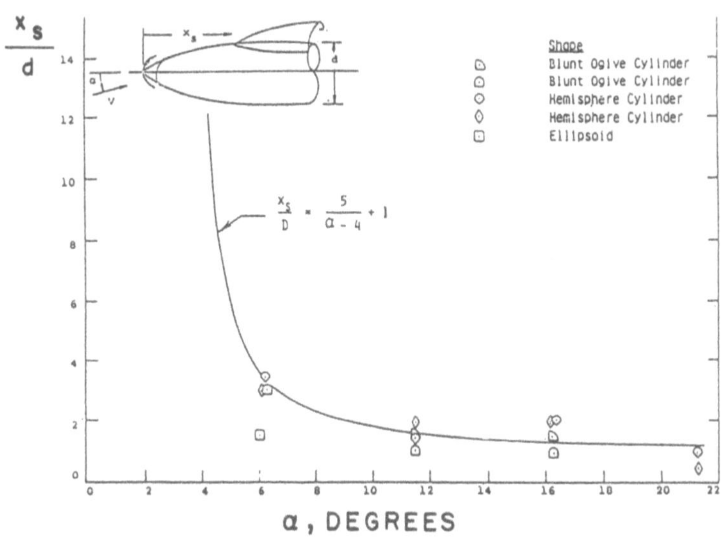

Fig. 7.31b. Axial location of separation for blunt-nosed bodies

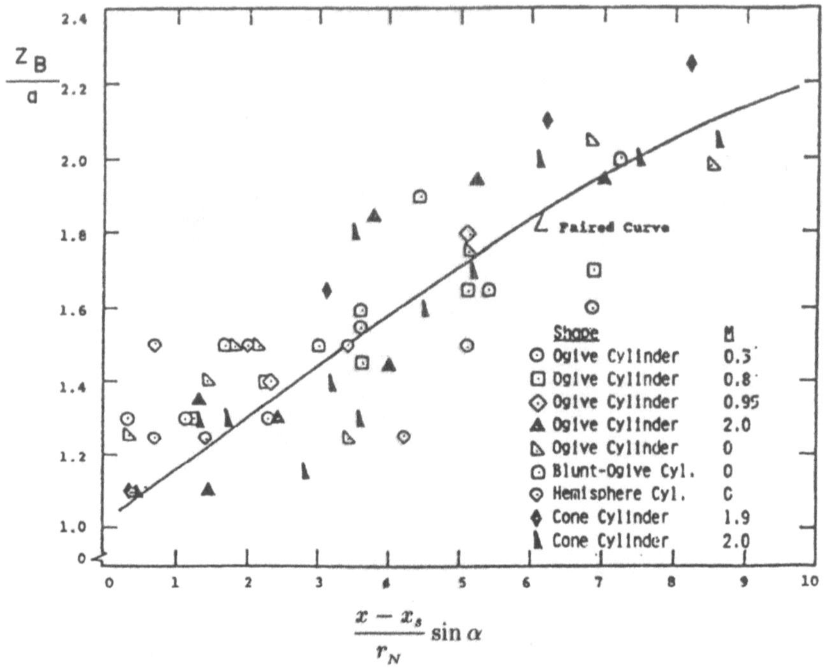

Fig. 7.32a. Vertical location of concentrated vortex over bodies

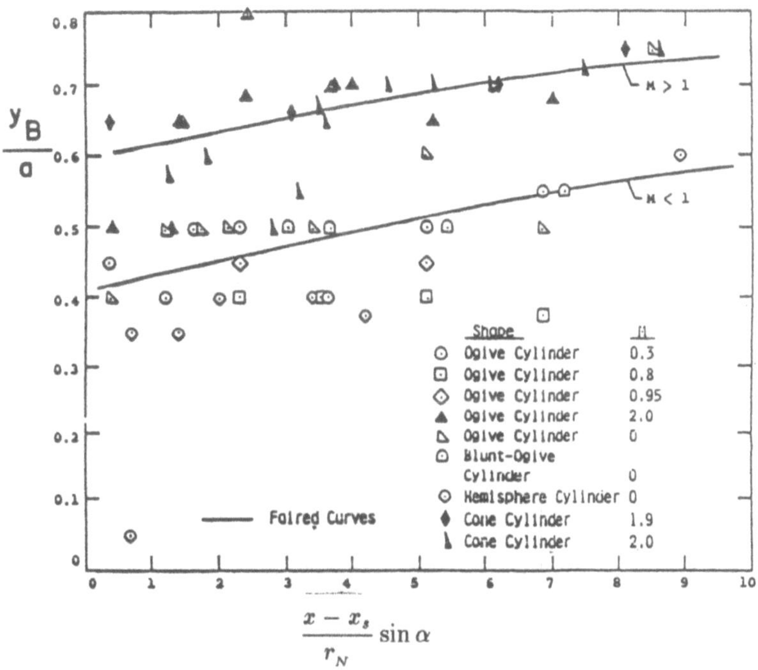

Fig. 7.32b. Horizontal location of concentrated vortex over bodies

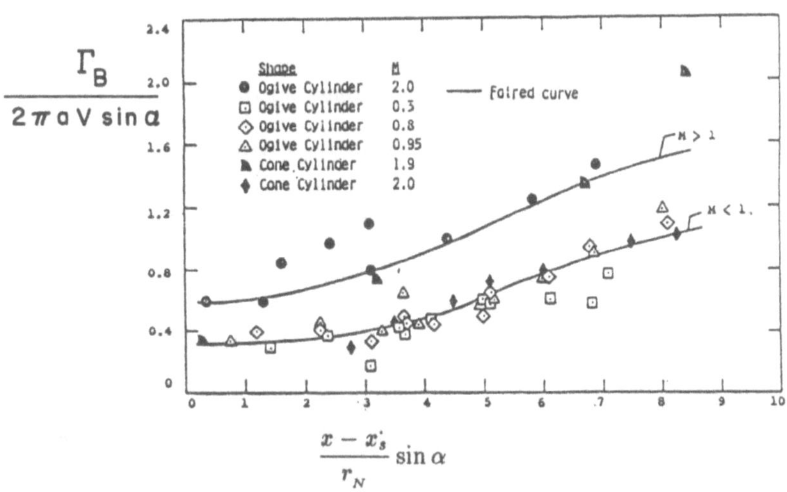

Fig. 7.33. Nondimensional strength of concentrated vortex over body

Experimental results as to the transition from the symmetric vortex separation to the asymmetric separation are presented in Fig. 7.34. For moderately slender bodies, asymmetric separation can already occur at angles of attack above 18° to 20°.

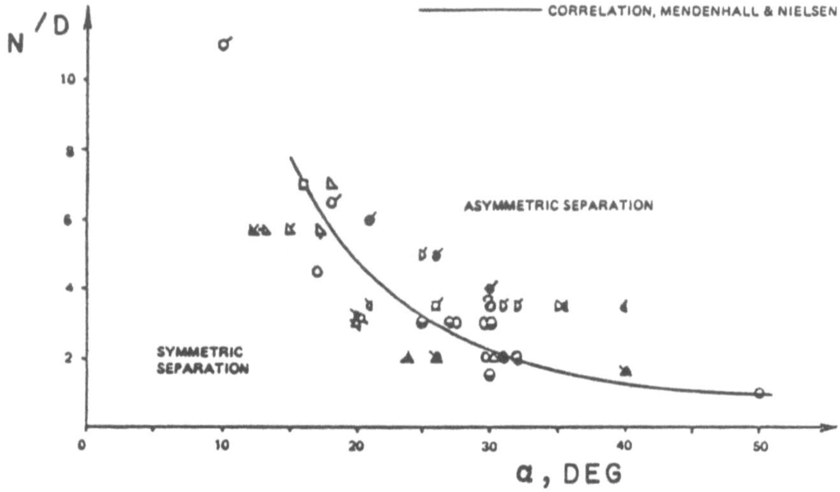

Fig. 7.34. Transition from symmetric to asymmetric vortex separation over bodies

7.1.4.2 Calculation of the Aerodynamic Characteristics for Symmetric and Asymmetric Vortex Separation over Elongated Bodies.

7.1.4.2.1 Symmetric Vortex Separation. The calculation must start by determining the line of separation on the body or by selecting a criterion for determining that line. On a flattened body, such as the elongated ellipsoid shown in Fig. 7.28b and the Shuttle configuration shown in Fig. 7.28c, the separation line can be assumed along the panels of the leading-edges of these flattened bodies. In the case of a thick body, like the Ogive-Cylinder at an angle of attack, the line of separation can be taken from experimental data or, alternatively, a criterion for separation can be introduced into the computation, so that those panels that are on the positions of minimum pressure can be taken as defining the line of separation, as indicated in Figs. 7.35, where both the experimental line and the line of minimum pressures are indicated.

Once the points of release of the trailing vortices are thus determined, it is possible to calculate by the iterative procedure the vortex trajectories, as shown in Fig. 7.36 in a side view and in Fig. 7.37 in various cross-section plans, where the symmetric roll-up of these free vortices is seen.

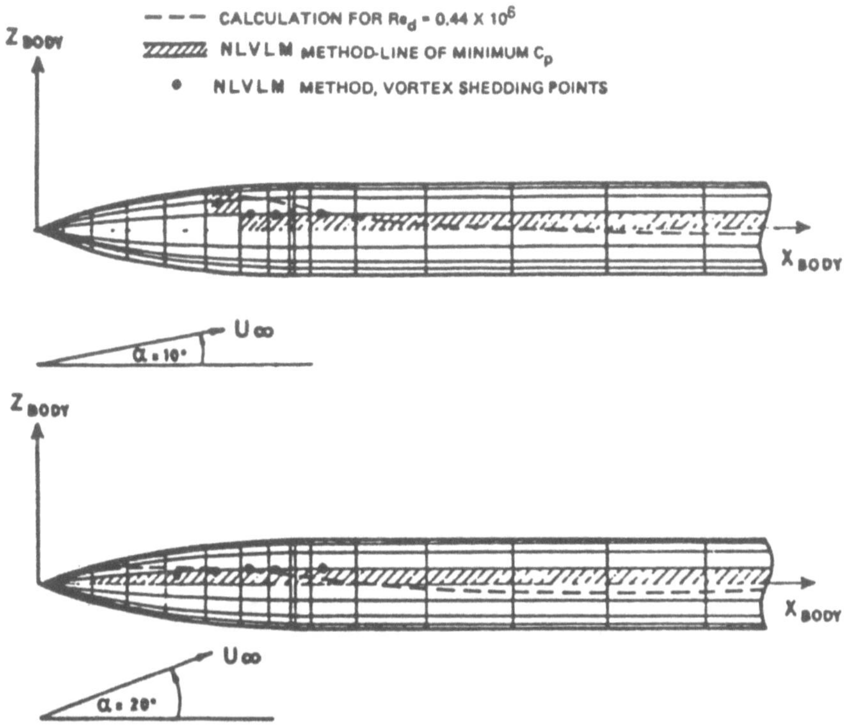

Fig. 7.35. Lines of separation on the Ogive-Cylinger body (Rom and Almosnino 1980)

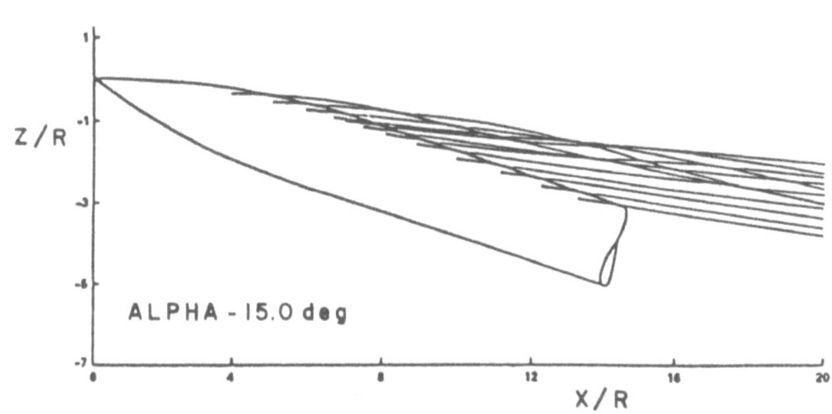

Fig. 7.36. Calculated trajectories of vortices shed from the Ogive-Cylinder body at $\alpha = 15°$ – Side view (Rom and Almosnino 1980)

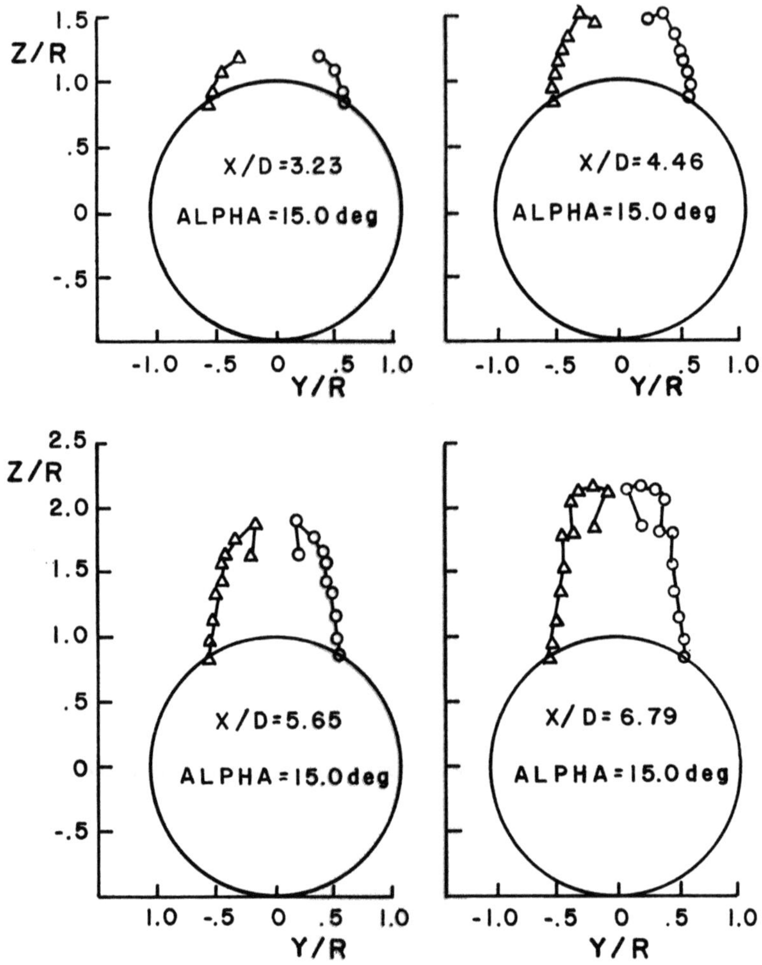

Fig. 7.37. The roll-up of the vortex trajectories over the Ogive-Cylinder body (Rom and Almosnino 1980)

The calculated pressure distributions at various cross-sections over the Ogive-Cylinder body are presented in Fig. 7.38, where in Fig. 7.38a results are presented at $\alpha = 15^o$, and in Figs. 7.38b and 7.38c at $\alpha = 20^o$ a various axial cross-section positions.

Examination of the experimental results shows some effect of Reynolds number on the circumferential pressure distribution, particularly in the leeward parts. The calculated results accord relatively well with the experimental data, although the calculations with no separation do not follow the experimental results beyond the circumferential angle of $\theta = 90^o$. As seen in Fig. 7.38b, there is good agreement between the nonseparated calculation values and the calculated results from the source-panel theory, while the

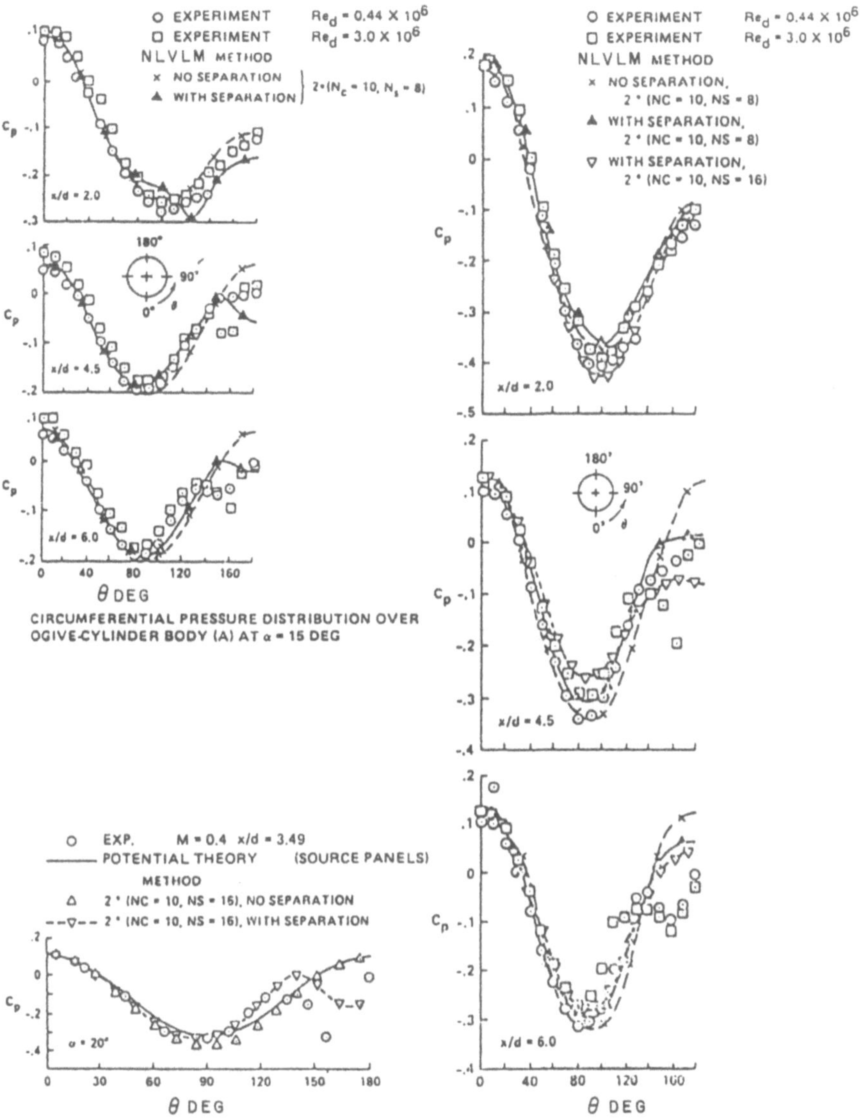

Fig. 7.38. Circumferential pressure distributions over Ogive-Cylinder body at various angles of attack (Almosnino and Rom 1983)

calculations that include the separation effects are in better agreement with the experimental results over the complete circumference, although there are some differences at the rear of the body in the range of θ between 150° to 180°.

The pressure distribution on a three-dimensional body, such as the Shuttle configuration (Fig. 7.28c), as calculated by the NLVLM over the upper centerline of the nose and canopy is presented in Fig. 7.39, where it is compared with experimental data as well as with results from an Euler code computation. The NLVLM results are in about the same reasonable agreement with the experimental data as the Euler code results.

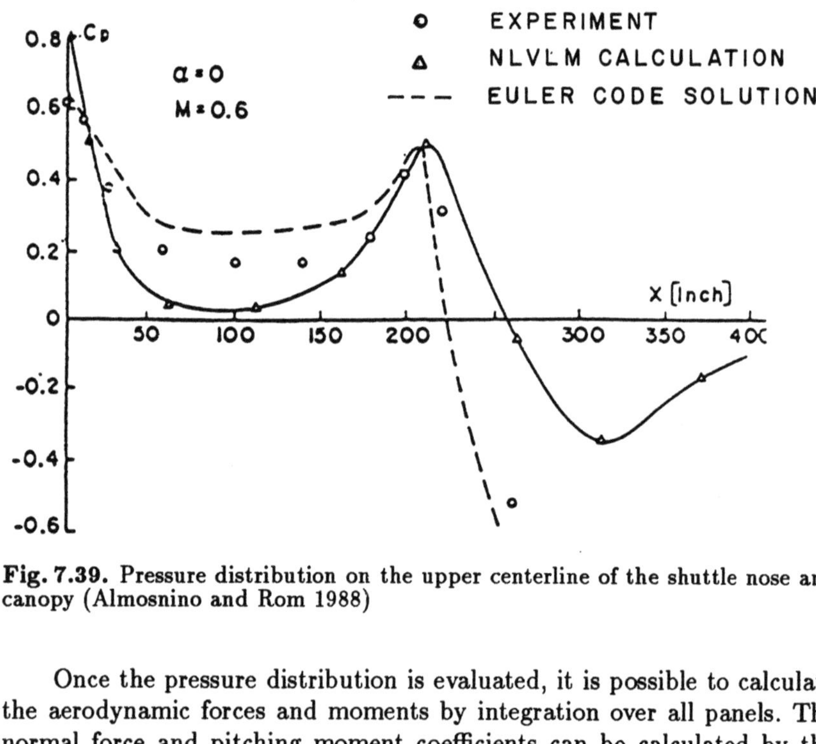

Fig. 7.39. Pressure distribution on the upper centerline of the shuttle nose and canopy (Almosnino and Rom 1988)

Once the pressure distribution is evaluated, it is possible to calculate the aerodynamic forces and moments by integration over all panels. The normal force and pitching moment coefficients can be calculated by the NLVLM, assuming a given separation position. As the separation points move forward, the effect of the vortices becomes stronger and the nonlinear lift force increases. This effect is seen in Fig. 7.40, in which the normal force and pitching moment are calculated for two values of separation points and are compared with experimental measurements at two Reynolds numbers. The agreement of the calculated results with the experimental values is rather good.

The capability of the NLVLM to evaluate the aerodynamic coefficients of a more complex body configuration is demonstrated in Figs. 7.41a to 7.41c, where the calculated variations of the lift, the induced drag and

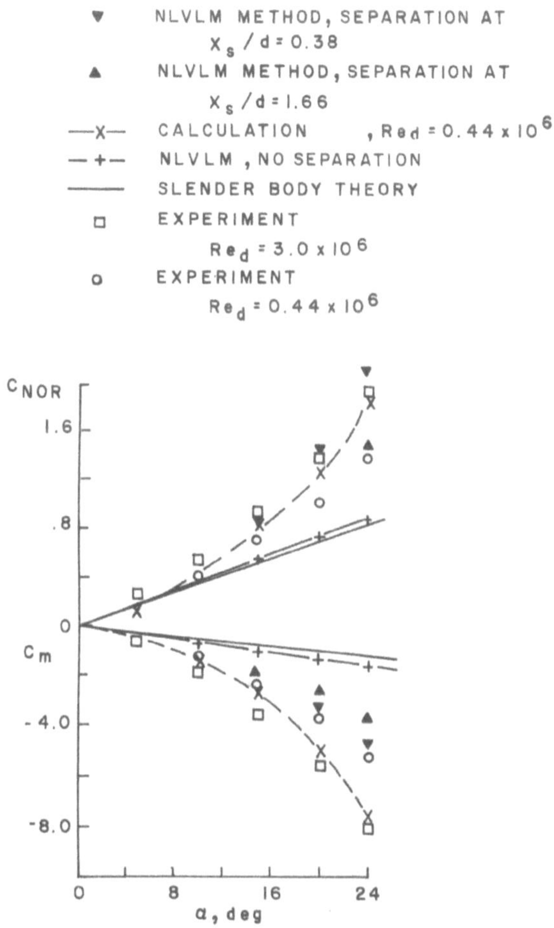

Fig. 7.40. Variation of normal force and pitching moment coefficients as a function of angle of attack for an Ogive-Cylinder body (Almosnino and Rom 1983)

the pitching moment coefficients, respectively, as a function of the angle of attack are compared with the experimental data. Excellent agreement is found for the lift and induced drag variations, while that for the pitching moment is more qualitative.

7.1.4.2.2 Asymmetric Vortex Separation. An asymmetric flow structure can be obtained on an elongated body by arbitrarily imposing delay in separation of the vortex on one side of the model, as sketched in Fig. 7.42. This results in the shift in the positions of the vortices and therefore, an asymmetry in the vortices which roll-up over the body, as can be seen in Fig. 7.42. This asymmetry also affects the pressure distribution around the circumference. By having the pressures on the right side lower than those

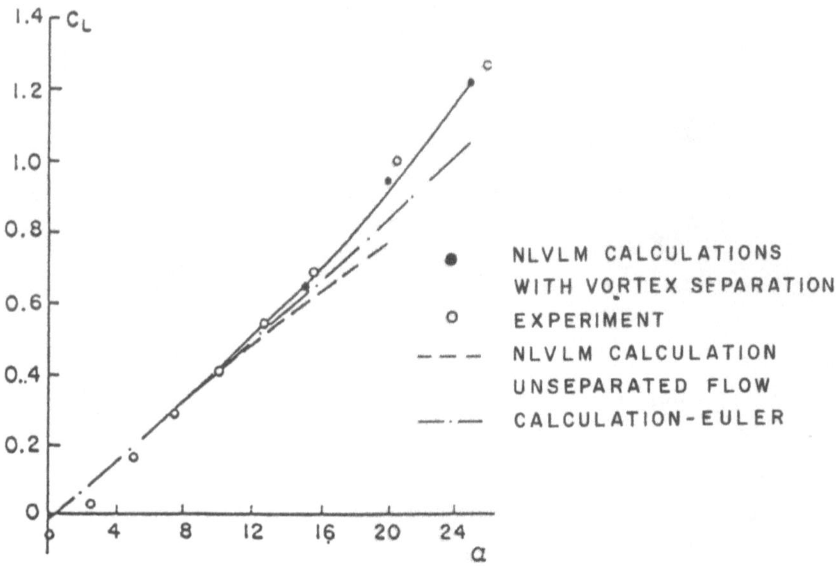

Fig. 7.41a. Variation of the lift coefficient as a function of angle of attack for the shuttle configuration (Almosnino and Rom 1988)

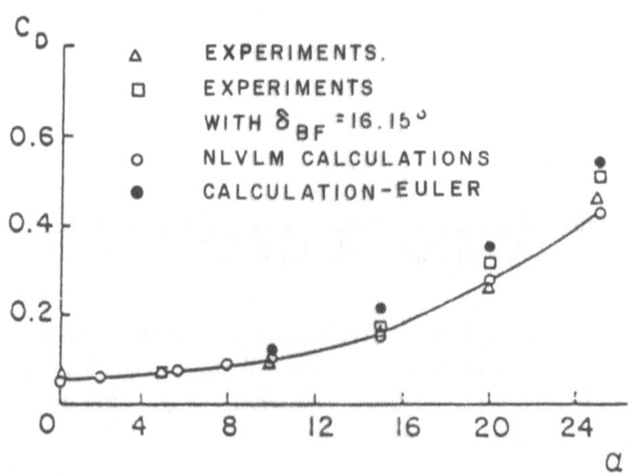

Fig. 7.41b. Variation of the drag coefficient as a function of angle of attack of the shuttle configuration (Almosnino and Rom 1988)

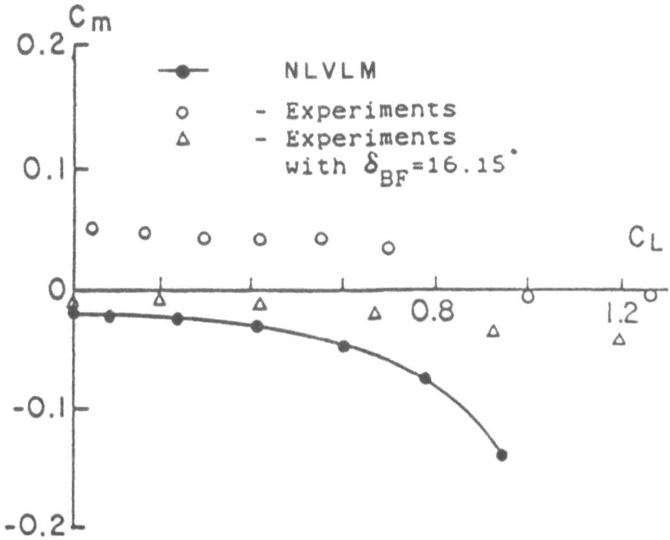

Fig. 7.41c. Variation of the pitching moment as a function of the angle of attack for the shuttle configuration (Almosnino and Rom 1988)

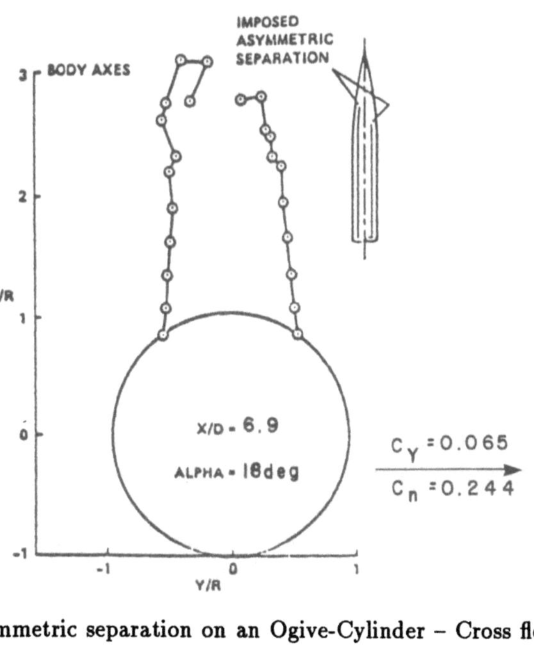

Fig. 7.42. Asymmetric separation on an Ogive-Cylinder – Cross flow view (Almosnino 1985)

on the left side, the results are a net side force and a side moment, which can be calculated using the NLVLM, as seen in Fig. 7.43.

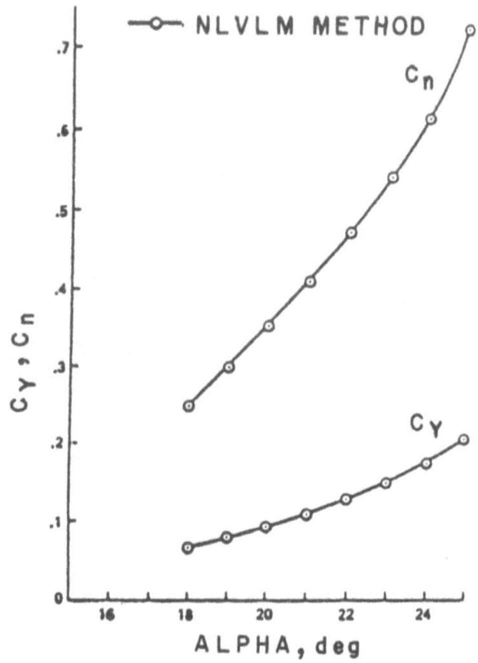

Fig. 7.43. Side force and yawing moment variation with angle of attack due to asymmetric separation on an Ogive-Cylinder (Almosnino 1985)

7.2 The Free Vortex Sheet (FVS) Method for Subsonic Flows

The Free Vortex Sheet (FVS) method is developed by considering a flow model based on the assumption that the three-dimensional vortical flow over the lifting surfaces and bodies is dominated by the separated thin shear layers and slender vortex cores which are embedded in the otherwise irrotational flow. It is assumed that the thin sheer layers are separated as vortex sheets from known separation lines. In many cases, the separation lines are the sharp edges of the configuration. The method then employs the three-dimensional inviscid flow equations which are solved by distributing singularities on the surfaces of the configuration, on the separated vortex sheets which roll-up into concentrated vortices, and on the surfaces of the wake. The quadratic doublet and the linear source distributions are used to fulfill the potential flow equation and to satisfy the proper boundary conditions. Since the shapes and positions of the rolled-up vortex sheets over and in the wake of the configuration are not known a priori, the strengths of the singularity distributions and the shapes and positions of the vortex sheets are computed by an iterative process, starting with an assumed initial sheet geometry. It is then possible to calculate the aerodynamic forces and moments as well as the pressure distributions. This chapter will present the theoretical basis for this method, discuss the numerical programs which were developed to enable converging solutions, and present the capabilities of this method to evaluate the aerodynamic characteristics of configurations of various shapes in subsonic flows.

7.2.1 The Theoretical Model

The flow field is represented by a combination of the flows over the following elements: the configuration surfaces (wing, body, additional surfaces); the trailing sheet (wake); the sheet emerging from the wing leading-edge and tip (free sheet); and the rolled-up core-spiral region (fed sheet) which is fed by the leading-edge sheet, as shown in Fig. 7.44.

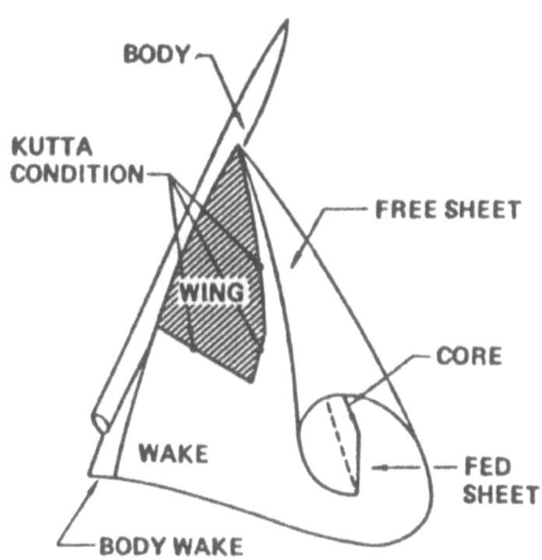

Fig. 7.44. Definition of the vortex sheet method flow field model

The governing equation for the velocity perturbation function in the steady inviscid subsonic compressible flow, is given by the Prandtl-Glauert equation, where $\beta^2 = 1 - M_\infty^2$

$$\beta^2 \phi_{xx} + \phi_{yy} + \phi_{zz} = 0 \tag{7.7}$$

Introducing a mass flux vector $\mathbf{W} = \rho_\infty \mathbf{V}_\infty + \mathbf{w}$, where $\mathbf{w}$ is the perturbation mass flux vector defined by $\mathbf{w} = (\beta^2\phi_x, \phi_y, \phi_z)$. In terms of the mass flux vector, the surface impenetrability can be expressed as $(\mathbf{W}\cdot\mathbf{n})=0$, where $\mathbf{n}$ is the vector normal to the surface. In this case, the force is defined by the relation

$$F = -\iint_S [\mathbf{V}(\mathbf{W}\cdot\mathbf{n}) + p_2\mathbf{n}]\, dS \tag{7.8}$$

where $p_2 = p_\infty - [\rho_\infty(\mathbf{V}_\infty \cdot \mathbf{v}) + 1/2(\mathbf{v}\cdot\mathbf{w})]$. Then, by the use of Green's theorem, it can be shown that the solution of the Laplace equation at a point $P(x, y, z)$ may be expressed as the potential induced by the superposition of source singularities σ and doublet singularities μ distributed on the surface B, as follows:

$$\phi(P) = -\iint_B \sigma\left(\frac{-1}{4\pi R}\right) dS + \iint_B \mu\mathbf{n}\cdot\beta^2 A^{-1}\nabla\left(\frac{1}{4\pi R}\right) dS \tag{7.9}$$

Then, the perturbation velocity vector, $\mathbf{v}(\mathbf{P})$, at a point $\mathbf{P}(x, y, z)$ can be expressed by the relation

$$\begin{aligned}\mathbf{v}(\mathbf{P}) = &\iint_B \sigma\nabla_P\left(\frac{-1}{4\pi R}\right) dS + \iint_B \zeta\times\beta^2 A^{-1}\nabla\left(\frac{1}{4\pi R}\right) dS + \\ &+ \int_L \Delta\mu\beta^2 A^{-1}\nabla\left(\frac{1}{4\pi R}\right)\times d\mathbf{l}\end{aligned} \tag{7.10}$$

Here, $\zeta = \mathbf{n}\times\nabla\mu$, $R = [\mathbf{R}\cdot A\cdot\mathbf{R}]^{1/2}$ and A is the matrix

$$A = \begin{pmatrix} 1 & 0 & 0 \\ 0 & \beta^2 & 0 \\ 0 & 0 & \beta^2 \end{pmatrix}$$

and L is any curve on the surface B across which μ has a discontinuity, $\Delta\mu$.

The boundary conditions which are imposed on the various elements in the flow field, as shown in Fig. 7.44, are that the velocity normal to the configuration surface must be zero, that the free sheet and the wake cannot support any pressure difference, and that the velocity normal to these surfaces must also be zero. The fed sheet (indicated in Fig. 7.44) is an extension of the free sheet and feeds vorticity to the vortex core which is assumed to be a line vortex. The proper boundary condition for the fed sheet, governing its size and core orientation, is that the total force induced on the fed sheet and core by the rest of the configuration be parallel to the core. The size of the fed sheet is chosen initially by experience or from the conical flow results of Smith (1966). Along all sharp edges (leading, side and trailing) of the lifting surfaces from which the free sheets are emanating, Kutta conditions are imposed.

The mathematical implementation of the boundary value problem and the development of the numerical solutions have been presented by Johnson and Rubbert (1975), Weber et al. (1976) and Johnson et al. (1980). The unknowns are the source and doublet strengths distribution on all surfaces and the positions and shapes of the free, fed and wake sheet surfaces. These unknowns can be calculated from the simultaneous solution of the set of the equations formulated in the potential flow and the appropriate boundary conditions.

On thin surfaces bounded on both sides by fluid, such as free sheet, wake and zero thickness wing, the source strength must be zero, so that on such surfaces only doublet singularity is required. The no penetration condition must be fulfilled, so that the boundary condition is

$$\mathbf{W}_a \cdot \mathbf{n} = 0 \tag{7.11}$$

where the subscript a denotes the average value across the sheet.

On the free sheet and wake, the pressure difference must be zero. The pressure difference is evaluated by the relation

$$\Delta p_2 = -\beta^2 A^{-1}\mathbf{n} \cdot (\mathbf{W}_a \times \zeta)/[\mathbf{n} \cdot (\beta^2 A^{-1}\mathbf{n})] = 0 \tag{7.12}$$

and the force difference is then

$$\Delta\mathbf{F} = -\iint [\mathbf{W}_a \times \zeta - \mathbf{V}_a \rho_\infty \sigma]\, dS + \int \mu \mathbf{W}_a \times d\mathbf{l} \tag{7.13}$$

It can be seen that Eqs. 7.12 and 7.10 define the doublet strength on these surfaces while Eq. 7.11 defines the vector normal to the surface which enables the calculation of the surface geometry. An approximate value for $\mathbf{W}_a$ in the far wake is $\mathbf{W}_a = \rho_\infty \mathbf{V}_\infty$; however, this approximation is not valid in the near wake.

The Kutta condition at the junction between the wing surface and the vortex sheet requires special attention, since on the wing the boundary condition of Eq. 7.11 is applicable, while on the free vortex Eq. 7.12 holds. However, in order to ensure finite velocities at the junction, it is required that

$$\Delta\zeta \cdot \mathbf{l} = 0, \tag{7.14}$$

where $\Delta\zeta = \zeta_{\text{vortex}} - \zeta_{\text{wing}}$ and $\mathbf{l}$ is the unit vector along the junction.

The fed sheet is needed in order to condense the free sheet vorticity into a line vortex core, thereby terminating the free sheet roll-up. Hence the fed sheet is chosen to be a doublet sheet whose strength is equal to the doublet value at the junction with the free sheet. Only the size and position of the fed sheet remain to be determined from the boundary conditions, which in this case require that the force normal to the core be zero

$$\mathbf{l} \times \Delta\mathbf{F} = 0, \tag{7.15}$$

where $\mathbf{l}$ is the unit vector along the line vortex core and $\Delta\mathbf{F}$ is given by Eq. 7.13.

It can be shown that the thickness effect of the thick wing or body can be simulated by a source distribution without any doublet strength $\mu = 0$, while a zero thickness lifting surface can be simulated by a doublet distribution without any sources $\sigma = 0$.

The integral equation for the perturbation velocity, Eq. 7.10, and the boundary conditions defined in Eqs. 7.11 to 7.15 pose a well-defined boundary value problem which can be solved numerically by the panel method proposed by Johnson and Rubbert (1975). The boundary conditions may be of the Neumann type (specification of $\partial\phi/\partial n$) or Dirichlet type (specification of ϕ) or of the mixed type. The boundary value problem with Neumann type boundary conditions is designated as an "analysis" problem and with Drichlet type boundary conditions as a "design" problem.

7.2.1.1 Method of Computation. The first step in the computation program is to divide the boundary surface into networks where a network is defined as a smooth portion of the boundary. Each network is divided into quadralinear panels, so that each panel is specified by a rectangular array of corner point coordinates. The panel surface is approximated by an analytical function. On each panel, linear source and/or quadratic doublet distributions are defined, accompanied by well posed boundary conditions to be fulfilled at specified control points in each panel. This is accomplished by assigning discrete values of singularity strength to certain standard points on each network. These values are interpolated by source and doublet splines which on each panel are assumed to be defined by linear and quadratic distributions, respectively. Then, certain points on each panel are assigned as control points at which the boundary conditions are applied. These points include panel center points as well as edge abutment points in the case of doublet networks. The latter serve to impose standard aerodynamic edge conditions, such as the Kutta condition, zero potential jump at thin edges, and continuity of singularity strength across abutting network. Now, the influence coefficients are evaluated to conform to the potential and velocity integral equations, Eqs. 7.9 and 7.10. The arrangement of the singularities and control points for various network types is presented in Fig. 7.45.

The panelling of a configuration comprising a thin delta wing and a body is shown in Fig. 7.46 and a breakdown of this configuration into the appropriate networks is shown in Fig. 7.47.

The position and shape of the free and fed sheets are determined during the iterative calculation procedure. Their shape and position are, therefore, defined by a set of parameters which determine their kinematics and which are evaluated by the calculations, as shown in Fig. 7.48.

The numerical procedures and some of the improvements developed to overcome certain difficulties encountered in the computation programs and improve the convergence and accuracy of the calculations are discussed by Johnson et al. (1980). They also treat the various verification computations

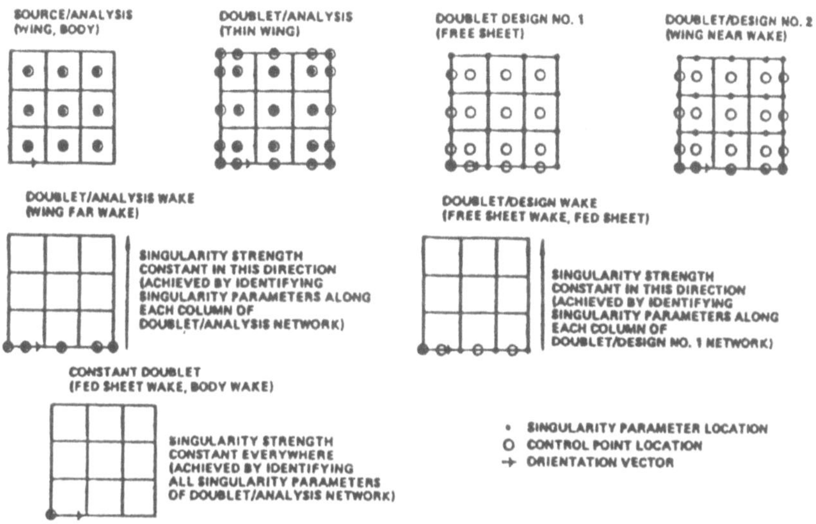

Fig. 7.45. Location of discrete singularity parameters and control points for various network types (Johnson et al. 1980)

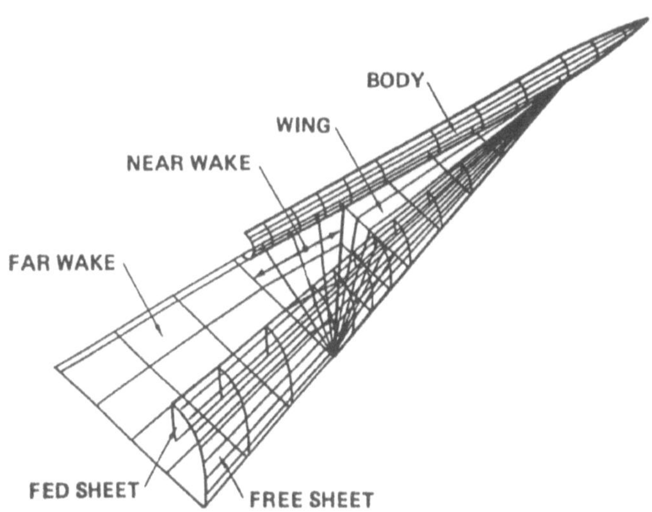

Fig. 7.46. Panelling of a Thin Delta Wing and Body Configuration (Johnson et al. 1980)

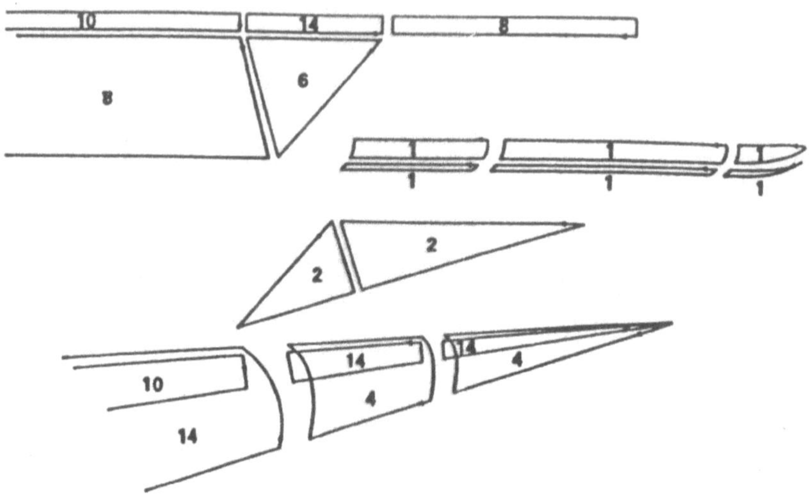

Fig. 7.47. View of the networks for the calculations of the thin delta wing and body configuration (Johnson et al. 1980)

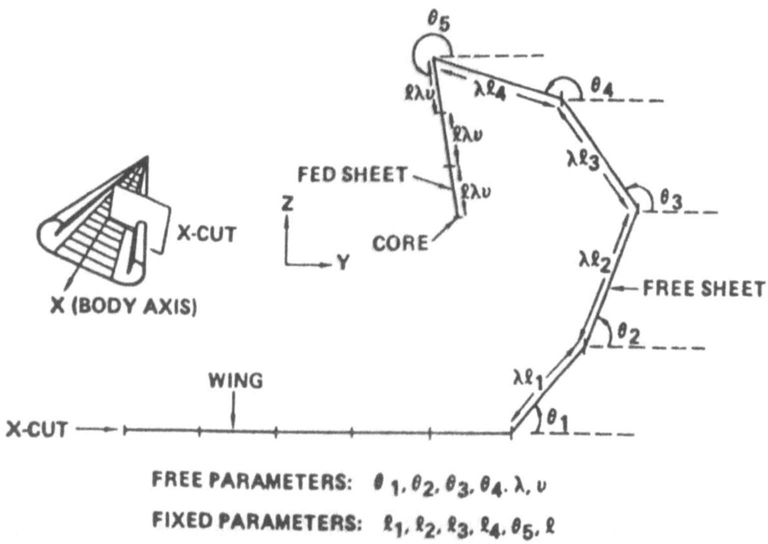

Fig. 7.48. The parameters defining the kinematics of the free and fed sheets (Johnson et al. 1980)

to determine the effects of the following: panel density and layout, vortex sheet roll-up and vortex kinematics, and initial shape of the free and fed sheet. It was shown that by selecting the proper panel geometry and singularity distributions, it is possible to obtain converged solutions for the flow field, the pressure distributions and the aerodynamic forces and moments on wings of various planforms and wing/body configurations. It is of interest to compare the results of these calculations with experimental measurements and with results of other computational methods.

7.2.2 Results of the FVS Method Calculations

7.2.2.1 Delta Wings. The aerodynamic characteristics of delta wings with AR varying from 0.25 up to 2 were calculated using the panel system with 30 and 60 panels on the wing. The networks and panels divisions on the delta wing are presented in Fig. 7.49.

The lift coefficients for delta wings of Aspect Ratios from 0.25 to 2.0 at an angle of attack of $\alpha = 20^o$ are presented in Fig. 7.50. The calculated values by the FVS method with simple wake and with the near wake networks are presented and are compared with the calculated results of the Polhamus Suction Analogy and with those of the NLVLM. These calculated lift coefficients are compared with the experimental measurements of Peckham (1961), Ticono and Yoshihara (1979), Wentz (1972), Kirpatrick (1970) and Fox and Lamar (1974). The Vortex Sheet calculations indicate the important effect of the near wake networks on the results. The inclusion of the near wake networks by this method is clearly seen to improve the agreement of the calculated lift values with the experimental data for both the low and the higher Aspect Ratio delta wings. With the near wake configuration, the FVS results accord well with the experimental data for delta wings with Aspect Ratios between about 0.5 to 1.5, but the calculated results are too low for AR=0.25 and too high at AR=2.0. The FVS results are in good agreement with the Suction Analogy results for delta wings of AR above about 1. The Suction Analysis calculated results accord well with the experimental measured values for AR between 0.5 to 1.5, but at AR=0.25, the Suction Analogy results are higher than the experimental data and are also too high at AR=2.0. It is interesting to note that the NLVLM results are in good agreement over the complete range of the delta wings examined here, from AR=0.25 to 2.0. The variation of the lift coefficient with angle of attack for the AR=0.25 delta wing is presented in Fig. 7.51.

The examination of the variation of the lift coefficient with angle of attack for AR=0.25, presented in Fig. 7.51a, indicates that the NLVLM calculated results accord well with the experimental data up to about $\alpha = 20^o$; the FVS calculated values are in good agreement up to $\alpha = 15^o$ and are too low at higher angles, while the Suction Analogy calculations result in too high values for α above 15°. The variation of the pitching moment as a

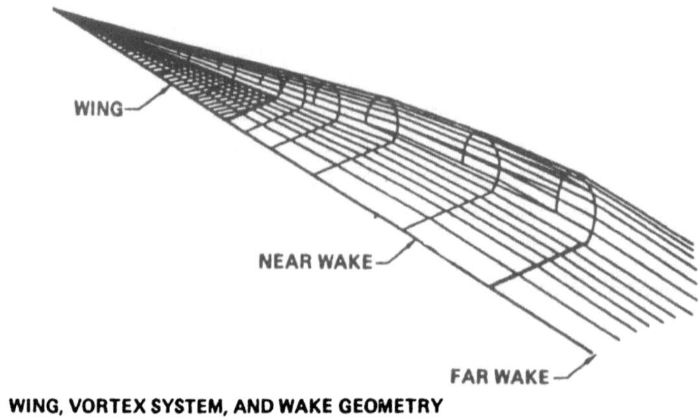

Fig. 7.49. Panelling of delta wing with wake (Johnson et al. 1980)

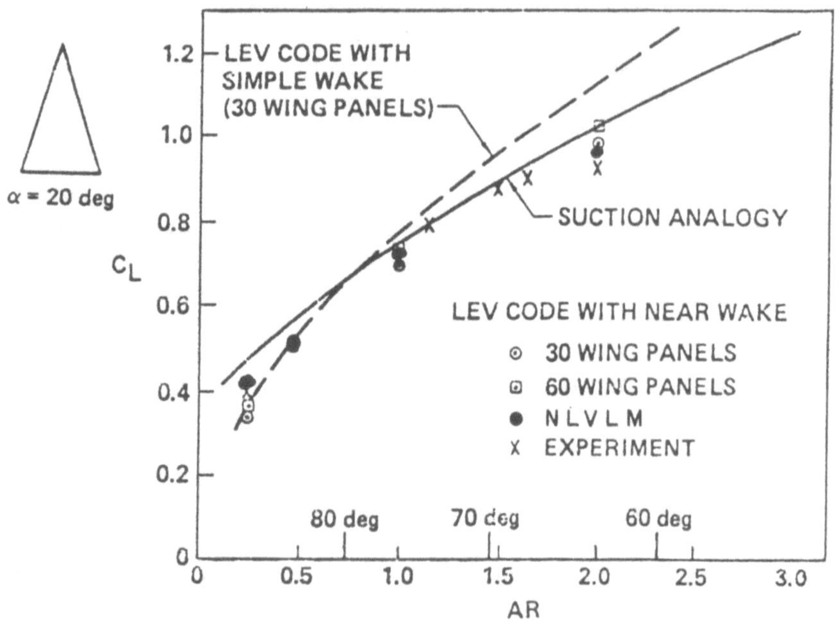

Fig. 7.50. Comparison of the measured and calculated lift coefficient variation as a function of Aspect Ratio for delta wings

function of lift coefficient is presented in Fig. 7.51b. The pitching moment evaluated by the Suction Analogy and by the FVS method results are in good agreement with the experimental data, while the NLVLM results indicate somewhat higher (less negative) pitching moment values. The induced drag coefficient variation with the lift coefficient is presented in Fig. 7.51c. The calculated induced drag by the FVS method and by the NLVLM are in reasonably good agreement over the complete range of measurements.

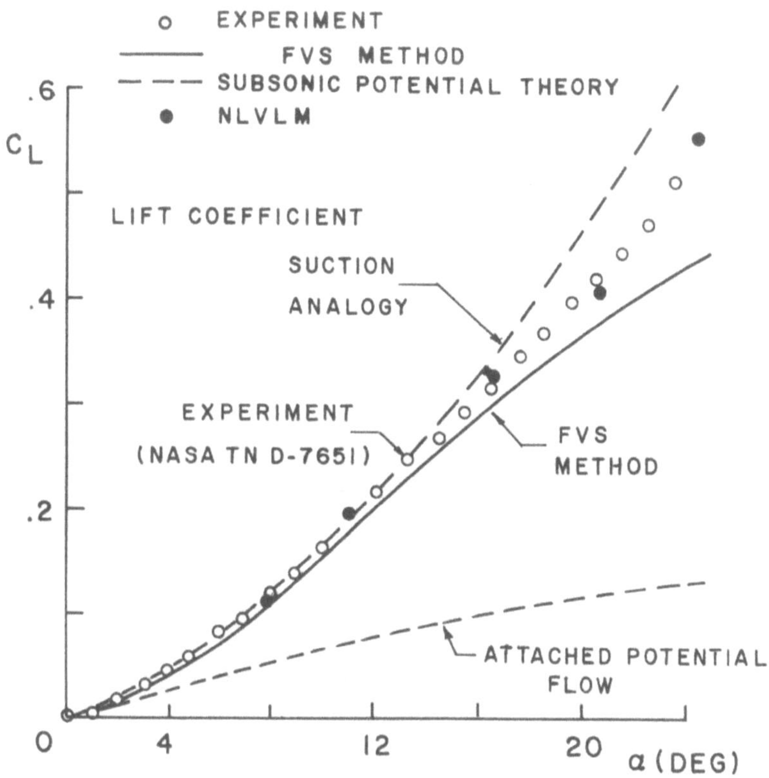

Fig. 7.51a. Variation of lift coefficient with angle of attack

One of the important advantages of the FVS method is its success in computing the pressure distributions over delta wings, havig relatively good agreement with experimental measurements. Comparison of the calculated pressure distributions with the measured values for delta wings of AR=1.0 and 1.46 are presented in Fig. 7.52a and 7.52b, respectively.

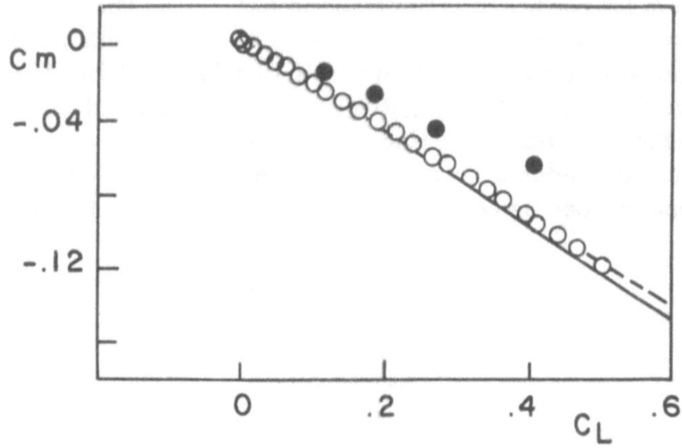

b. Variation of pitching moment as a function of lift coefficient

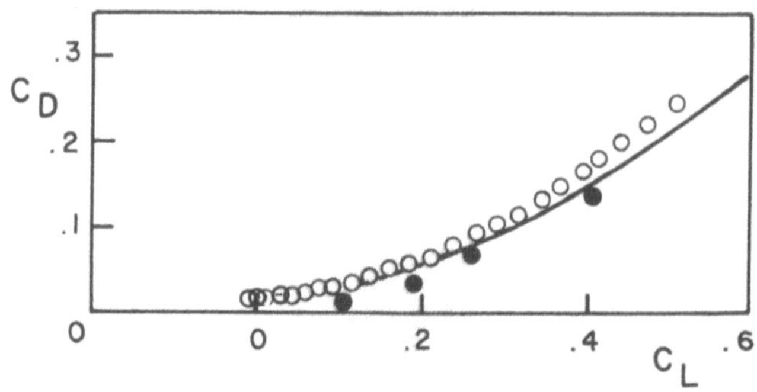

c. Variation of the induced drag coefficient as a function of lift coefficient

Fig. 7.51. Comparison of the measured and calculated lift, induced drag and pitching moment coefficients variation with angle of attack for delta wing of AR=0.25

For the AR=1 delta wing (Fig. 7.52a), the pressure distributions calculated by the FVS method are in very good qualitative and reasonably quantitative agreement with the experimental data. The calculated pressure distributions on the delta wing with the higher AR=1.46 (Fig. 7.52b) show only one peak (as expected from a model representing the roll-up as a single straight line vortex core), while the experimental data indicate a double peak which may be due to the effect of a secondary vortex.

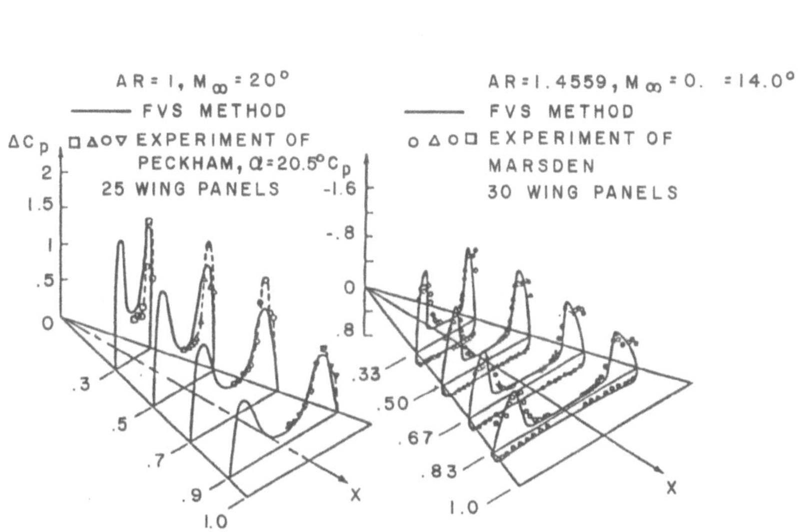

a. AR=1.0, $\alpha = 20^o$

b. AR=1.46, $\alpha = 14^o$

Fig. 7.52. Pressure distributions over delta wings (Johnson et al. 1980)

The differences in the aerodynamic characteristics due to camber were measured by Wentz (1972) where he compared the values for a flat delta wing of AR=1.15 and with a cambered wing of an identical planform. The cambered wing had a conical camber at the leading-edges as shown in Fig. 7.53.

The pressure distributions on the and cambered wings were measured, and these results are compared with the calculated values by the FVS method in Fig. 7.54. The main effect of the camber on the pressure distribution is to change the pressure near the leading-edge of the lower surface, as seen in Fig. 7.54. The calculated pressure distributions on both the upper surface (showing very little effect there due to camber) and on the lower surface accord well with the measured results.

The aerodynamic characteristics of the cambered delta wing were compared with those of the flat delta wing, by making both experiments and calculations using the FVS method and the NLVLM. Comparison of the drag polar measured by Wentz (1972) with the calculations of Johnson et al. (1980) and of Gordon and Rom (1985) is presented in Fig. 7.55. The ex-

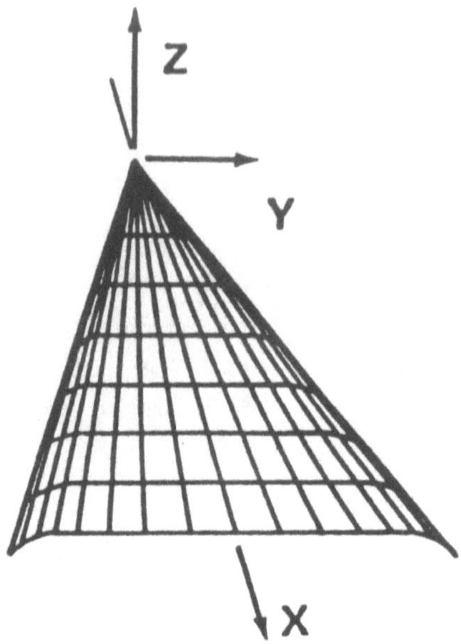

Fig. 7.53. Delta wing of AR=1.15 with conical camber

perimental data show that there is a drag reduction at a fixed lift coefficient due to the camber of about 7%. The calculated results of the FVS method also indicate such reduction, but the lift coefficient values are about 10% higher than the experimental data; the NLVLM results are in very good agreement with the data as shown in this figure. The excellent results of the NLVLM in calculating the variation of the lift and pitching moment with angle of attack are presented in Figs. 7.56a and 7.56b, respectively.

The comparison of the calculated pressure distributions by the FVS method with the experimental measurements on a slender arrow wing is presented in Fig. 7.57. It can be seen as another example of the ability of this method to evaluate the pressure distributions on lifting surfaces on which a single leading-edge vortex separation is established. The experimental data indicates the existence of a second peak (probably due to the secondary vortex) which cannot be obtained in the FVS method calculations.

Further refinements of the FVS method were developed by Hoeijmakers (1989). These developments include applications to the calculations of quasi two-dimensional vortex wakes generated behind lifting surfaces and the calculations of quasi two-dimensional vortex flows about slender configurations, such as delta type planforms, including an option of introducing a secondary vortex sheet separated from underneath the primary vortex

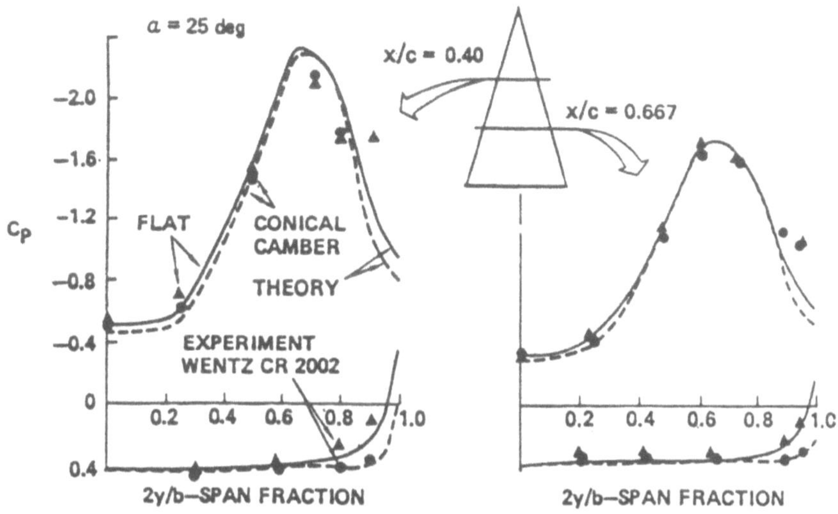

Fig. 7.54. Pressure distributions over a flat and cambered delta wing of AR=1.15 at $\alpha = 25°$ (Johnson et al. 1980)

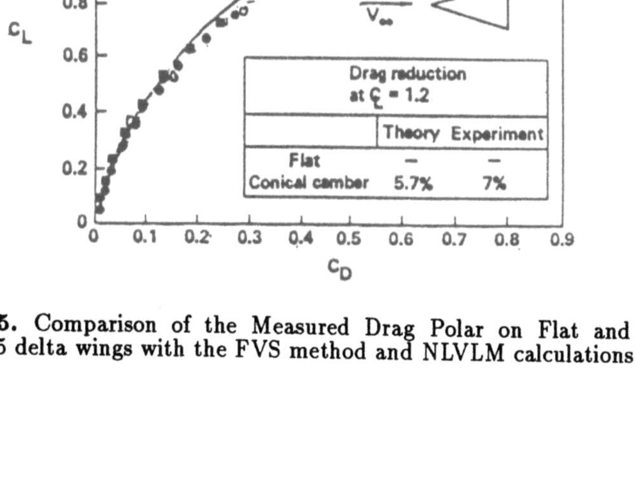

Fig. 7.55. Comparison of the Measured Drag Polar on Flat and cambered AR=1.15 delta wings with the FVS method and NLVLM calculations

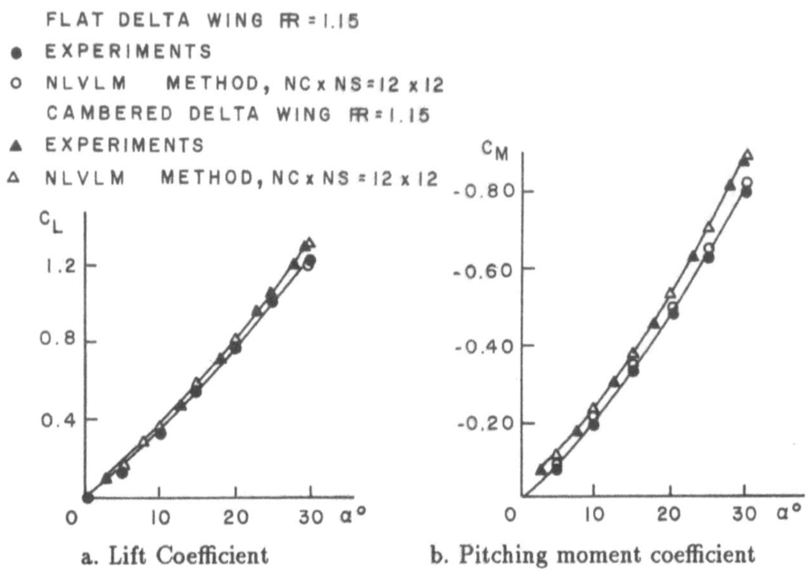

Fig. 7.56. Variation of the lift and pitching moment coefficients as a function of angle of attack for flat and cambered delta wings (Gordon and Rom 1985)

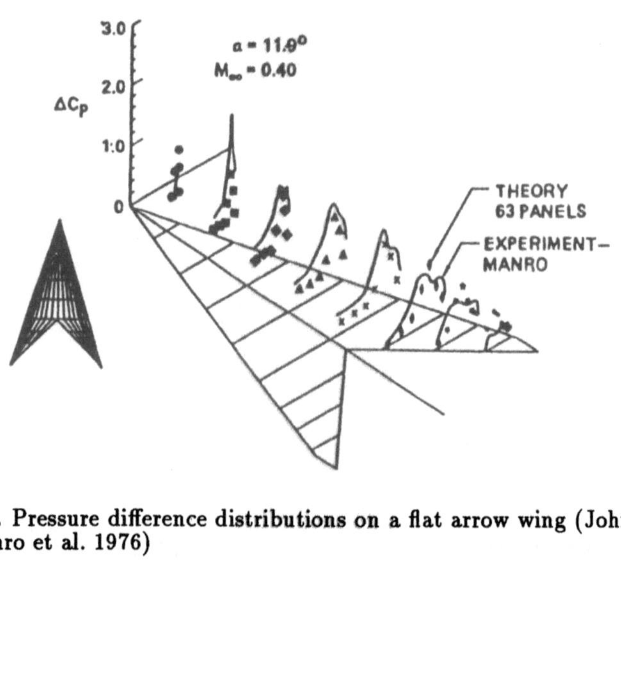

Fig. 7.57. Pressure difference distributions on a flat arrow wing (Johnson et al. 1980, Manro et al. 1976)

core in order to improve the pressure distribution calculations. The FVS method was also applied for the calculations of symmetric and asymmetric flows about slender cones. Three-dimensional separated vortex flows were computed using the FVS method. For thin slender configurations for which the leading-edges' vortex roll-up is approximately conical, reasonably good agreement is obtained between the computations and the results of experiments.

7.3 The Nonlinear Vortex Lattice Method for Supersonic Flow

The supersonic linear Vortex Lattice Method, described in Section 4.4.2, is based on utilizing the elementary solutions of the supersonic source and supersonic vortex while assuming fixed free vortex wake geometry to calculate the aerodynamic characteristics of a given configuration. The induced velocities, due to the singularities in the supersonic flow, affect the strength and position of the free vortices shed into the wake in a somewhat analogous way to that of the subsonic flow. It may be expected that by allowing the vortex wake to re-form due to the interactions with the flow field affected by the supersonic vortex, singularities will result in a nonlinear variation of the aerodynamic characteristics, i.e. a supersonic Nonlinear Vortex Lattice Method.

The supersonic NLVLM, first developed by Tennenbaum (1990), is presented here both for completeness of the coverage of the subject matter and as a further illustration of the utility of the NLVLM for aerodynamic design calculations This supersonic flow NLVLM complements the subsonic flow NLVLM calculations presented in Section 7.1.

7.3.1 Method of Calculations

The mathematical formulation of the equations for the perturbation potential and the boundary conditions on the lifting surfaces and the wake are similar to that used for the linear VLM. The appropriate equation is the hyperbolic potential wave equation $\beta^2\phi_{xx} - \phi_{yy} - \phi_{zz} = 0$ where $\beta^2 = M^2 - 1$. The solution of this equation can be constructed by superposition of the elementary solutions of the supersonic source and vortex singularities. The characteristics of these supersonic characteristics are discussed in Section 4.4.2, so it is of importance to distinguish the following properties of these supersonic singularities: 1. the existence of domains of influence in the supersonic flow field. 2. the bound supersonic vortex induces velocity disturbances only in the Mach cones issuing from its ends. 3. the singular behavior due to the supersonic vortex is distributed only on the Mach cone issuing from its ends.

In the linear supersonic VLM, like in the linear subsonic VLM, the vortex wake is assumed to separate from the trailing edge and remain a flat vortex sheet in the plane of the wing. Applying the method devised for the subsonic NLVLM to the supersonic nonlinear case; the vortices are allowed to separate from the leading, side and trailing edges and their trajectories are calculated by an iterative procedure in which both the trajectories of the free vortices and the strengths of the bound vortices are simultaneously calculated.

The first step requires the division of the lifting surfaces into a large number of small panels. Various schemes of divisions are possible, as discussed by Miranda et al. (1978) for the linear case and Tennenbaum (1990) for the nonlinear case. A horseshoe vortex is introduced into each panel. The bound vortex induces velocity components, given by Eq. 4.50 (repeated here for completeness, using the notation of 4.4.2),

(7.16a) $$u(x_o, y_o, z_o) = \frac{\Gamma}{2\pi} \frac{z_o}{x^{*2} + (t^2 - \beta^2) z_o^2} \left[\frac{t x_1 - \beta^2 y_1}{R_{H_1}^{1/2}} - \frac{t x_2 - \beta^2 y_2}{R_{H_2}^{1/2}} \right],$$

(7.16b) $$v(x_o, y_o, z_o) = \frac{\Gamma}{2\pi} \frac{t z_o}{x^{*2} + (t^2 - \beta^2) z_o^2} \left[\frac{t x_1 - \beta^2 y_1}{R_{H_1}^{1/2}} - \frac{t x_2 - \beta^2 y_2}{R_{H_2}^{1/2}} \right],$$

(7.16c) $$w(x_o, y_o, z_o) = \frac{\Gamma}{2\pi} \frac{x^*}{x^{*2} + (t^2 - \beta^2) z_o^2} \left[\frac{t x_1 - \beta^2 y_1}{R_{H_1}^{1/2}} - \frac{t x_2 - \beta^2 y_2}{R_{H_2}^{1/2}} \right],$$

and for a point in the plane of the panel an additional normal velocity component must be added

(7.17) $$w^* = -\frac{\gamma \cos \Lambda}{2} [\beta^2 - t^2]^{1/2}$$

There is no significance to any particular position for placing the bound vortex in the panel in supersonic flow; however, maintaining uniformity with the subsonic geometry the bound vortex is placed at the 1/4 chord and the control point is positioned at the center of the 3/4 chord line of the panel. From the tips of each bound vortex, two semi-infinite trailing vortices are added to complete the horseshoe vortex of that panel. The velocity components at every point in the flow field can be calculated by summing up the velocities induced by all the panels; for example the velocity component in the x direction, $u(x, y, z)$ can be expressed by (similar expressions can be written for v, and w)

(7.18) $$u(x, y, z) = \sum_{i=1}^{N} (u_{ib} + u_{il} + u_{ir})$$

where N is the total number of panels, u_{ib} is the velocity due to the bound vortex and u_{il} and u_{ir} are the velocities induced by the left and right trailing vortices, respectively. These velocity components are evaluated by Eq. 7.16 with the addition of the component w^* from Eq. 7.17 for points in the plane of the panel.

The condition of tangency of the velocity,

$$(U_\infty + u) \cdot \mathbf{i} + v \cdot \mathbf{j} + w \cdot \mathbf{k}) \cdot \mathbf{n} = 0 \tag{7.19}$$

is imposed at every control point. This leads to the set of linear algebraic equations which can be expressed in the matrix equation

$$\left[A_{ij}\right] \cdot \left[\Gamma_i\right] = \left[V_{ni}\right] \tag{7.20}$$

where A_{ij} is the influence matrix representing the velocity induced by the vortex at panel j on panel i, evaluated by Eq. 7.16 and when $i = j$ the velocity represented by Eq. 7.17 is added. V_{ni} is the velocity normal to the panel $V_{ni} = V_\infty \cdot \mathbf{n}_i$. The matrix equation, Eq. 7.20, can be solved for the strengths of the bound vortices Γ_i, and the pressure distribution, aerodynamic forces and moments can be then evaluated.

In the linear VLM, the trailing vortices leaving the lifting surface at the trailing edge remain as straight lines in the plane of the trailing edge. In the NLVLM, the trailing vortices are allowed to separate from the leading, side and trailing edges (or from any of the panels on the surface) and their shape is calculated by dividing the trailing vortex to segments, where each is aligned with the local velocity vector since these free trailing vortices cannot support any pressure difference. In supersonic flow, since the zone of influence eliminates the possibility of induced velocities in the upstream direction, there is no need for the inner iteration cycle for determining of the free vortices positions as was required in the subsonic case. It is possible to calculate the positions of all free trailing vortex segments at each calculation step so as to evaluate the three-dimensional vortex wake in the supersonic flow. With this new vortex wake shape, it is possible to recalculate the influence matrix and solve for a new vortex strength distribution. This process is repeated until both the strength of the vortices and their position converge within a specified error for successive iterations. Tennenbaum (1990) developed a computer program for such calculation, a flow diagram is presented in Fig. 7.58. This supersonic program is different from the subsonic one because the supersonic vortices have a limited zone of influence as well as a surface of singularity on the Mach cones issuing from the edges of each of the bound vortices. Therefore, the first iteration using the initial shape of the free vortex wake, is equivalent to the linear supersonic VLM, so that its results can be used as the reference linear values. For the nonlinear option, the calculation proceeds to evaluate the new free vortices trajectories which are influenced by the interactions with the bound and free vortices, with the limitations of the proper zones of influence. This process is repeated in further iterations until the force coefficients and the vortex trajectories converge to the required accuracy.

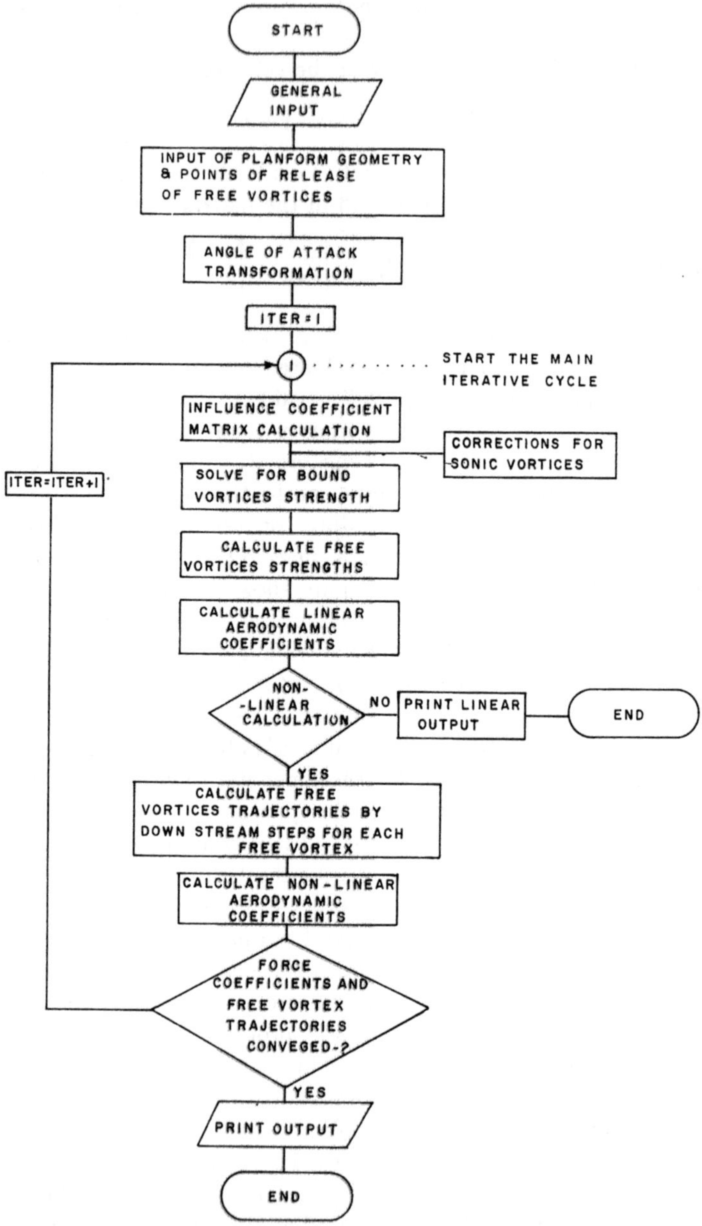

Fig. 7.58. Flow Chart for the supersonic NLVLM (Tennenbaum 1990)

References

1. Almosnino, D. (1977), "A Method for the Calculation of Longitudinal Aerodynamic Coefficients of Wings and Lifting surface Configurations in Incompressible Flow and at High Angles of Attack", Thesis submitted in partial fulfillment of the requirements for the degree of M.Sc., Department of Aeronautical Engineering, Technion, I.I.T., Haifa Israel.

2. Almosnino, D., Zorea, C. and Rom. J. (1978) "A Method for Calculating Longitudinal Characteristics of Wings and Multiple Lifting Surfaces in Subsonic Flow and at High angles of Attack", Israel Journal of Technology, Vol. 16, pp. 66–74.

3. Almosnino, D. and Rom, J. (1980), "Alleviation of the Lateral Forces and Moments acting on a Slender Body at High Angles of Attack using Jet Injection at Subsonic and Transonic Speeds", AIAA Paper 80-1558-CP.

4. Almosnino, D. (1981), "A Method for the Calculation of the Longitudinal Aerodynamic Coefficients of Wing-Body Configurations in Subsonic Flow and at High Angles of Attack", Thesis Submitted in Partial fulfillment of the requirements for the degree of D.Sc., Department of Aeronautical Engineering, Technion, I.I.T., Haifa, Israel.

5. Almosnino, D. and Rom, J. (1981), "Lateral Forces on a Slender Body and their Alleviation at High Incidence", Journal of Spacecraft and Rockets, Vol. 18, No. 5, pp. 393–400.

6. Almosnino, D. and Rom, J. (1983), "Calculation of Symmetric Vortex Separation Affecting Subsonic Bodies at High Incidence", AIAA Journal, Vol. 21, No. 3, pp. 398–406.

7. Almosnino, D. (1985), "High Angles of Attack Calculations of the Subsonic Flow on Slender Bodies", AIAA Journal, Vol. 23, No. 8, pp. 1150–1156.

8. Almosnino, D. and Rom, J. (1988), "Subsonic Aerodynamic Prediction of Shuttle Like Configurations Using Non-Linear Vortex Lattice Method", ICAS Paper 88-4.4.2.

9. Belotserkovskiy, S.M. (1967), *Theory of Thin Wings in Subsonic Flow*, Pleneum Press, New York.

10. Belotserkovskiy, S.M. (1969), "Calculation of the Flow Around Wings of Arbitrary Planforms in a Wide Range of Angles of Attack", NASA TT F-12, 291.

11. Brennenstuhl, U. and Hummel, D. (1982), "Vortex Formation over Double Delta Wings", ICAS Paper 82-6.6.3, pp. 1133-1146.

12. Chigier, N.A. and Corsiglia, V.R. (1972), "Wind Tunnel Studies of Wing Wake Turbulence", AIAA Paper 72-41.

13. Deffenbaugh, F.D. and Koerner, W.G. (1976), "Asymmetric Wake Development and Associated Side Force on Missiles at High Angles of Attack", AIAA Paper 76-364.

14. Del Frate, J.H. and Zuniga, F.A. (1990), "In-Flight Flow Field Analysis on the NASA F-18 High Alfa Research Vehicle With Comparisons to Ground Facility Data", AIAA Paper 90-0231.

15. Garner, H.C. and Lehrian, D.E. (1964), "Non-Linear Theory of Steady Forces on Wings with Leading Edge Flow Separation", R.A.E. R&M 3375.

16. Gordon, R. and Rom, J. (1985), "Calculation of Non-Linear Subsonic Characteristics of Wings with Thickness and Camber at High Incidence", AIAA Journal, Vol. 23, No. 6, pp. 817–825.

17. Gordon, R. (1990), "Numerical Simulation of Vortical Flows Over a Strake-Delta Wing and a Close Coupled Delta-Canard Configuration", AIAA Paper 90-3002.

18. Harloff, G.J. (1987), "High Angle of Attack Hypersonic Aerodynamics", AIAA Paper 87-2548.

19. Hummel, D. and Oelker, H.C. (1986), "Vortex Interference Effects on Close-Coupled Canard Configurations in Incompressible Flow" Proceedings of the International Vortex Flow Experiments on Euler Code Validation", Stockholm, Sweden, pp. 47–61.

20. Johnson, F.T., Lu, P., Ticono, E.N. and Epton, M.A. (1980), "An Improved Panel Method for the Solution of Three-Dimensional Leading-Edge Vortex Flows", NASA CR-3278.

21. Johnson, F.T. and Rubbert P.E. (1975), "Advanced Panel Type Influence Coefficient Methods Applied to Subsonic Flows", AIAA Paper 75-50.

22. Kandil, O.A., Mook, D.T. and Nayfeh, A.H. (1976), "Nonlinear Prediction of the Aerodynamic Loads on Lifting Surfaces", Journal of Aircraft, Vol. 13, pp. 22-28.

23. Kandil, O.A., Mook, D.T. and Nayfeh, A.H. (1977), "A Numerical Technique for Computing Subsonic Flow Past Three-Dimensional Canard Wing Configuration with Edge Separation" AIAA Paper 77-1.

24. Lamont, P.J. (1982), "The complex Asymmetric Flow over a 3.5 Ogive Nose and Cylindrical Afterbody at High Angles of Attack", AIAA Paper 82-0053.

25. Legendre, R. (1952), "Ecoulement au Voisinage de la Pointe avant d'une Ail à Forte Flèche aux Incidences Moyennes", La Recherche Aéronautique (O.N.E.R.A), No. 30, Nov.-Dec., pp. 3–8, and No. 31, Jan.-Feb., 1953, pp. 3–6.

26. Manro, M. E., Manning, K. J. R., Hallstaff, T. H. and Rogers, J. T. (1976), "Transonic Pressure Measurements and Comparison of Theory to Experiments for an Arrow-Wing Configuration – Summary Report", NASA CR 2610.

27. Marshall, F.J. and Deffenbaugh, F.D. (1974), "Separated Flow over Bodies of Revolution Using an Unsteady Discrete Vorticity Cross Wake", NASA CR 2414.

28. Mendenhall, M.R. and Nielsen, J.M. (1975), "Effect of Symmetrical Vortex Shedding on the Longitudinal Aerodynamic Characteristics of Wing-Body-Tail Combinations", NASA CR 2473.

29. Oelker, H.C. and Hummel, D. (1988), "Investigation on the Vorticity Sheets of a Close-Coupled Delta Canard Configuration", ICAS Paper 88-5.4.1, pp. 649–662.

30. Peake, D.J., Owen, K.F. and Higuchi, H. (1978), "Symmetrical and Asymmetrical Separations about a Yawed Cone", AGARD CP 247, pp. 16-1 to 16-27.

31. Quick, A.W. (1961), "Möglichkeiten der Luft und Raumfahrtforschung in der Bundesrepublik Deutschland", Z. Flug-Wiss., Vol. 10, pp. 307–333.

32. Rehbach, C. (1973), "Etude Numérique de Nappes Turbillonaires Issues d'une Ligne de Découlement Près du Bord d'Attaque", O.N.E.R.A. Rech. Aéro. 1973-6, pp. 325–330.

33. Rom, J. and Zorea, C. (1973), "The Calculation of the Lift Distribution and the Near Vortex Wake Behind High and Low Aspect Ratio Wings in Subsonic Flow", TAE Report 168, Department of Aeronautical Engineering, Technion-I.I.T., Haifa, Israel.

34. Rom, J., Portnoy, H. and Zorea, C. (1974), "Investigation into the Formation of Wing Tip Vortices", TAE Report 199, Department of Aeronautical Engineering, Technion, I.I.T., Haifa, Israel.

35. Rom, J., Zorea, C. and Gordon, R. (1974), "On the Calculation of Non-Linear Aerodynamic Characteristics and the Near Vortex Wake", ICAS Paper 74-27.

36. Rom, J., Portnoy, H. and Zorea, C. (1976), "Investigation of the Rolling-Up of the Vortex Wake, and the Calculation of the Non-Linear Aerodynamic Characteristics of Wings", TAE Report 277, Department of Aeronautical Engineering, Technion, I.I.T., Haifa, Israel.

37. Rom, J., Almosnino, D. and Zorea, C. (1978), "Calculation of the Non-Linear Aerodynamic Coefficients of Wings of Various Shapes and their Wakes, including Canard Configurations", Collection of Papers of the 11th Congress of ICAS, pp. 333–344.

38. Rom, J. and Almosnino, D. (1980), "Investigation of the Non-Linear Aerodynamic Characteristics of Slender Bodies at High Incidence in Subsonic and Transonic Speeds", TAE Report 431, Department of Aeronautical Engineering, Technion, I.I.T., Haifa, Israel.

39. Rom, J., Almosnino, D. and Gordon, R. (1981), "Method for the Calculation of the Non-Linear Aerodynamic Characteristics of Thick Wings and Bodies Including Symmetric Vortex Separation in Subsonic Flows", TAE Report 437, Department of Aeronautical Engineering, Technion, I.I.T., Haifa, Israel.

40. Rom, J., Almosnino, D. and Gordon, R. (1987), "High Angles of Attack Subsonic Non-Linear Vortex Flow Calculations", AIAA Paper 87-2275.

41. Rom, J. and Gordon, R. (1988), "High Angle of Attack Non-Linear Vortex Flow Calculations of Canard-Wing", AIAA Paper 88-0484.

42. Rom, J., Er-El, J. and Gordon, R. (1989), "Measurements of the Aerodynamic Characteristics of Various Wing-Canard Configurations and Comparison with NLVLM Results", AIAA Paper 89-2217.

43. Rom, J., Melamed, B. and Almosnino, D. (1990), "Comparison of Experimental Results with the Non-Linear Vortex Lattice Method Calculations For Various Wing-Canard Configurations", ICAS Paper 90-3.3.4.

44. Rusak, Z., Wasserstrom, E. and Seginer, A. (1983), "Numerical Calculation of Nonlinear Aerodynamics of Wing-Body Configurations", AIAA Journal, Vol. 21, No. 7, pp. 929–936.

45. Rusak, Z. and Seginer, A. (1985), "An Analytical Parametric Investigation of Numerical Nonlinear Vortex-Lattice Methods", Proceedings of the 27th Israel Conference on Aviation and Astronautics, pp. 13–22.

46. Rusak, Z., Seginer, A. and Wasserstrom, E. (1985), "Convergence Characteristics of a Vortex-Lattice Method for Nonlinear Configuration Aerodynamics", Journal of Aircraft, Vol. 22, No. 9, pp. 743–749.

47. Scallion, W.I. et al. (1985), "Study of Several Factors Affecting Crew Escape Trajectories from the Space Shuttle", NASA TM 86303.

48. Scholz, N. (1949), "Kraft und Druckverteilungmessungen an Tragflächen Kleiner Steckung", Forsch. Ingenieurwes., Vol. 16, 3, pp. 85–91.

49. Szema, K.Y., Chakravarthy, S.R., Pan, D. and Bihari, B.L. (1988), "The Application of a Unified Marching Technique for Flow Over Complex Three-Dimensional Configurations Across the Mach Number Range", AIAA Paper 88-0276.

50. Tinling, B.E. and Allen, C.Q. (1962), "An Investigation of the Normal Force and Vortex Wake Characteristics of an Ogive-Cylinder Body at Subsonic Speeds", NASA TN D-1297.

51. Weber, J.A., Brune, G.W., Johnson, F.T., Lu, P. and Rubbert, P.E. (1976), "Three-Dimensional Solution of Flow over Wings with Leading-Edge Vortex Separation", AIAA Journal, Vol. 14, No. 4, pp. 519–525.

52. Wickens, R.H. (1967), "The Vortex Wake and Aerodynamic Load Distribution of Slender Rectangular Wings", Canadian Aeronautical Space Journal, Vol. 13, pp. 247–260.

Chapter 8

Solutions of the Euler Equations for Flows over Configurations at High Angles of Attack

Given the greater efficiency of algorithms, coupled with ever-faster computers with increased memory capacity, it is becoming more practical to calculate flow over complex configurations using direct solutions of the differential equations (Euler equations for inviscid flows, Navier-Stokes for viscous flows) representing mass, momentum and energy conservation laws. The solutions must comply with the appropriate initial and boundary conditions. The availability of large vector computers has enabled researchers to attempt the modeling of increasingly more complex geometries and the calculation of flow at higher angles of attack by solving Euler or Navier-Stokes equations. The recent development of parallel computers may facilitate further increases in the speed and complexity of such calculations. This chapter will be devoted to the discussion of the methods of solving Euler equations and their application to studies of flows at high angles of attack. The solutions of the Navier-Stokes equations and their application to computations of high angles of attack flows will be discussed in Chapter 9.

The concept that three-dimensional separation flows evolving from free rolled-up vortices can be modeled by an essentially inviscid potential flow field by introducing vortices along specified boundaries (thus confining the vortical flow to limited regions in the essentially inviscid flow field) has been the basis of the various linear and nonlinear potential panel methods discussed in Chapters 4 to 7. The solution of the Euler equations has the advantage over the potential equation solution of providing the correct Rankin-Hugoniot shock wave jump conditions. Furthermore, the Euler equations are able to capture regions of vortical flows generated at certain parts of the aerodynamic configuration. The generation of vorticity in the inviscid flow can be explained by applying Crocco's theorem which equates the vorticity to the gradient of entropy in inviscid steady flow. Therefore, it is reasonable to assume that the complete flow field, except for isolated zones of vortex generation, can be calculated by the solution of the inviscid Euler equations. In this case, it seems plausible that the direct numerical solution of the Euler equations is a reasonable approximation for the calculation of flows with three-dimensional separations and free vortices. This last statement is still controversial because of the uncertainty as to the capability of the Euler solutions to correctly determine the vortical flow regions, and the effects of "numerical viscosity" and grid system on the results of these calculations. Still, the solutions of the Euler equations are basically solutions of the exact nonlinear partial differential equations which apply over the complete Mach number range, including the nonlinear transonic regime. In this respect, the Euler solutions have a unique advantage over the panel methods which present approximate solutions for linearized small perturbation potential equations, applicable only in the subsonic or supersonic ranges, and not applicable in the transonic speed range, as noted in previous chapters. We will now proceed to discuss the methods of solutions of the Euler equations and the results obtained on various configurations. It is hoped that this discussion will enable us to better understand the significant advantages as well as the problems associated with the use of these Euler solutions for aerodynamic calculations, in particular at high angles of attack.

8.1 The Euler Equations

The conservation of mass, momentum and energy for a compressible inviscid fluid in thermodynamic equilibrium, can be represented by the partial differential equations – the Euler equations – which can be written in the following form

$$\frac{\partial}{\partial t}\left\{\begin{matrix}\rho\\ \rho u\\ \rho v\\ \rho w\\ e\end{matrix}\right\}+\frac{\partial}{\partial x}\left\{\begin{matrix}\rho u\\ \rho u^2+p\\ \rho uv\\ \rho uw\\ (e+p)u\end{matrix}\right\}+ \tag{8.1}$$
$$+\frac{\partial}{\partial y}\left\{\begin{matrix}\rho v\\ \rho uv\\ \rho v^2+p\\ \rho vw\\ (e+p)v\end{matrix}\right\}+\frac{\partial}{\partial z}\left\{\begin{matrix}\rho w\\ \rho uw\\ \rho vw\\ \rho w^2+p\\ (e+p)w\end{matrix}\right\}=0$$

The equation for the state of an ideal gas in thermodynamic equilibrium can be written as

$$\frac{p}{\rho}=(\gamma-1)\left[e-\frac{u^2+v^2+w^2}{2}\right] \tag{8.2}$$

and $h = e + p/\rho$.

The integral form of the Euler equation is useful and is used in various methods of computation.

$$\frac{\partial}{\partial t}\iiint_V \mathbf{q}\,\mathrm{dvol} + \iint_S \mathbf{H}\cdot\mathbf{n}\,ds = 0 \tag{8.3}$$

where

$$\mathbf{q} = \begin{Bmatrix} \rho \\ \rho u \\ \rho v \\ \rho w \\ \rho e \end{Bmatrix} \quad \text{and } \mathbf{H} = \begin{pmatrix} \rho & u\rho & v\rho & w \\ \rho & u^2 + p\rho & uv\rho & uw \\ \rho & uv\rho & v^2 + p\rho & vw \\ \rho & uw\rho & vw\rho & w^2 + p \\ \rho & uh\rho & vh\rho & wh \end{pmatrix}$$

The first integration is performed over the volume V, representing the time variation of the elements of $\mathbf{q}$ within this control volume, while the second integration is performed over a surface S that bounds the volume V with normal unit vector $\mathbf{n}$ representing the net flux of mass, momentum and energy across the surface S.

These sets of equations, the conservation relationships in the differential and integral forms, Eqs. 8.1 and 8.3, respectively, and the equation of thermodynamic equilibrium, Eq. 8.2, represent a well posed mathematical problem which in principle, can be solved with the appropriate initial and boundary conditions.

8.2 Numerical Methods of Solution of the Euler Equations

An outline of the procedure for the calculation of flow over a complex aerodynamic configuration using the numerical solution of the Euler equations is presented in Fig. 8.1. This procedure first requires input of the configuration geometry, construction of a grid system, definition of the grid topology, generation of surface and field grids (including graphical editor for visualization of the surface grid), and then solution of the Euler equations by the proper algorithm for the differencing scheme.

A very short description of some of the various grid systems which are being used will be included here, as well as a short description of the algorithms being developed and used in the numerical solutions of the Euler equations. A general introduction to these numerical methods can be found in various textbooks such as in "Computational Fluid Mechanics and Heat Transfer" by Anderson, Tannehill and Pletcher (1984), while specific details can be found in the extensive reference list cited at the end of this chapter.

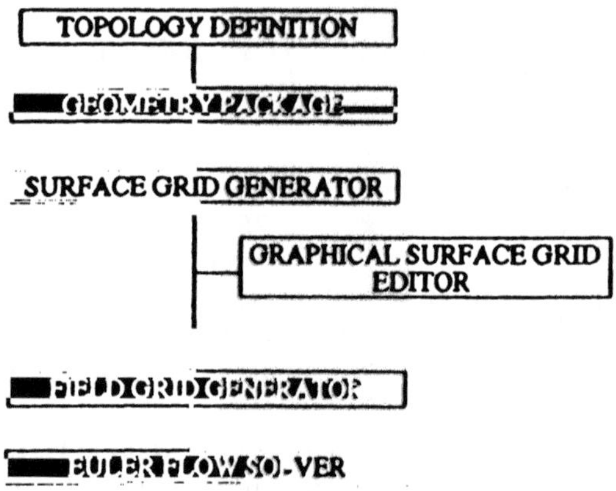

Fig. 8.1. Euler equations solution procedure

8.2.1 Methods for Grid Generation

It is well known that the numerical solution of a system of differential equations – in our case, the Euler equations – can be greatly simplified by the choice of a suitable grid system. An improper choice of grid point locations may lead to unsatisfactory results involving apparent instabilities or lack of convergence. However, a correct choice of grid system may enable greater numerical accuracy, a better fit to configuration surface shape and a great savings in computer resources. This is particularly important for the calculation of the flow over complex aircraft configurations.

The general practice is to devise methods for generating grids aligned with the body shape. The grid generation involves the mapping of the physical plane onto a computational domain. In such a mapping, the body surface can be selected as a boundary in the computational plane, permitting relatively easy application of the boundary conditions on the body. There are a number of conditions required for proper mapping; the mapping must be one-to-one, the grid lines should be smooth to provide continuous transformation derivatives, grid points should be closely spaced in regions of rapid changes (without which large numerical errors can be expected), and excessive grid skewness should be avoided. Given a certain geometry, grids can be generated by various techniques which may be classified in three categories: analytical transformation methods based on complex variables (restricted to two-dimensional cases), methods based on algebraic transformations, and methods based on solutions of differential equations. For the calculation of flows over three-dimensional aircraft and missile configurations at high angles of attack, grid generation by the algebraic and differential equations methods is generally used.

The grid system can be composed of quadrilateral cells in two-dimensional geometry or hexahedra in three-dimensions – such grid systems are called "structured" grids. Alternatively, grid systems can be composed of triangular cells for two-dimensions, or tetrahedra for three-dimensions – such grid systems are called "unstructured" grids. These latter type of grids are so named since neighboring nodes are in general, not directly addressable, in contrast to cases with structured grids which are directly addressable due to the fact that they are composed of quadrilateral cells (in 2-D) or hexahedra (in 3-D).

In the case of a complicated geometry, it is advisable to use different grid arrangements over various portions of the flow field – this can be accomplished by using combinations of structured and unstructured grids on different parts of the configuration. Furthermore, the grid system can be divided into many blocks, each block having its own divisions, and the solutions can be matched at block interfaces. Adaptive grid methods can also be advantageously applied in certain flow calculations. These various grid generation methods may be applied in either a multigrid system and multiblock system, or in the combinations of schemes, such as multigrid inside multiblock or multiblock inside multigrid.

A different way of solving the grid problem is the use of Cartesian grids. Cartesian grids enable a simple and good numerical approximation to the governing equations in the flow. As they are not body surface aligned, they eliminate much of the work incurred in generating body aligned grids. However, the use of Cartesian grids introduces the serious problem of correct implementation of boundary conditions on the configuration boundary. The application of Cartesian grids and local refinement to the solution of the three-dimensional Euler equations on aircraft configurations was developed at the Israel Aircraft Industries (IAI) and is presented by Epstein et al. (1989).

8.2.1.1 Grid Topology. Grid topology can be defined and classified according to various types of mapping, useful for aerodynamic configurations. The H, C and O types of mapping (over a wing, for example) are shown in Fig. 8.2, and the characteristics of a O-C type mapping for a wing-body configuration is shown in Fig. 8.3. In some cases, it is useful to use a C-H mapping or to introduce an unstructured grid into parts of the configuration, as can be seen in the various cited references. It should be noted that the mapping process may involve the appearance of singular points which must be dealt with in the numerical schemes.

8.2.1.1 Body Aligned Grid Systems. The determination of the grid system involves the important and often difficult task of surface grid generation as well as field grid generation which must be matched to the surface grid. The surface grid must adapt its structure to the particular shape of the configuration surfaces. An example of surface grid over an aircraft configuration is presented in Fig. 8.4.

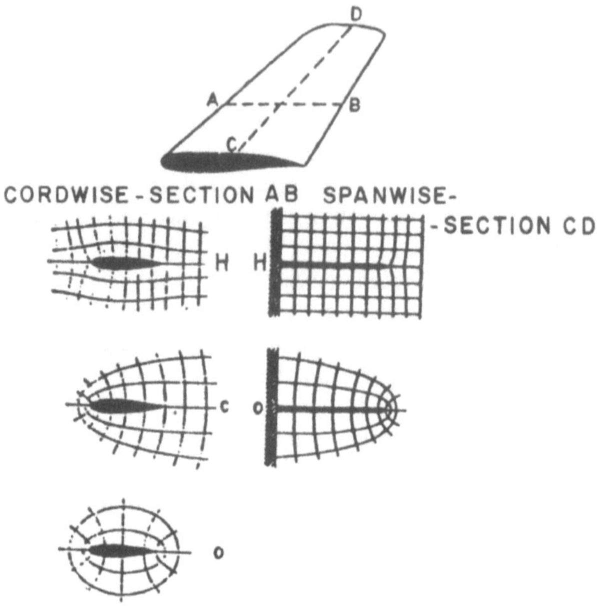

Fig. 8.2. Basic mapping types for a wing

In general, the grid system for a complex aircraft configuration is composed of many blocks matched at their interfaces, each one containing its own division of structured or unstructured meshes, with the possibility of having adaptive meshes in certain regions of the flow field. The large number of options and combinations for the grid generation methods available to the interested user are presented in the many publications listed in the references of this chapter. Once the grid is established over both the configuration surfaces and the flow field, it is necessary to determine the various methods of solutions of the Euler equations by using the appropriate numerical scheme.

An excellent overview of several numerical methods available for the solution of the Euler equations was presented by Jameson (1987). He states that some of the requirements for designing satisfactory algorithms can be summarized by the following guidelines: the conservation laws of fluid mechanics should be satisfied in discrete form by numerical approximation; shock waves and contact discontinuities should be automatically captured by the differencing scheme; in steady flow calculations, the final steady state must be independent of the time-stepping scheme; invariant quantities in the flow field, such as entropy upstream of shock wave or total enthalpy in

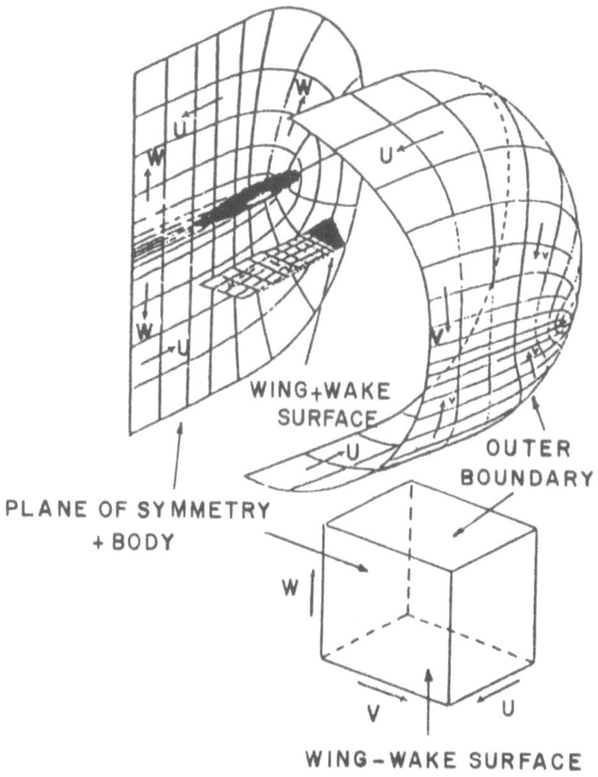

Fig. 8.3. Mapping characteristics of grid type O-C

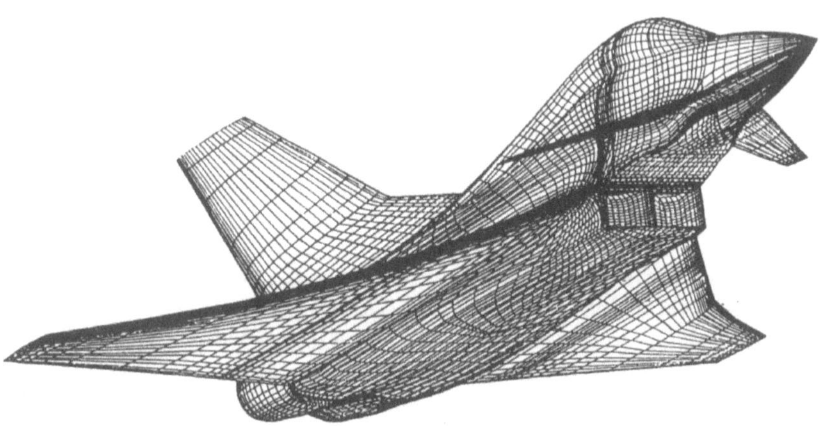

Fig. 8.4. Surface grid over an aircraft configuration (Bocci 1990)

steady flow, should be invariant in the numerical scheme and uniform flow should of coarse be an exact solution of the difference equations on any arbitrary mesh. The steady state solution of the Euler equations can be obtained by introducing a space discretization procedure, which reduces the problem to the solution of a large number of coupled nonlinear algebraic equations by iterative methods. Another very useful approach is the solution of time-dependent Euler equations by time-stepping, hence reaching the steady state solutions. Jameson (1987) presents the advantages of using this strategy as follows: it is simple; the same computer program can be used to calculate steady and unsteady flows; the time-dependent program provides a natural framework for the design of nonoscillatory shock capturing schemes reflecting the physics of wave propagation; and algorithms can be devised for concurrent computations on vector, pipelined or parallel processors, either through the use of an explicit time-stepping scheme or through the use of an iterative procedure at each time-step of an implicit scheme.

The numerical methods for the solution of the Euler equations using body aligned grids can be classified into the following categories: finite volume methods, finite difference methods and finite element methods. In addition, a method utilizing multigrid calculations with equally meshed Cartesian (not necessarily rectangular) grids and local refinements is available.

8.2.2 Finite Volume Methods

Various finite volume numerical methods suitable for the solution of the Euler equations for flows over complex aircraft configurations in the transonic, as well as in the subsonic and supersonic speed ranges, have been developed. One of the efficient computer codes used widely for calculations of flow characteristics over aircraft configurations is the finite volume method of Jameson, denoted as FLO-57. The method has been presented by Jameson, Schmidt and Turkel (1981), and various modifications and applications have been presented by Agarwal and Deese (1983), by Volpe, Siclari and Jameson (1987a, 1987b), and in other papers listed in the references. Another finite volume method was developed by Rizzi and Eriksson (1981, 1984, 1985), Rizzi (1982) and Eriksson (1982).

8.2.2.1 The Finite Volume Method of Jameson. This method is based on the integration of the time-dependent Euler equations until a steady state is reached. The integration is done using a Runge-Kutta scheme on a set of discretized equations which are obtained first by dividing the flow field into hexahedral cells and then by applying the conservation laws in integral form for each cell. Space discretization is accomplished by using a single block H-O grid which is recommended for low Aspect Ratio aircraft configurations. The grid is generated by the union of two-dimensional meshes obtained in planes normal to the longitudinal axis of the aircraft. Planes can be spaced as required to fit the configuration. Each of these two-dimensional meshes is obtained by mapping the local cross-section of the wing-body into a near circle. The resulting polar-like coordinate grid in the transformed plane corresponds to a grid in the physical plane that not only has a good concentration of points in regions of high curvature but is also body-conforming, a highly desirable feature since it simplifies the application of boundary conditions. The resulting three-dimensional grid is of the O-type, if viewed in the normal planes, and of the H-type, if viewed in the planes parallel to the longitudinal axis.

The finite volume discretization is performed on the integrated conservation equations. The integral form of the mass, momentum and energy conservation laws for three-dimensional flows in a volume **V** enclosed by a surface S can be expressed in the integral form of the Euler equations

$$\frac{\partial}{\partial t}\iiint_V W^{(i)} d\Omega + \iint_S \mathbf{F}^{(i)} d\mathbf{S} = 0 \tag{8.4}$$

where $W^{(i)}$ is a conserved quantity and $\mathbf{F}^{(i)}$ is its corresponding flux term. For the Euler equations, the conservation terms are

Mass	$W^1 = \rho$	and $\mathbf{F}^1 = (\rho u, \rho v, \rho w)$
Momentum	$W^2 = \rho u$	and $\mathbf{F}^2 = (\rho u^2 + p, \rho uv, \rho uw)$
	$W^3 = \rho v$	and $\mathbf{F}^3 = (\rho uv, \rho v^2 + p, \rho vw)$
	$W^4 = \rho w$	and $\mathbf{F}^4 = (\rho uw, \rho vw, \rho w^2 + p)$
Energy	$W^5 = \rho E$	and $\mathbf{F}^5 = (\rho Hu \rho Hv, \rho Hw)$

The integral equation, Eq. 8.4, can be applied to any subvolume $\mathbf{V}_N$ of **V**, thereby dividing the volume **V** into a large number of hexahedral cells and applying the Euler conservation integral to each one of these results in discrete approximation for the total conservation integral, Eq. 8.4, over the complete volume **V**. The flow variables, $W^{(i)}$, can be assumed to be stored at the center of the cell or at the nodes of the mesh. In the latter case, the control volume for each node (i, j, k) is given by the union of the eight cells meeting at that node. Applying the conservation integral equation to each of these eight cells for the control volume enclosing the node (i, j, k)

$$\frac{d}{dt}\left(\sum_{n=1}^{8} V_n\right) W_n^{(i)} + \sum_{n=1}^{8} Q_n^{(i)} = 0 \tag{8.5}$$

where the summation runs over the eight cells meeting at the node (i, j, k), V_n is the volume of the n-th cell and $Q_n^{(i)}$ is the flux balance for the n-th cell, approximated by

$$Q_n^{(i)} = \sum_l \mathbf{F}_l^{(i)} \cdot S_l \tag{8.6}$$

where S_l is the directed face area of the l-th face (of the n-th cell) and $\mathbf{F}_l^{(i)}$ is an estimate of the mean flux vector across that face. Clearly, contributions to the flux balance through the control volume are given only by its external, faces since the fluxes across internal faces cancel.

8.2.2.1.1 Artificial Dissipation. Dissipation terms must be added over the entire volume in order both to prevent oscillations in the flux balance in each interior cell and to eliminate oscillations near shock waves and near contact discontinuities.

Several dissipation schemes for scalar conservation laws have been developed using the concept of Total Variation Diminishing (TVD) Difference and have been extended to the treatment of the Euler equations by splitting the flux into components based on the characteristic speeds and by applying the TVD construction separately to each component. Flux splitting allows the construction of terms which introduce just enough dissipation to prevent oscillations. This, in turn, permits the clear capture of shock waves spread out over only one or two cells.

Another approach is to use adaptive coefficients for the dissipation terms. This approach entails the use of low level dissipation terms over the entire volume and an increase in dissipation terms values in the neighborhood of a shock wave. It has been found that the second difference of the pressure provides an effective measure for detecting locations where these additional dissipation levels are required.

The adaptive coefficients scheme is implemented by substituting for $Q_n^{(i)}$ in Eq. 8.5 by a term $Q_n^{(i)} - D_n^{(i)}$, where $D_n^{(i)}$ is the dissipative flux constructed in such a way as to preserve the conservative form of the equations and is the sum of three dissipative terms constructed for each of the three coordinate directions. At a node (i, j, k), the dissipation term in the i (or x) direction is given by

$$D_n^{(i)} = d_{i+1,j,k} - d_{i,j,k} \tag{8.7}$$

where

$$d_{i,j,k} = r\,(\,\varepsilon^{(2)} - \varepsilon^{(4)}\delta_x^2\,)\,(\,W_{i,j,k}^{(i)} - W_{i-1,j,k}^{(i)}\,) \tag{8.8}$$

and δ_x^2 is a second difference central operator, $\varepsilon^{(2)}$ and $\varepsilon^{(4)}$ are adaptive coefficients and r is a scaling factor proportional to an estimate of the maximum local wave speed. In the explicit scheme, r can be made proportional to $1/\Delta t^*$, where the local time step Δt^* is a measure of the time it takes for the fastest wave to cross a mesh interval. $\varepsilon^{(4)}$ provides the background dissipation, which is of third order in smooth regions of the flow. $\varepsilon^{(2)}$ controls the dissipation near shock waves and is made proportional to a normalized second difference of the pressure

$$\varepsilon^{(2)} = \nu_{i,j,k} = \left| \frac{p_{i+1,j,k} - 2p_{i,j,k} + p_{i-1,j,k}}{p_{i+1,j,k} + 2p_{i,j,k} + p_{i-1,j,k}} \right| \tag{8.9}$$

This part of the dissipation is, therefore, of third order everywhere except near regions of steep pressure gradients. The dissipation terms in the j (or y) and k (or z) directions are constructed in a similar manner. The low order dissipative terms effectively introduce upwind biasing. Instead of blending low and high order dissipative terms, Jameson (1985a) proposed a nonoscillatory scheme by eliminating entirely the low order terms and introducing flux limiters into the high order terms. This approach leads to a class of TVD schemes.

Other schemes of artificial dissipation are being used by various investigators. Detailed investigation of various modifications to the dissipation terms is presented by Swanson and Turkel (1987) as well as in many other publications listed in the references to this chapter.

8.2.2.1.2 Time-Stepping. Since the mesh used in this method is fixed with time, Eq. 8.5 can be re-written in the following form

$$\frac{d}{dt}W_{i,j,k}^{(i)} + R(W^{(i)}) = 0 \tag{8.10}$$

where the residual $R(W^{(i)})$ is given by

$$R(W^{(i)}) = \frac{1}{V_{i,j,k}}(Q_{i,j,k} - D_{i,j,k}) \tag{8.11}$$

and $V_{i,j,k}, Q_{i,j,k}$ and $D_{i,j,k}$ are the control volume, the total flux and total dissipative flux through the control volume, respectively. The steady state solution is obtained by integrating Eq. 8.10 using a five stage Runge-Kutta scheme with a locally varying time-step Δt. This scheme has been selected by Jameson (1985b) for its stability and damping properties rather than its accuracy in tracking the transient solution, since the objective is the attainment of steady state as rapidly as possible. The stability properties of multi-stage schemes, including this scheme, are discussed in detail by Jameson (1985b).

The time-step advance of each variable W from time (t) to time $(t+\Delta t)$ can be presented in the following relations, using a superscript notation to denote each particular stage

$$W^{(0)} = W(t) \tag{8.12a}$$

$$W^{(1)} = W(t) - \alpha_1 \Delta t R[W^{(0)}] \tag{8.12b}$$

$$W^{(2)} = W(t) - \alpha_2 \Delta t R[W^{(1)}] \tag{8.12c}$$

$$W^{(3)} = W(t) - \alpha_3 \Delta t R[W^{(2)}] \tag{8.12d}$$

$$W^{(4)} = W(t) - \alpha_4 \Delta t R[W^{(3)}] \tag{8.12e}$$

(8.12f)
$$W^{(5)} = W(t) - \alpha_5 \Delta t R[W^{(4)}]$$

(8.12g)
$$W(t + \Delta t) = W^{(5)}$$

with $\alpha_1 = 1/4$, $\alpha_2 = 1/6$, $\alpha_3 = 3/8$, $\alpha_4 = 1/2$. Jameson (1975a) notes that, in practice, it has been possible to freeze the artificial dissipation terms at the values computed at the second stage, a practice which noticeably reduces the computational effort and also increases the stability margin for the dissipative terms.

8.2.2.1.3 Methods of Accelerating Convergence to Steady State. Acceleration to the steady state solution is obtained by maximizing the time-step, by implicit smoothing of the residuals, by enthalpy damping and by using an efficient multigrid system.

The stability limit on the local Courant number sets restrictions on the maximum permissible time-step. Jameson and Baker (1983) indicated that smoothing of the residuals increases the support of the numerical scheme, thus relaxing the restriction on the maximum time-step. Smoothing is obtained by replacing the residual at each point by an average of its neighbors. This average is computed implicitly by solving

(8.13)
$$(1 - \epsilon_x \delta_x^2)(1 - \epsilon_y \delta_y^2)(1 - \epsilon_z \delta_z^2)\overline{R} = R$$

where ϵ_x, ϵ_y, ϵ_z are smoothing coefficients and δ_x, δ_y, δ_z are second difference operators.

It was shown (Jameson 1975b) that stability can be maintained for any Courant number λ if the smoothing parameter is chosen to be

(8.14)
$$\varepsilon \geq \frac{1}{4}\left[\left(\frac{\lambda}{\lambda^*}\right)^2 - 1\right]$$

where λ^* denotes the stability limit in the unsmoothed scheme. It has been found that a faster rate of convergence is usually obtained by setting $\lambda \cong 3\lambda^*$ and by using the smallest possible values of the smoothing parameters which still maintain stability.

Enthalpy damping is an additional method for accelerating convergence. In a homoenthalpic flow, it can be assumed that the time rate of change of each dependent variable W is proportional to the difference between the instantaneous local value of the enthalpy and the steady state constant value of the enthalpy at infinity. Thus, let

(8.15)
$$\frac{\partial W}{\partial t} + \beta W(H - H_\infty) = 0$$

where β is a user-defined constant. This condition can be applied implicitly at the end of each time-step by using the following relation

$$W - \overline{W} + \beta \Delta t W(\overline{H} - H_\infty) = 0 \tag{8.16}$$

where W is the new estimate of the value of the variable and is the computed value of W at the end of the time-step. Each dependent variable can be evaluated by Eq. 8.15, with only a slight modification for (ρE), as follows

$$\rho E - \bar{\rho}\bar{E} + \beta \Delta t[\rho E + \bar{p} - \bar{\rho} H_\infty] = 0 \tag{8.17}$$

The multigrid scheme can be used as well to accelerate the convergence. This is accomplished by transferring part of the effort of tracking the evolution of a system of discrete equations to a sequence of successively coarser meshes. On the coarser grids, the use of larger control volumes permits the tracking of larger scales. With the Runge-Kutta time-stepping scheme, this is made possible by using larger time steps on the coarser grids without violating the stability bounds. In addition, the computational effort is greatly reduced on the coarser meshes.

The coarser meshes can be generated by eliminating alternate points in each coordinate direction. This simplifies the task of transferring information through several mesh levels. Passing from a grid level k to a coarser grid level $k+1$, the first step of the Runge-Kutta operation, described by Eq. 8.12a, is therefore modified to

$$W_{k+1}^{(0)} = T_{k+1,k} W_k \tag{8.18}$$

where $T_{k+1,k}$ denotes a transfer operator. The modified residuals in the Runge-Kutta operation are then

$$W_{k+1}^{(1)} = W_{k+1}^{(0)} - \alpha_1 \Delta t[R_{k+1}^{(0)} + P_{k+1}] \tag{8.19a}$$

$$W_{k+1}^{(2)} = W_{k+1}^{(0)} - \alpha_2 \Delta t[R_{k+1}^{(1)} + P_{k+1}] \tag{8.19b}$$

Similarly, the relations for $W_{k+1}^{(3)}$ to $W_{k+1}^{(5)}$ can be written. R_{k+1} is the residual computed on the mesh $k+1$, and

$$P_{k+1} = Q_{k+1,k} R_k(W_k) - R_{k+1}^{(0)} \tag{8.20}$$

where $Q_{k+1,k}$ is another transfer operator. In this scheme $[Q_{k+1,k} R_k(W_k)]$ is a weighted average of the residuals at the 27 points nearest the coincident point on mesh level k (including the point itself).

Typically, one time-step is performed on each mesh level until the coarsest (desired) mesh is reached. Then, the end result of this Runge-Kutta step

$$W_{k+1}(t + \Delta t) = W_{k+1}^{(5)} \tag{8.21}$$

is transferred back to grid level k to provide an improved value of W. This improved value at grid level k is set to

$$W_k^+ = W_k + I_{k,k+1}(W_{k+1}^+ - W_{k+1}^{(0)}) \tag{8.22}$$

where W_{k+1}^+ is the value of W_{k+1} obtained from both the value of correction calculated by Eq. 8.21 at the end of the time-step, and any correction transferred from a still coarser grid $k+2$, and $I_{k,k+1}$ is an interpolation operator. At points that are coincident on mesh levels k and $k+1$, the value of W is simply transferred. At intermediate points, the correction is obtained by a trilinear interpolation. This is usually performed by a V-cycle, in which correction data is transferred up through the meshes without invoking a time-integration step. Better convergence rates have sometimes been obtained with a W-cycle, in which time-steps are performed on each of the meshes on the way up.

8.2.2.1.4 Boundary Conditions. In the case of an aircraft configuration which is symmetric about a vertical plane in the longitudinal direction and with zero yaw angle, symmetry-type boundary conditions can be imposed along one face of the computational cube. Flow tangency is enforced at node points lying on the aircraft, and continuity is enforced on the slit in the mesh past the wing. These boundary conditions are enforced during each stage of the Runge-Kutta time-step. At the same time, the value of all flow variables at each node on the degenerate axis ahead of the aircraft is replaced by the average of the values computed for the nodes sharing a common point. In the far field, appropriate inflow or outflow boundary conditions are imposed using Riemann invariants to attenuate the spurious reflection of outgoing waves into the far field, as discussed by Jameson and Baker (1983).

8.2.2.1.5 Mesh Generation Around Aircraft Configuration. A useful method for generating the three-dimensional mesh using O-H topology around a typical aircraft configuration is to generate two-dimensional meshes around preselected cross-section planes normal to the longitudinal axis of the aircraft, as shown by Volpe, Siclari and Jameson (1987a). Upstream and downstream of the aircraft, the mesh is a continuation of the first and last sections on the aircraft, respectively. A cross-section plane may include the fuselage body alone, the fuselage body and wing, or even the wing alone with or without an intervening slit between it and the plane of symmetry. Each of these cross-sections, $z = x + iy$, is mapped into a near circle by sequentially removing all singularities at corners of the cross-section by a Karman-Trefftz transformation. Once all the singularities at the corners are removed, the cross-section is a near-circle in the final mapped plane, and a polar-like coordinate grid can be generated around the wing-body shape. Grid lines wrapping around the body are gradually distorted into perfect circles at a predetermined distance denoting the outer boundary of the mesh. Radial grid lines emanating from a point equidistant from the vertical and horizontal extremes of the mapped cross-section will then give a second family of lines for the mesh. Since the mapped cross-section is nearly circular, the mesh will usually automatically provide adequate clustering near the wing tips. Additional clustering in the azimuthal and radial directions can be obtained by the use of stretching functions.

The distance to which the mesh in each two-dimensional plane extends is kept constant and is typically set to 10 times the body length. Upstream of the aircraft nose, the grid is generated by extending upstream the first mesh plane at the aircraft nose to a predetermined distance with a gradually increased spacing between planes. Since the aircraft nose eventually comes to a point, the points on the configuration cross-section degenerate into a single point at the nose. The grid will, therefore, contain a degenerate axis upstream of the aircraft nose, a topological feature that requires special attention in the solution of the discretized equations. The grid past the last station of the aircraft is similarly extended downstream of the last two-dimensional mesh plane of the aircraft. This last mesh usually wraps around a slit, since the last cross-section typically includes the wing trailing edge and/or its wake.

A representative grid is shown in Figs. 8.5 to 8.7. Typically, the grid extends downstream with one quarter of the planes; one half of the planes are stationed between the nose and the last section of the aircraft, and one quarter are downstream. The planes are clustered near the nose and the tail. These figures depict the grid in the symmetry plane along with the surface grid, in representative cross-sectional planes at the leading-edge (Fig. 8.5a) and at the wing-body station (Fig. 8.5b). Fig. 8.6 shows the grid surface wrapping around the body, and a surface running along the body of the aircraft is shown in Fig. 8.7.

8.2.2.2 The Finite Volume Method of Rizzi and Eriksson. Rizzi and Eriksson (1981, 1984, 1985) developed a method of solving the Euler equations by an explicit time-marching finite volume procedure using transfinite interpolation to construct mesh networks of either the C-O or O-O types.

8.2.2.2.1 Grid Generation. The construction of a three-dimensional boundary conforming grid for a wing-body configuration is done by using a direct algebraic method based on the concept of transfinite or multivariant interpolation. The procedure for the grid generation is presented in detail by Rizzi and Eriksson (1981) and by Eriksson (1982), and only a short outline will be included here.

The generation of an O-O type grid, (as shown in Fig. 8.8) is accomplished by the following steps:

1. definition of the configuration surface corresponding to the $w = w_1$ surface of the parametric box.
2. generation of out-of-surface derivatives on the configuration surface.
3. definition of the outer boundary corresponding to the $w = w_2$ surface of the parametric box.
4. definition of blending functions for the interpolation in the w direction between the configuration surface and outer boundary.

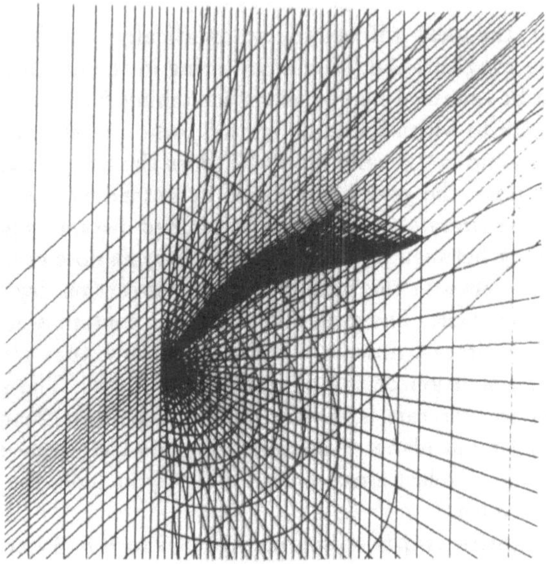

a. Plane at leading-edge of aircraft

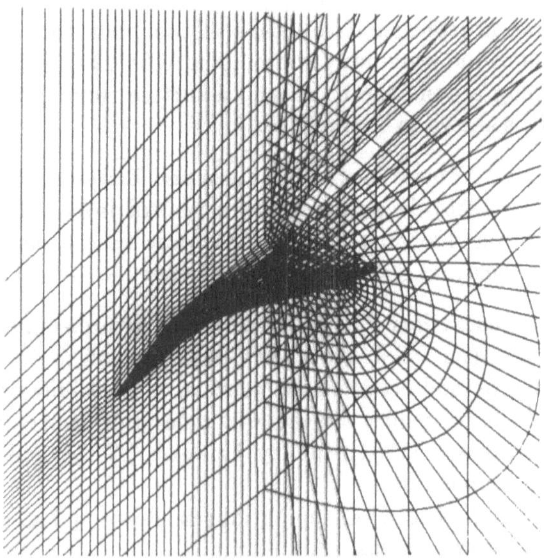

b. Plane through the wing

Fig. 8.5. Grid lines in symmetry plane, on wing/fuselage/wake surface and in cross-sectional plane (Volpe et. al 1987a)

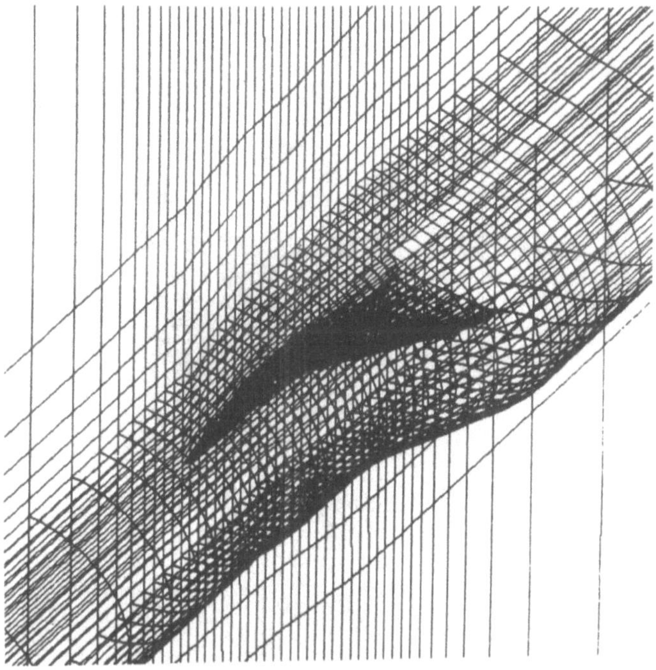

Fig. 8.6. Grid lines in symmetry plane, on wing/fuselage/wake surface and on the shell around the aircraft (Volpe et. al 1987a)

5. interpolation in the w direction (first step of the generalized transfinite interpolation procedure).
6. definition of the combined plane-of-symmetry and body surface corresponding to the $v = v_1$ surface of the parametric box.
7. definition of the blending functions for the interpolation in the v direction.
8. interpolation in the v direction (second, and in this case, final step of the generalized transfinite interpolation procedure).

An example of the O-O type grid, generated by this method for the RAE "W_A/B_2" wing-body configuration, is presented in Fig. 8.9. The grid dimensions are $65 \times 11 \times 25$ for the parameters u, v and w, respectively. It can be seen in this figure that, although the grid conforms to the body surface, it is not aligned with the line separating the body from the plane-of-symmetry.

8.2.2.2.2 Numerical Procedure. This method proposes to solve the Euler equations in their integral form by a pseudo-unsteady approach that marches a hyperbolic system of equations forward in time, without strict concern for time accuracy until a steady state is reached. The integral form of the Euler equations is written in Eq. 8.3 and is solved by differencing this

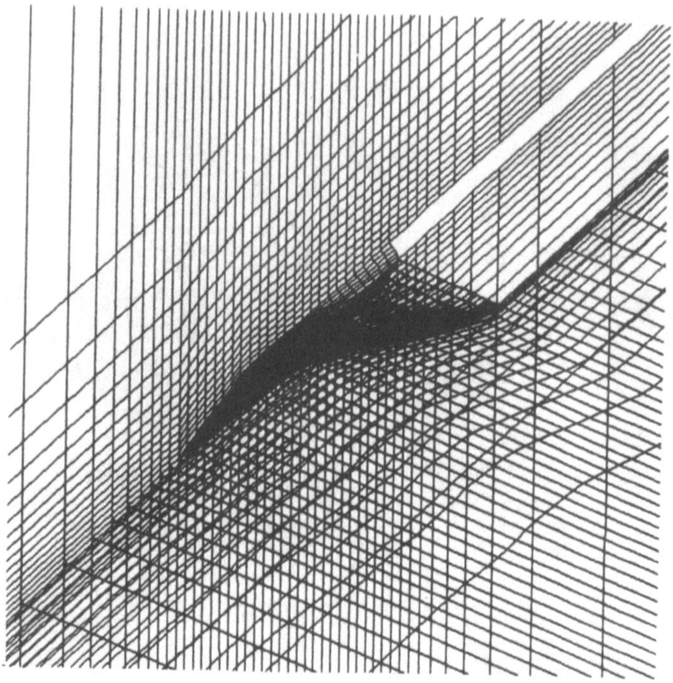

Fig. 8.7. Grid lines in symmetry plane, on wing/fuselage/wake surface and in longitudinal plane touching the wing tip (Volpe et. al 1987a)

integral equation directly, i.e. the finite-volume approach. The details of this numerical procedure have been described by Rizzi and Eriksson (1981) and by Rizzi (1982), so only an outline of the method will be presented here.

The flow field is discretized using the transfinite grid generation technique. An example of the O-O type grid is shown in Fig. 8.8. The mesh generating parameters u, v, w that trace out the grid lines can be renamed by I, J, K and scaled to attain integer values i, j, k at the centers and half integers at the vertices of the cells. The edges of the cells define the nonorthogonal coordinate directions X_I, X_J, X_K in physical units and the unit vectors n_I, n_J, n_K are normal to the grid surfaces on which I, J, K, respectively, are equal to constants. Since the integral equation, Eq. 8.3, is valid for any arbitrary volume, it holds also locally for each cell i, j, k, thus

$$\frac{\partial}{\partial t}\int_{V_{i,j,k}}\iint q_{i,j,k}\, dvol + \iint_{S_{i,j,k}} \mathbf{H}\cdot\mathbf{n}\, ds = 0 \tag{8.23}$$

where the bounding surface $S_{i,j,k}$ now consists of the family of the three coordinate surfaces $\mathbf{S} = (\mathbf{S}_I, \mathbf{S}_J, \mathbf{S}_k)$ that delineate the hexahedronal mesh cell.

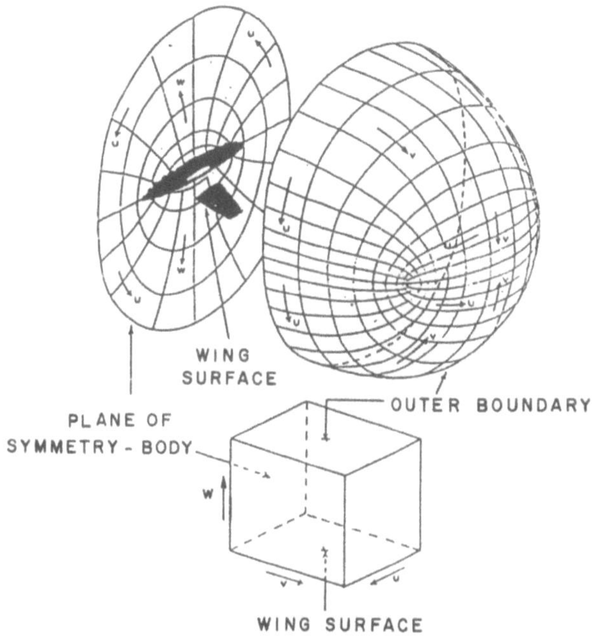

Fig. 8.8. Topology of grid type O-O

The discrete approximation which is used for the solution of this continuum equation uses only a single point evaluation per cell for the dependent variable q so that by the mean value theorem, Eq. 8.23 becomes

$$V_{i,j,k}\frac{\partial q_{i,j,k}}{\partial t} + \delta[\mathbf{H}(q)\cdot S]_{i,j,k} = 0 \tag{8.24}$$

where $q_{i,j,k}$ is now interpreted as a volumetric average located at the center of the cell and $\mathbf{H}(q)$, and S is the corresponding flux evaluated at the surfaces S. The total central difference operator $\delta\psi_{i,j,k}$ equals

$$\begin{aligned}(\delta_i + \delta_j + \delta_k)\psi_{i,j,k} = (\psi_{i+1/2,j,k} - \psi_{i-1/2,j,k}) + \\ +(\psi_{i,j+1/2,k} - \psi_{i,j-1.2,k}) + (\psi_{i,j,k+1.2} - \psi_{i,j,k-1/2})\end{aligned}$$

which expresses the net gain of flux into the cell.

For this finite-volume method, no global coordinate transformation needs to be specified. The coordinate geometry of the mesh system which is transmitted to the method is the three Cartesian coordinates of the eight vertices of every cell. The required metric quantities are the three components of each of the three surface areas S_i, S_j, S_k of a cell as well as the volume of the cell. Since $q_{i,j,k}$ is located in the center of the cell while $\mathbf{H}(q)$ is expressed at its surfaces, an interpolating function and quadrature that defines the specific spatial difference scheme is chosen:

a)

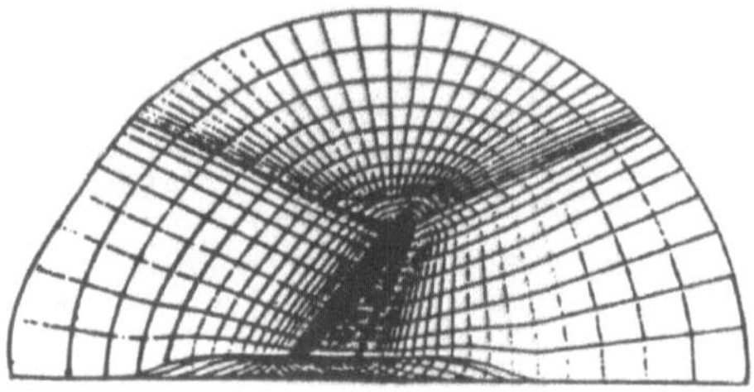

b)

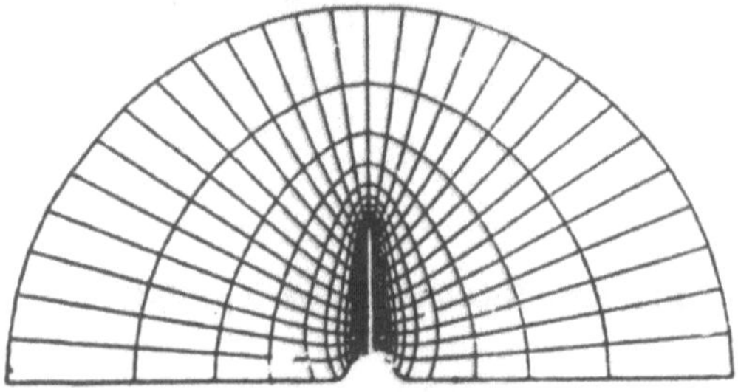

a. Planar view of the upper half, grid dimensions: 65 × 11 × 25

b. Planar view of a spanwise grid surface showing the cross-section of the wing and fuselage

Fig. 8.9. O-O type grid for the RAE W_A/B_2 wing-body configuration (Eriksson 1982)

(8.25a) $$[\mathbf{H}\cdot S]_{i,j,k} = [\mathbf{H}(\mu_i q_{i,j,k})]\cdot S_i + [\mathbf{H}(\mu_j q_{i,j,k})]\cdot S_j + [\mathbf{H}(\mu_k q_{i,j,k})]\cdot S_k$$

where μ is the averaging operator

$$\mu_i \psi_{i,j,k} = 1/2(\psi_{i+1/2,j,k} + \psi_{i-1/2,j,k}).$$

An alternative form of Eq. 8.25a is presented by Rizzi and Eriksson (1984) where it is suggested to compute the flux separately for each of the neighboring dependent variables and then average the two results

(8.25b) $[\mathbf{H}\cdot S]_{i,j,k} = [\mu_i \mathbf{H}(q_{i,j,k})]\cdot S_i + [\mu_j \mathbf{H}(q_{i,j,k})]\cdot S_j + [\mu_k \mathbf{H}(q_{i,j,k})]\cdot S_k$

Eq. 8.24 combined with Eq. 8.25a or Eq. 8.25b leads to a spatial difference operator, completely centered in all three coordinate directions, which is second order accurate in space if the variable in mesh size is reasonably smooth. These three point differences lead to a relatively simple program structure requiring no logic to decide whether to skew the differencing to one side or the other, a feature making it amenable to a large degree of computer vectorization.

The temporally dissipative explicit multistage scheme, devised by Rizzi and Eriksson (1981) and Rizzi (1982), is presented by the sequence (with the spatial indices i, j, k suppressed) of successive iterates $\bar{q}^{(r)}$ at each time-step Δt, whose dissipation can be increased by increasing θ. With $\theta > 1/2$, it is time accurate to first order and stable for sequence numbers $r \geq 2$ and Courant numbers $\lambda < 2$. Second order time accuracy can be attained by using $\theta = 1/2$, but then stability is lost for $r = 2, 5, 6, 9, 10\ldots$ irrespective of λ.

$$\bar{q}^{(0)} = q^n$$

$$\bar{q}^{(1)} = q^n - \frac{\Delta t}{2V}[\delta(\mathbf{H}\cdot S)^n + \delta(\overline{\mathbf{H}}\cdot S)^{(0)}]$$

$$\ldots$$

$$\ldots$$

$$\bar{q}^{(r)} = q^n - \frac{\Delta t}{V}[(1-\theta)\delta(\mathbf{H}\cdot S)^n + \theta\delta(\overline{\mathbf{H}}\cdot S)^{(r-1)}]$$

$$q^{n+1} = \bar{q}^{(r)} \tag{8.26}$$

For this calculation, it is beneficial for the overall convergence to set the parameter θ to a value of greater than or equal to one in order to increase the damping of the medium-to-long wave transients since these waves typically take the longest time to die out in time-marching methods, as presented by Rizzi (1982).

Although the time-stepping scheme, Eq. 8.26, effectively damps out the medium-to-long wavelength components, it only slightly dissipates short waves, a feature representing the drawback of this scheme. Various oscillations which are due to the short waves can be controlled, if not completely eliminated, by a digital filter, such as the Shuman filter with switch (Rizzi 1982). However, when shocks are to be captured in the nonlinear flow field represented by the Euler equations, it is imperative to introduce a small dissipation term to the convective differences by the so-called artificial viscosity model.

The introduction of the artificial viscosity to this finite-volume method is presented by Rizzi and Eriksson (1984). This is accomplished by introducing a dissipation term, $F_D(q)$, to the convective difference operator $F_C(q_{i,j,k}) = -\delta(\mathbf{H}\cdot S)_{i,j,k}/V_{i,j,k}$ in Eq. 8.24, which can then be written

$$\frac{d}{dt}q_{i,j,k} = F_C(q_{i,j,k}) + F_D(q_{i,j,k}) = F(q_{i,j,k}) \tag{8.27}$$

The discrete dissipative operator, $F_D(q_{i,j,k})$, is expressed as the sum of linear and nonlinear terms, $F_D(q_{i,j,k}) = Dq_{i,j,k} + g(q_{i,j,k})$, where D is a constant matrix, and the nonlinear term $g(q_{i,j,k})$ is designed to provide dissipation at discontinuities. Expressions for the evaluation of the linear and nonlinear artificial viscosity terms in the flow field and on its boundaries are given by Rizzi and Eriksson (1984).

In this formulation of the artificial viscosity, the time integration of the equations and the convergence to steady state are obtained by a three-stage two-step scheme, as follows (where $\theta = 1/2$ or 1, and the spatial indices i, j, k are suppressed):

$$q_0 = q^n$$

$$q' = q_0 + \Delta t F(q_0)$$

$$q'' = q_0 + \Delta t[(1-\theta)F(q_0) + \theta F(q')]$$

$$q^{n+1} = q_0 + \Delta t[(1-\theta)F(q_0) + \theta F(q'')] \tag{8.28}$$

The solution for a particular flow field is determined by the conditions imposed upon it at its boundaries. The boundary conditions for the case of the O-O type mesh conforming to a wing-body configuration are discussed by Rizzi (1982) and Rizzi and Eriksson (1984). The boundary conditions are imposed at the six outer surfaces of the computational space, as indicated in Fig. 8.10. There are three distinct types of boundary conditions which are discussed: 1. flow into and out of the far field, 2. periodic conditions across coordinate cuts, and 3. conditions on solid walls.

8.2.2.2.3 The Trailing-Edge Vorticity Generation – Kutta Condition. Any solution of the potential flow equations past an aerodynamic configuration, exact analytical or any numerical finite-difference approximation, *cannot* admit distributed vorticity in the flow field. Therefore, numerical methods based on solutions of the potential equations must introduce vortex lines or vortex sheets which are shed from predetermined positions on the boundaries (e.g. the trailing edge of a wing) in order to model the vorticity generated by the lifting surfaces, so that nonzero lift and induced drag can be obtained. In this way, the Kutta condition fixes the lift when applied to a wing with sharp trailing edges. These arbitrarily introduced vortex lines or sheets can be planar and rigid, as in the case of linear aerodynamic coefficients, or can be allowed to roll up following the streamlines of the flow, in which case the nonlinear coefficients are obtained. It has been observed that solutions of the Euler equations of flows over airfoils did evaluate the correct lift coefficient, even though no explicit procedure to introduce vorticity into the solution method was implemented. This was further enforced by the results of calculations on three-dimensional wing

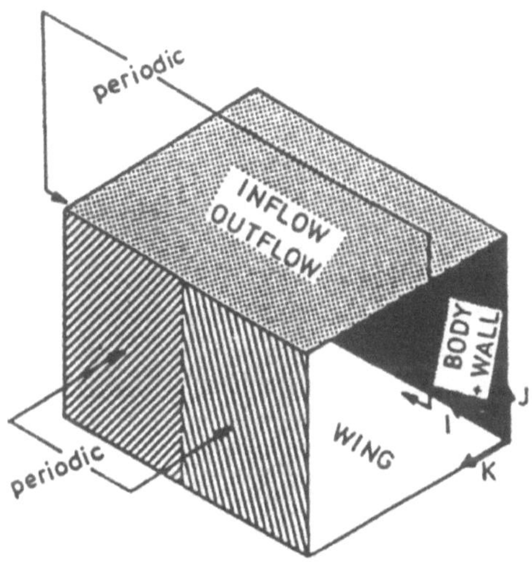

Fig. 8.10. Types of boundary conditions on the six outer surfaces of the computational space IJK resulting from the O-O type grid mapping (Rizzi and Eriksson 1984)

configurations, where reasonable lift coefficients were obtained without explicitly imposing the Kutta condition. It is argued by Rizzi (1982) and Rizzi and Eriksson (1984) that the Euler equations, unlike the potential equation, admit vorticity and vortex sheets as genuine and weak solutions, respectively. Consequently, if vortex sheets are somehow introduced into the flow, they can be "captured" in the same sense that shock waves are captured, and a Kutta condition need not be imposed explicitly.

However, there remains the question as to how a separating vortex sheet can be initiated in the flow field governed by the Euler equations without imposing the Kutta condition. There is no conclusive answer to this problem. Some plausible explanations were proposed by Rizzi (1982), attributing the initiation of the vortex to an unsteady mechanism in the Euler equations. These explanations are inconclusive, particularly since the discrete solution of the Euler equations introduces numerical errors that depend on the mesh size. Lacking a definite explanation as to the fixing of the vortex separation at the sharp trailing edge of the wing and the capturing of the free rolled-up vortices in the flow field by the Euler equations solution, it can only be stated that, in practice, numerical solutions of the Euler equations do fulfill the Kutta conditions at sharp trailing edges and are capable of presenting regions of vorticity in the flow field, such as concentrated leading-edge vortices above low aspect ratio wings.

8.2.3 Finite Differences Methods

Finite difference numerical schemes can be developed using the time-dependent Euler equations when they are expressed in terms of (characteristic type) extended Riemann variables in a streamline coordinate system (s, n, m). The remaining dependent variables are entropy and the two flow angles for three-dimensional flows. The streamline coordinate system allows upwind differencing with only small logic penalty. The boundary conditions are consistent with the characteristic formulation and are, therefore, very simple to implement. The advantages of this formulation are in the unrestricted Mach number range of applicability which can be easily extended to supersonic and hypersonic speeds, and, furthermore, no extrapolation is required to obtain the dependent variables' values on the boundaries at the configuration surfaces. This finite difference numerical scheme does not require the introduction of artificial dissipation or various smoothing techniques needed in the finite-volume methods.

The three-dimensional Euler equations are written in the QAZID formulation introduced by Verhoff and O'Neil (1984). The solution of these Euler equations using a second order accurate finite difference scheme was presented by Agarwal et al (1988) for transonic speed range and by Verhoff and O'Neil (1987) for supersonic and hypersonic flows.

8.2.3.1 The Governing Equations. The three-dimensional Euler equations in the extended Riemann variables formulation written in the streamline coordinates s, n, m are: (note that the variables p, ρ, q, and a are nondimensionalized with respect to their free stream values)

$$\begin{aligned}\frac{\partial Q}{\partial t} + (q+a)\frac{\partial Q}{\partial s} = &-\frac{\gamma-1}{2}a\left(S-\frac{2}{\gamma-1}\right)\frac{\partial}{\partial s}\left(q-\frac{2}{\gamma-1}a\right) - \\ &-\frac{\gamma-1}{2}qaS\left(\frac{\partial\theta}{\partial n}+\cos\theta\frac{\partial\phi}{\partial m}\right)\end{aligned} \tag{8.29a}$$

$$\begin{aligned}\frac{\partial R}{\partial t} + (q-a)\frac{\partial R}{\partial s} = &+\frac{\gamma-1}{2}a\left(S-\frac{2}{\gamma-1}\right)\frac{\partial}{\partial s}\left(q+\frac{2}{\gamma-1}a\right) + \\ &+\frac{\gamma-1}{2}qaS\left(\frac{\partial\theta}{\partial n}+\cos\theta\frac{\partial\phi}{\partial m}\right)\end{aligned} \tag{8.29b}$$

$$\frac{\partial S}{\partial t} + q\frac{\partial S}{\partial s} = 0 \tag{8.29c}$$

$$\frac{\partial\theta}{\partial t} + q\frac{\partial\theta}{\partial s} = -\frac{a^2}{\gamma q}\frac{\partial(\ln p)}{\partial n} \tag{8.29d}$$

(8.29e)
$$\frac{\partial \phi}{\partial t} + q\frac{\partial \phi}{\partial s} = -\frac{a^2}{\gamma q \cos\theta}\frac{\partial(\ln p)}{\partial m}$$

The extended Riemann variables are defined by the expressions

(8.30)
$$Q = q + aS \quad \text{and } R = q - aS$$

where q is the magnitude of the velocity and S is a modified entropy defined by the expression

(8.30)
$$S = \frac{1}{\gamma(\gamma-1)}[2\gamma - \ln(p\rho^{-\gamma})]$$

The coordinate directions as well as the definitions of the flow angles θ and ϕ are shown in Fig. 8.11.

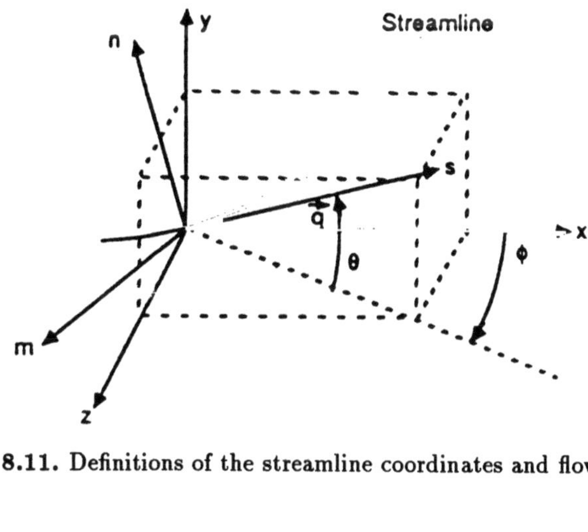

Fig. 8.11. Definitions of the streamline coordinates and flow angles

8.2.3.2 Numerical Solution Methods. The numerical method for the solution of Euler equations expressed in natural streamline coordinates, Eqs. 8.29a to 8.29e, utilizes the procedure of flux vector splitting, and its details are presented by Agarwal et al. (1988). The QAZID equations, Eq. 8.29a to 8.29e, can be written in matrix form

(8.31)
$$\frac{\partial}{\partial t}\{F\} + [A]\frac{\partial}{\partial s}\{F\} + [B]\frac{\partial}{\partial n}\{F\} + [C]\frac{\partial}{\partial m}\{F\} = 0$$

where $\{F\} = \{QR\theta\phi\}^T$ and [A] is the diagonal matrix of characteristic velocities which are $q \pm a, q, q$. In matrix [A] the sign of the eigenvalues along the diagonal clearly distinguish between upstream and downstream characteristics. In order to distinguish directions of propagation normal to the local streamline direction along the n coordinate direction, the matrix $[B]$ is split into $[B] = [B^+] + [B^-]$ where

$$[B^+] = \begin{bmatrix} a/4 & -a/4 & 0 & qa/2 \\ -a/4 & a/4 & 0 & qa/2 \\ 0 & 0 & 0 & 0 \\ a/4q & -a/4q & 0 & a/2 \end{bmatrix} \tag{8.32a}$$

$$[B^-] = \begin{bmatrix} -a/4 & a/4 & 0 & qa/2 \\ a/4 & -a/4 & 0 & -qa/2 \\ 0 & 0 & 0 & 0 \\ a/4q & -a/4q & 0 & -a/2 \end{bmatrix} \tag{8.32b}$$

Similarly, in order to distinguish directions of propagation along the m coordinate direction, the matrix $[C]$ is split into

$$[C] = [C^+] + [C^-]$$

where

$$[C^+] = \begin{bmatrix} a/4 & -a/4 & 0 & qa\cos\theta/2 \\ -a/4 & a/4 & 0 & -qa\cos\theta/2 \\ 0 & 0 & 0 & 0 \\ a/4q\cos\theta & -a/4q\cos\theta & 0 & a/2 \end{bmatrix} \tag{8.33a}$$

$$[C^-] = \begin{bmatrix} -a/4 & a/4 & 0 & qa\cos\theta/2 \\ a/4 & -a/4 & 0 & -qa\cos\theta/2 \\ 0 & 0 & 0 & 0 \\ a/4q\cos\theta & -a/4q\cos\theta & 0 & -a/2 \end{bmatrix} \tag{8.33b}$$

The Euler equations in the natural stream line coordinates, Eqs. 8.29a to 8.29e can also be written by similarly splitting the directions of the characteristics as follows:

(8.34a)

$$\begin{aligned} \frac{\partial Q}{\partial t} + (q+a)\frac{\partial Q}{\partial s} = -\frac{a}{4}\Big[\frac{\partial}{\partial n}(Q-R)^+ - \frac{\partial}{\partial n}(Q-R)^- + \\ +\frac{\partial}{\partial m}(Q-R)^+ - \frac{\partial}{\partial m}(Q-R)^-\Big] - \\ -\frac{qa}{2}\left[\left(\frac{\partial\theta^+}{\partial n} + \frac{\partial\theta^-}{\partial n}\right) + \cos\theta\left(\frac{\partial\phi^+}{\partial m} + \frac{\partial\phi^-}{\partial m}\right)\right] \end{aligned}$$

$$\begin{aligned} \frac{\partial R}{\partial t} + (q-a)\frac{\partial R}{\partial s} = \frac{a}{4}\Big[\frac{\partial}{\partial n}(Q-R)^+ - \frac{\partial}{\partial n}(Q-R)^- + \\ +\frac{\partial}{\partial m}(Q-R)^+ - \frac{\partial}{\partial m}(Q-R)^-\Big] + \\ +\frac{qa}{2}\left[\left(\frac{\partial\theta^+}{\partial n} + \frac{\partial\theta^-}{\partial n}\right) + \cos\theta\left(\frac{\partial\phi^+}{\partial n} + \frac{\partial\phi^-}{\partial n}\right)\right] \end{aligned} \tag{8.34b}$$

$$\frac{\partial\theta}{\partial t}+q\frac{\partial\theta}{\partial s}=-\frac{a}{4q}\left[\frac{\partial}{\partial n}(Q-R)^{+}+\frac{\partial}{\partial n}(Q-R)^{-}\right]-\frac{a}{2}\left(\frac{\partial\theta^{+}}{\partial n}-\frac{\partial\theta^{-}}{\partial n}\right) \tag{8.34c}$$

$$\frac{\partial\phi}{\partial t}+q\frac{\partial\phi}{\partial s}=-\frac{a}{4q\cos\theta}\left[\frac{\partial}{\partial m}(Q-R)^{+}+\frac{\partial}{\partial m}(Q-R)^{-}\right]-\frac{a}{2}\left(\frac{\partial\phi^{+}}{\partial n}-\frac{\partial\phi^{-}}{\partial n}\right) \tag{8.34d}$$

where $\frac{\partial}{\partial n}(\,)^{+}$, $\frac{\partial}{\partial n}(\,)^{-}$ are one-sided derivatives along n and $\frac{\partial}{\partial m}(\,)^{+}$, $\frac{\partial}{\partial m}(\,)^{-}$ are one-sided derivatives along m, corresponding to the positive and negative eigenvalues, respectively.

The splitting of the matrices allows numerical differencing normal to the stream line direction, simulating the physical direction of wave propagation. It is found that this aids in the convergence of the solution and also acts to inhibit any odd-even decoupling of the solution (saw tooth pattern), which is sometimes observed with central differencing procedures. In the finite-volume methods, the inhibition of this odd-even decoupling is accomplished by introducing artificial dissipation. In the case of the splitting of the matrices, dissipation is produced by truncation errors which have similar effects on the artificial dissipation terms. However, in this case, these dissipation values are not determined arbitrarily unlike in the case of the methods using artificial dissipation.

8.2.3.3 Mesh Generation and Boundary Conditions. For the subsonic-transonic flow solution presented by Agarwal et al. (1988), a body conforming C-H type mesh is generated about the wing configuration using a simple shearing and conformal mapping method, similar to that of Jameson and Baker (1983).

For the supersonic-hypersonic application, Verhoff and O'Neil (1987) used an H-O type grid topology which was generated by solving an elliptical system of partial differential equations. Selecting the proper forcing functions allows strong localized control of node spacing and grid orthogonality on and near grid boundaries. This grid control near boundaries allows multiple grid blocks to be generated independently and matched smoothly at their common boundaries. This feature is particularly useful for hypersonic vehicle geometries.

For the H-O topology used by Verhoff and O'Neil, the three-dimensional grid is constructed by successive generation of two-dimensional grids at stations positioned along the longitudinal axis of the configuration. The two-dimensional grid code automatically divides the complex domain into a number of simpler domains on the basis of singular geometry points, such as sharp internal or external corners. A high quality grid is then generated for each of the simpler sections. The sections are matched smoothly at their common boundaries. The procedure is automatic, not user interactive.

The attached oblique shock waves in the supersonic flows can be approximated with relatively reasonable accuracy in these computations by assuming isentropic shocks. For hypersonic flows with stronger shock waves, the Rankin-Hugoniot relations can be used to fit accurately the shock waves.

In the supersonic-hypersonic cases, the solution is initiated by assuming a conical flow near the sharp nosed configuration and by guessing a starting plane solution and shock geometry. An iterative procedure is used to evaluate the conditions at the starting plane. In the case of a blunt nosed body, an iterative procedure must again be used to evaluate the position and conditions of the starting plane at a location where the flow has become supersonic.

The applicable boundary conditions for the subsonic flow case are as follows: zero normal velocity at surfaces (ensuring flow tangency at the solid surfaces); symmetry conditions on all variables (except the flow angle ϕ at the plane of symmetry); and nonreflecting boundary conditions at the far-field at large distances from the wing.

8.2.4 Finite Element Method

Unstructured assemblies of tetrahedra can be used to represent three-dimensional domains of arbitrary shape. Furthermore, the unstructured grid can be easily adapted to incorporate the methods of adaptive mesh for the solution of the Euler equations. These advantages can be used in the application of the finite element method for the solution of Euler equations in two- and three-dimensions. The development of the finite element methods for the solution of the Euler equations is presented in various publications, such as Peraire et al. (1987), Angrand et al. (1985) and others listed in the references of this chapter.

8.2.4.1 Governing Equations. The Euler equations for the flow of a perfect inviscid gas can be written in the conservation form by Eqs. 8.1 and 8.2. A solution vector $\mathbf{V}(x, y, z, t)$ must be found which satisfies the Euler equations inside a given spatial domain Ω and for $t > 0$, together with the appropriate initial and boundary conditions. The domain Ω is subdivided into tetrahedral finite elements, and the approximation $\mathbf{V}^*$ to the solution vector $\mathbf{V}$ is interpolated using C^0 linear shaped functions. A semi-discrete system with time as the only independent variable can be written by using the Galerkin weighted residual statement

$$\int_{\Omega} \left(\frac{\partial \mathbf{U}}{\partial t} + \frac{\partial F_i}{\partial x_i}\right) N_j \, d\Omega = 0 \tag{8.35}$$

where

$$\mathbf{U} = \begin{bmatrix} \rho \\ \rho u_1 \\ \rho u_2 \\ \rho u_3 \end{bmatrix} \qquad F_i = \begin{bmatrix} \rho u_i \\ \rho u_1 u_i \\ \rho u_2 u_i \\ \rho u_3 u_i \end{bmatrix}$$

and N_j is the linear finite element shape function associated with node j. Making use of the Gauss divergence theorem and the linear finite element approximation for $\mathbf{U}$ and F_i results in the relation

$$M_{jk} \frac{d\mathbf{U}_k^*}{dt} = \int_\Omega F_i^* \frac{dN_j}{\partial x_i} \, d\Omega - \int_\Gamma F_i^* n_i N_j \, d\Gamma \tag{8.36}$$

where M_{jk} is the jk-th component of the standard finite element mass matrix defined as

$$M_{jk} = \int_\Omega N_j N_k \, d\Omega \quad \text{and } \mathbf{U}^* = \mathbf{U}_k^* N_k; \quad F_i^* = F_i(\mathbf{U}_k^*) N_k$$

and n_i is the i-th component of the outward unit normal to the surface Γ which is the boundary of Ω.

Performing time discretization on Eq. 8.36 by approximating the time derivatives by a central difference expression gives

$$M_{jk} \frac{\mathbf{U}_k^* \,|^{n+1} - \mathbf{U}_k^* \,|^n}{\Delta t} = \int F_i^* \,|^{n+\frac{1}{2}} \frac{\partial N^j}{\partial x_i} \, d\Omega - \int F_i^* \,|^{n+\frac{1}{2}} n_i N_j \, d\Gamma \tag{8.37}$$

where the superscript n denotes an evaluation at time level n and the time-step is given by $\Delta t = t^{n+1} - t^n$.

The evaluation of the fluxes in the right hand side of Eq. 8.37 is done explicitly starting from the value of $\mathbf{U}^*$ at time level n in two steps: first, an approximation to $\mathbf{U}^*$ at time level $n + 1/2$ is computed, and then the constant element value for variation of $\mathbf{U}$ in element e is computed which is given by

$$\delta \mathbf{U}_e^* = -\frac{\Delta t}{2} \frac{\partial F_i^*}{\partial x_i} \bigg|_e^n$$

where $F_i^* \,|^n = F_i(\mathbf{U}_k^* \,|^n) N_k$. Using the computed value of $\delta \mathbf{U}^*$ and the linear distribution of $\mathbf{U}^* \,|^n$, a piecewise linear discontinuous approximation to $\mathbf{U}^* \,|^{n+1/2}$ can be evaluated by

$$\mathbf{U}^* \,|^{n+1/2} = \mathbf{U}^* \,|^n + \delta \mathbf{U}^* \tag{8.38}$$

The values of $F_i^* \,|^{n+1/2}$ in Eq. 8.37 are then linearly interpolated separately in each element for element e

$$F_i^* \,|^{n+1/2} = F_i^*(\mathbf{U}_k^* \,|^n + \delta \mathbf{U}_e^*) N_k$$

All the integrations in Eq. 8.37 can be performed exactly and lead to an equation system of the form $M_{jk}\Delta\mathbf{U}_k^* = \text{RHS}_j$, where RHS_j is the vector of element contributions to the node j and $\Delta\mathbf{U}_k^* = \mathbf{U}_k^* \mid^{n+1} - \mathbf{U}_k^* \mid^n$. In order to maintain the explicit character of the algorithm, this equation system is solved iteratively using the relation

$$M^L \Delta\mathbf{U}^* \mid^r = \text{ RHS } - \epsilon(M - M^L)\Delta\mathbf{U}^* \mid^{r-1}$$

starting with $\Delta\mathbf{U}^* \mid^0$, where M^L is the lumped (diagonal) mass matrix and $\varepsilon = 1$. This process converges in three iterations and can be used for computations of transient solutions.

The boundary conditions are evaluated by using the calculated value of $\mathbf{U}^* \mid^{n+1/2}$ from Eq. 8.38, which is corrected by applying a linearized characteristic analysis, and then applied to the calculations of the conditions at the boundary from Eq. 8.37. This scheme was developed by Peraire et al. (1987) and can be viewed as a two-step implementation of the one-step Euler-Taylor-Galerkin scheme for convective transport problems.

The explicit character of this algorithm requires the use of a Courant-type stability criterion. It is recommended by Peraire et al. to choose a time-step Δt so that the following expression is satisfied for all elements

$$\Delta t \leq \frac{\alpha\beta h_o}{(u_i u_i)^{1/2} + c} \tag{8.39}$$

where c is the local speed of sound, $\alpha = (3)^{-1/3}, \beta$ is a factor generally taken to be 0.8 - 0.9 and h_o is a representative element length calculated as the arithmetic average of the four heights of the tetrahedra.

The number of iterations required to obtain convergence to steady state can be reduced by the following operations: by lumping the mass matrix M_{jk}, by local time-stepping, and by residual averaging (Peraire et al. 1987).

8.2.4.1.1 Artificial Dissipation Model. As in most Euler solutions, it is necessary also here to introduce artificial dissipation terms to prevent oscillations in the vicinity of discontinuities. Therefore, the computed values of $\mathbf{U}^* \mid^{n+1}$ need to be smoothed before proceeding to the next step by the relation

$$\mathbf{U}_j^* \mid^{n+1}_{\text{smooth}} = \mathbf{U}_j^* \mid^{n+1} + \Delta t_j (M^L)_j^{-1} \left(\sum \frac{C_e S_e}{\Delta t_e} (M - M^L)_e \overline{\mathbf{U}} \right)_j \tag{8.40}$$

where $\overline{\mathbf{U}}$ is evaluated by modifying $\mathbf{U}^* \mid^{n+1}$ in such a way so that the smoothing contribution added to the energy equation is zero when the enthalpy is constant, C_e is a constant specified by the user, and S_e is the element pressure switch coefficient which is the mean of the element nodal values S_i given by the relation

$$S_i = \sum_e \frac{[(M - M^L)_e \mathbf{p}]_i}{\mid (M - M^L)_e \mathbf{p} \mid_i} \tag{8.41}$$

where | .. | denotes that absolute values of contributions from the elements which are summed, and **p** is the vector of the nodal pressures. The values of S_i vary from 0 to 1, where it can be shown that $S_i = 1$ when the pressure has a local extremum at node i.

An optional procedure to employ a flux corrected transport (FCT) algorithm to enhance the resolution of flow discontinuities in conjunction with the artificial dissipation scheme is described by Peraire et al. (1987).

8.2.4.2 Mesh Generation. The operation of the finite element method requires the automatic generation of unstructured meshes of tetrahedral elements over any arbitrarily shaped regions in three-dimensions. Peraire et al. (1987) present a detailed description of their three-dimensional mesh generator which starts from an initial triangulation of the boundaries of the solution domain and fills the domain with tetrahedra using an advancing front approach. In this method, the element size can be varied throughout the domain, and the directionality can be achieved by introducing stretching in certain prescribed directions. The local characteristics of the mesh are defined by mesh parameters. These parameters are, in general, a function of the space coordinates and must be specified as initial data before the generation process starts. A special feature of the mesh generation process is that nodal points and elements are simultaneously created. The efficient implementation of the mesh generation algorithms involved in this method is accomplished by the use of a specific data structure to solve various data management problems. By this method, it is possible to perform mesh generation of unstructured grids with an expense proportional to $N \cdot \log N$ where N is the total number of elements to be generated.

The computation program includes the capability of using the adaptive mesh methods. The adaptive mesh is regenerated by the use of the computed solution on the current mesh to provide the information for adjusting the mesh. The distribution of the mesh generation parameters involves spacing and stretching, enabling the generation of the new mesh which is based on using the current mesh as the background mesh in the generation process. In this adaptive program, the mesh is completely regenerated at each mesh adaptation. In this way, the solution quality can be improved in an optimal manner.

8.2.5 Multigrid Calculations with Cartesian Grids and Local Refinements

The use of Cartesian grids eliminates most of the difficult problems associated with the generation of body aligned grids needed for the finite volume, finite difference and finite element methods of calculations. However, the use of Cartesian grids introduces the problem of implementation of the proper boundary conditions on the three-dimensional surfaces of the aircraft or missile configuration. The development of a three-dimensional code for the solution of the Euler equations using uniform Cartesian (not necessarily rectangular) grids is presented by Epstein, Luntz and Nachshon (1989). Local grid refinement is used to achieve the needed resolution near the configuration surfaces in order not to overflood the computer memory with grid points. A multigrid approach is used mainly to hold all the local grids together in the field solution and to speed up the computation. A given aircraft configuration is treated as a Boolean sum of bodies. Discretization of the body boundary conditions is necessary for the the successful application of the method.

8.2.5.1 Numerical Scheme. The three-dimensional Euler equations in Cartesian coordinates and the equation of state are presented in Eqs. 8.1 and 8.2. Cartesian grids (rectangular or swept) are used for the discretization of the Euler equations

$$\frac{\partial \mathbf{F}_h}{\partial t} + \mathbf{E}_h = 0 \tag{8.42}$$

where $\mathbf{F}_h$ represents the discretized Euler variables vector

$$[\rho, \rho u, \rho v, \rho w, \rho e]_h$$

and $\mathbf{E}_h$ represents the discretized Euler spatial vector

$$[\ \mathrm{div}(\rho\mathbf{V}),\ \mathrm{div}(\rho u\mathbf{V}) + p_x,\ \mathrm{div}(\rho v\mathbf{V}) + p_y,\ \mathrm{div}(\rho w\mathbf{V}) + p_z,\ \mathrm{div}(\rho H\mathbf{V})]_h$$

Symmetrical schemes are used for the second order accurate approximation of the spatial derivatives. The Jameson, Schmidt and Turkel (1981) dissipative terms (second differences only) are introduced in the discretized equations. These dissipation terms are needed for reasons of numerical stability and consistent physical modeling as well as the elimination of the odd-even grid points decoupling of the solution. The discretized system is solved using the modified four stage Runge-Kutta scheme. After each Runge-Kutta stage, the calculated correction to the solution is smoothed, using factorized "residual smoothing" as presented by Jameson (1983).

Some details of this calculation process for the case of a rectangular grid with grid lines parallel to the global coordinate axes are presented in the following:

The four stage calculation is started by the calculation of the analog to

$$\Delta\tau = h/[\max_{G_h}(\mid u \mid + c)]$$

where the maximum is taken over all grid points of the grid box G_h.

The dissipative vector term, D_h , is defined at the grid point (i, j, k) for each of the Euler equations, as

$$\begin{aligned} D_{i,j,k} =& d_{i+1/2,j,k} - d_{i-1/2,j,k} + d_{i,j+1/2,k} - \\ & - d_{i,j-1/2,k} + d_{i,j,k+1/2} - d_{i,j,k-1/2} \end{aligned} \tag{8.43}$$

The values the $d_{i,j,k}$ are calculated by the relations, for the x coordinate as, for example,

$$d_{i+1/2,j,k} = [\epsilon_{i+1/2,j,k}(\eta_{i+1/2,j,k} - \eta_{i,j,k})]/\Delta\tau$$

where

$$\varepsilon_{i+1/2,j,k} = \max(\nu_{i+1/2,j,k}, \nu_{i,j,k}),$$
$$\nu_{i,j,k} == \max\left[\text{vs0}, \left|\frac{p_{i+1,j,k} - 2p_{i,j,k} + p_{i-1,j,k}}{p_{i+1,j,k} + 2p_{i,j,k} + p_{i-1,j,k}}\right|\right] \tag{8.44}$$

and η represents one of the variables in the Euler equations such as ρ, ρu, ρv, ρw, ρH.

The dissipative vector, D_h , is calculated with the first of the four Runge-Kutta stages for the discretized Euler equation

$$\partial \mathbf{F}_h/\partial t + \mathbf{E}_h - \text{vs2}\ D_h = 0 \tag{8.45}$$

and frozen for the following three stages. The terms vs0 and vs2 are dissipation parameters which are fixed in this calculation, so that vs0=0.01 and vs2=1.

The time step used in this calculation is

$$\Delta t = \Delta\tau C_{\text{cfl}}$$

In the calculation, the value of C_{cfl} is limited by 1 in the calculations not using residual smoothing, while with residual smoothing, the actual limit is up to 8.

8.2.5.2 Boundary Conditions. The outer boundary conditions, at the computational "infinity", are evaluated at the faces of the coarsest grid box using Riemann invariants. The velocity component normal to the face and the speed of sound are evaluated by the relations

$$V_n = \mathbf{V}\cdot\mathbf{n} = \frac{1}{2}(R_e - R_\infty)$$

$$c = \frac{1}{4}(\gamma - 1)(R_e - R_\infty)$$

where, at the outflow boundary, the velocity vector is defined by

$$\mathbf{V} = \mathbf{V}_e + (V_n - \mathbf{V}_e\cdot\mathbf{n})\mathbf{n}$$

and

$$R_e = \mathbf{V}_e\cdot\mathbf{n} + 2c_e/(\gamma - 1)$$

while at the inflow boundary, the velocity vector is

$$\mathbf{V} = \mathbf{V}_\infty + (V_n - \mathbf{V}_\infty\cdot\mathbf{n})\mathbf{n}$$

and

$$R_\infty = \mathbf{V}_\infty \cdot \mathbf{n} - 2c_\infty/(\gamma - 1)$$

The energy relation is then calculated by setting the entropy to the corresponding values at the inflow and outflow boundary conditions, respectively.

At the outer boundary of the local grid, Dirichlet conditions are used. These are reflecting boundary conditions and would lead to instability if used in a series of iterations for an isolated grid. However, only one iteration (in the fourth stage) is used at a time, for a given grid, in the multigrid process. Between every two such iterations, there is a coarse grid correction or a fine grid smoothing. This appears to be enough to provide the necessary damping of the reflection and insure the overall convergence.

It is clear that the most critical problem is the implementation of the boundary conditions on the body surfaces with grids which are not body aligned. Therefore, the successful use of the Cartesian grid depends on finding the appropriate method for fulfilling the body boundary conditions. The details of this scheme, as well as the local refinement in multigrid calculations, are described by Epstein et al. (1989).

8.2.5.3 Computer Code. The computer code consists of three preprocessors, PP1, PP2 and PP3; a Multigrid Euler computation code, MGE; and the postprocessor, PTP.

The preprocessors interpret the geometry of a given aircraft configuration in terms of the chosen computational grids. Their output may be used as input for several runs of the Euler code, MGE, and the postprocessor, PTP, for calculations at different flight conditions. The preprocessor PP1 deals with every computational grid separately. It determines the grid point structure and the interpolations required to define the points across the body surface for fulfilling the body boundary conditions. Preprocessor PP2 deals with relations between grids. It prepares the formulae to be used in the Euler code, MGE, for coarse-to-fine and fine-to-coarse levels interpolation, the exchange of information between overlapping grids and the multigrid fine-to-coarse operator correction.

The use of nonaligned grids requires special care in the calculations of the physical parameters, like C_p, over the configuration surfaces. The preprocessor PP3 is designed to prepare the formulae to be used in the postprocessor PTP for integration of final results. The construction of these formulae consists of a panelling of the surface and formulae of extrapolation to the surface. The postprocessor PTP also prepares the results of the calculations for graphical output.

8.2.6 Results of Euler Computations on Three-Dimensional Configurations at High Angles of Attack

There are many options for applying of the methods for the solution of the Euler equations described in the previous sections: the finite volume, finite difference and finite element methods, as well as the Cartesian grid method. Each one of these computational methods can be applied using various mesh generation methods: structured or unstructured grids, adaptive grids, multiblock and multigrid schemes. These can be used in various combinations in such a way as to provide a program capable of solving the Euler equations at certain flow conditions on a specified configuration. Such solutions which are available in open publications are listed in the references of this Chapter. Since it is of interest to examine the capabilities of these Euler solutions to compute the aerodynamic characteristics of complex configurations at high angles of attack, the relevant results will be discussed in the following sections.

8.2.6.1 Delta Wings with Sharp and Rounded Edges. The flow around a highly swept delta wing, features most of the interesting flow elements of the high angles of attack flow field, including the effects of the three-dimensional separation. The separation structure begins with the appearance of separation bubbles on the leeward side of the delta wing at low angles of attack. Then, at increasing angles of attack, the shear layers leave the surface and roll up into the leading-edge vortices with symmetric structure which may become asymmetric at higher angles. In addition, at higher angles of attack, vortex breakdown may affect the concentrated leading-edge vortices, moving from the trailing edge towards the tip as the angle of attack increases. In the case of rounded leading-edges, the rolled-up vortex structure may be followed by free shear layers separating from the surfaces.

The question as to whether the solutions of the Euler equations can capture the separating sheath generated at the sharp leading-edges and its rolling-up into concentrated vortices is very intriguing. Furthermore, seeing that the Euler solutions do feature this rolled-up structure, the question then arises as to how an inviscid solution can simulate the spiraling flow structure of the viscous shear layer. Crocco's theorem relates the vorticity in an inviscid steady flow to the entropy gradients in such flow. In the Euler codes, the introduction of vorticity may be explained by the effect of the artificial "numerical viscosity". Defining this numerical viscosity in the same direction as the physical viscosity can trigger such viscous effects as leading-edge separation and roll-up and even an effect that may be interpreted as vortex bursting (Lee and Brandt 1986).

However, even though the numerical viscosity in the Euler codes can activate these viscous effects qualitatively, a correct quantitative evaluation cannot be expected. In deference to the physical viscosity, the amount of numerical viscosity in the Euler code is determined by the numerical algorithm, by the grid topology and mesh size and by the variations of flow variables. In addition to these are the effects of the artificial dissipation terms introduced into the calculation scheme. In the case of sharp edge geometry, the separation point is fixed at the leading-edge, independent of the viscous characteristics of the flow represented by the Reynolds number. Therefore, just as the separation point at the sharp edge is independent of the real viscosity, it should also be insensitive to the values of the artificial viscosity introduced into the Euler solutions. This is validated by the results of computations which showed that the solutions of Euler equations include the means for inducing separation and generating a primary vortex. However, these solutions are incapable of producing secondary vortices generated by the viscous interactions on the leeward side of the wing. This disadvantage is tolerable since the comparisons of the Euler and the Navier-Stokes solutions, presented by Rizzetta and Chang (1986), show that the effects of the secondary vortices on the primary vortex flow structure and on the overall aerodynamic coefficients are small.

There are also specific effects of the grid system, as demonstrated by the numerical experiments of Lee and Brandt (1988). It was demonstrated that in nonuniform stretched grids in regions of expanding cells, the vortex flow can be deformed to different flow patterns for various grid meshes. These results illustrate some of the nonphysical mechanisms which may affect the Euler solutions. Still, it is interesting to examine the results of solutions to the Euler equations for subsonic, transonic and supersonic flows over delta wings.

The flow over a 55° cropped-delta wing was calculated by Eriksson and Rizzi (1984) using the time marching finite volume method to obtain steady-state solutions to the Euler equations. The wing planform and the O-O type mesh, constructed by transfinite interpolation, used in this calculation are shown in Fig. 8.12.

The capturing of the leading-edge vortex is seen in the velocity vector plot in the plane at 67.5% of the wing chord. The roll-up of the leading-edge primary vortex obtained from the solution of the Euler equation is clearly seen in Fig. 8.13.

Calculations of the aerodynamic characteristics of a sharp edge cropped-delta wing are presented by Raj et al. (1986). The computational method is based on the explicit finite volume, multistage Runge-Kutta pseudo-time-stepping algorithm for the solution of the three-dimensional Euler equations, designated as the TEAM code. The calculation initiated with the entire domain is at free stream conditions. This is equivalent to starting impulsively the configuration. Although no Kutta condition is explicitly applied, the leading-edge vortices are simulated.

The cropped-delta wing has an Aspect Ratio of 0.87, a leading-edge sweep of 63°, and a taper ratio of 0.4. The aerodynamic characteristics are calculated for a free stream of $M = 0.3$ at an angle of attack of 19.95^o. Since all edges are sharp, the separation is clearly fixed to these edges. The O-O type grid with 110,592 ($96 \times 48 \times 24$) cells is used. Calculation with a coarse grid of 24,576 ($64 \times 32 \times 12$) cells indicated only a small effect of the grid density on the flow parameters. The calculated cross-plane velocity field show the rolling-up of the leading-edge primary vortex, seen in Fig. 8.14.

The calculations of the pressure distributions at various chord stations by the Euler TEAM code are compared with experimental data as well

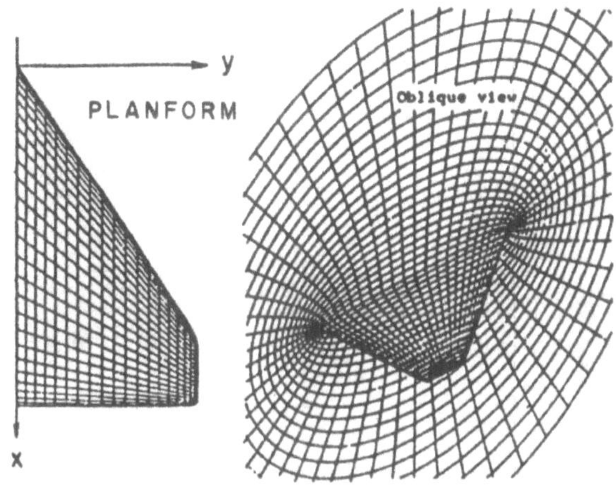

Fig. 8.12. The O-O grid around the 55° cropped-delta wing (Eriksson and Rizzi 1984)

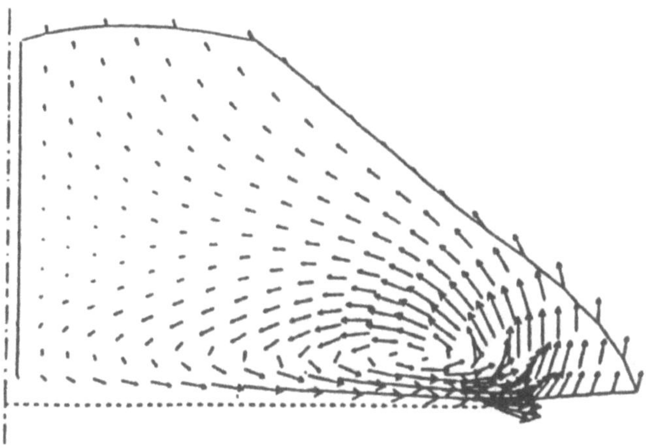

Fig. 8.13. Velocity vector plot in the 67.5% chord plane of the 55° cropped-delta wing for $M = 0.7$ at $\alpha = 10°$ (Eriksson and Rizzi 1984)

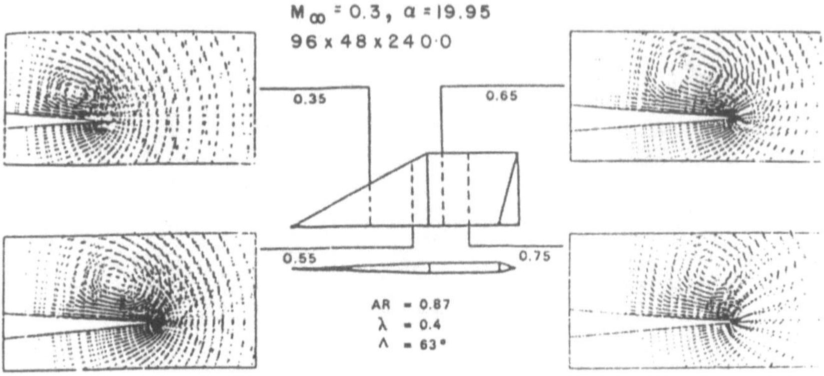

Fig. 8.14. The cross-plane velocity field for the cropped-delta wing calculated with the Euler TEAM code (Raj et al. 1986)

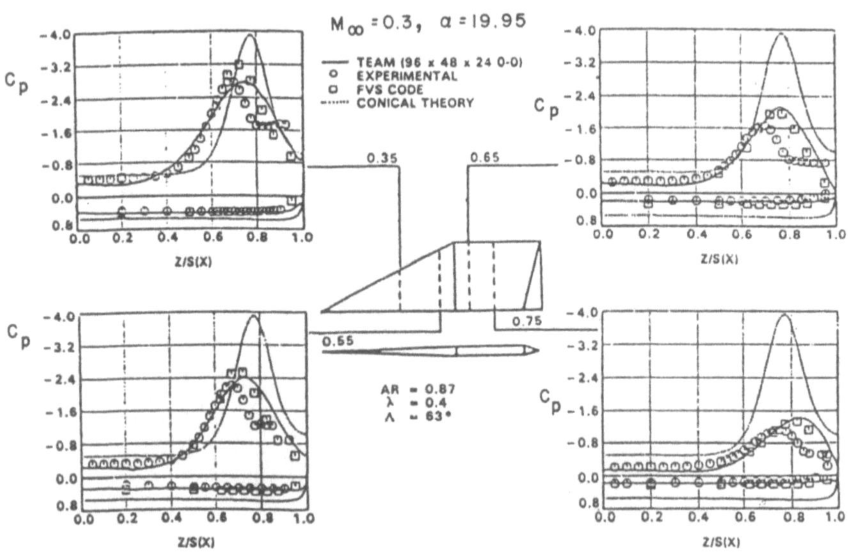

Fig. 8.15. Comparisons of pressure distributions calculated by Euler TEAM code and FVS method with experimental data (Raj et al. 1986)

as with the free-vortex-sheet (FVS) method calculations in Fig. 8.15. It is interesting to note that there is reasonably good agreement between the calculated pressure distributions by the Euler code and by the FVS method, but both overpredict the peak negative pressure in the rear part of the leeward side of the wing, while the Euler code accords somewhat better with experimental results in the forward part of the wing. The calculated results, both the Euler and FVS, indicate an outboard shift of the peak negative pressure in comparison with the experimental data.

Solutions of Euler equations for flow over a 65° delta wing at subsonic $M = 0.4$, transonic $M = 0.85$ and supersonic $M = 1.2$ conditions including comparisons with experimental data were presented by Sirbaugh (1987). The wing is a cropped-delta wing which has 65° leading-edges sweep with a taper ratio of 0.15. The sharp leading-edge wing is a biconvex airfoil over the forward 40% chord and a NACA 64A005 airfoil over the aft 60% chord. The rounded leading-edge airfoil is basically a NACA 64A005 airfoil with a 0.7% chord leading-edge radius. The grid is generated by constructing surface meshes, and the three-dimensional grid is generated using a fully three-dimensional elliptic method. The Euler solution algorithm is the finite volume method FLO57 code developed by Jameson et al. (1981), modified to permit arbitrary geometries through the use of multiple grid blocks. The aerodynamic coefficients, C_L, C_D and C_M and the pressure distributions were calculated for the three Mach numbers for angles of attack up to 25°.

The variation of the lift coefficient as the angle of attack is increased up to 25° on the wing with sharp edges for Mach numbers 0.4, 0.85 and 1.2 is presented in Fig. 8.16a. It is interesting to note that the lift coefficients for the subsonic to supersonic Mach numbers are almost the same, showing the very little effect of the Mach number, and the agreement between the Euler calculated lift and the experimental data is also very good. There is a very strong Mach number effect on the variation of the pitching moment coefficient with increased angles of attack, as seen in Fig. 8.16b, and here the agreement between the Euler calculations and the experimental data is reasonable but not as good as that found in the lift case. In the case of the wing with rounded leading-edges, the variation of the lift coefficients and the pitching moment coefficients with angle of attack for Mach numbers of 0.4 and 0.85 are shown in Figs. 8.17a and 8.17b, respectively. Here again there is very good agreement between the Euler calculations and the experimental lift coefficients, but less so for the pitching moment coefficients.

The calculated pressure distributions on the cropped-delta wing with the rounded leading-edges at $\alpha = 10^o$ and flow Mach numbers of 0.4, 0.85 and 1.2, and at $\alpha = 20^o$ and flow Mach numbers of 0.4 and 0.85, are compared with the measured distributions in Figs. 8.18 and 8.19, respectively. At $\alpha = 10^o$, there is a reasonable qualitative agreement between the calculated pressure and the measured data. However, a close examination of the 10° results indicate that the Euler solution fails to create the second peak due to the secondary vortex. The differences are even more accentuated in the comparison of the results at $\alpha = 20^o$ in Fig. 8.19.

The structure of the separated vortices generated on a sharp edge 75° swept delta wing at supersonic speed of $M = 1.2$ and at angles of attack of $10^o, 20^o$ and 30° are presented by Lee and Batina (1991). Using an unstructured grid of triangles, the conical Euler equations are solved for the

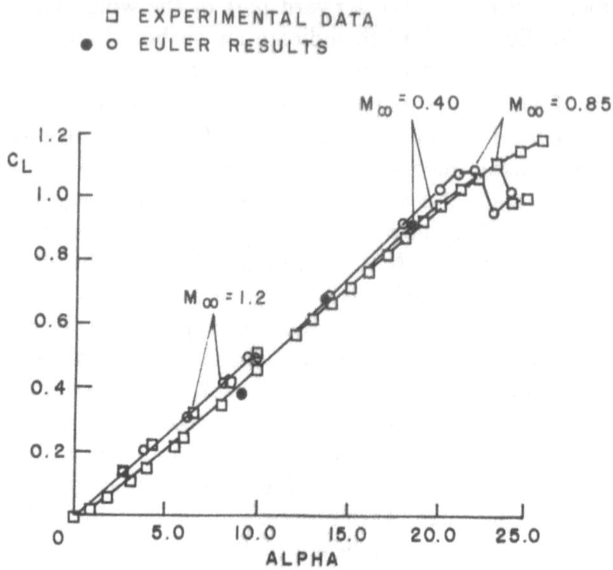

Fig. 8.16a. Variation of lift coefficients as a function of angles of attack for the cropped-delta wing with sharp leading-edges (Sirbaugh 1987)

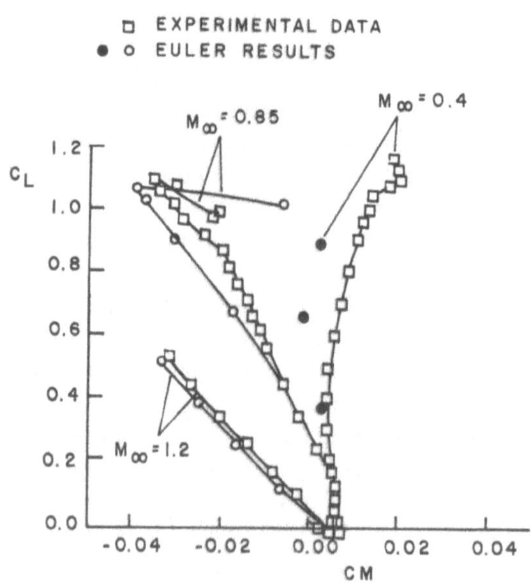

Fig. 8.16b. Variation of pitching moment coefficients as a function of angles of attack for the Cropped-Delta wing with sharp leading-edges (Sirbaugh 1987)

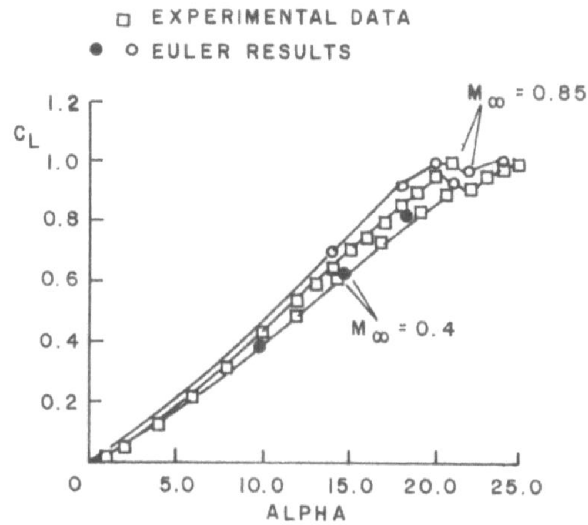

Fig. 8.17a. Variation of lift coefficients as a function of angles of attack for the cropped-celta wing with rounded leading-edges (Sirbaugh 1987)

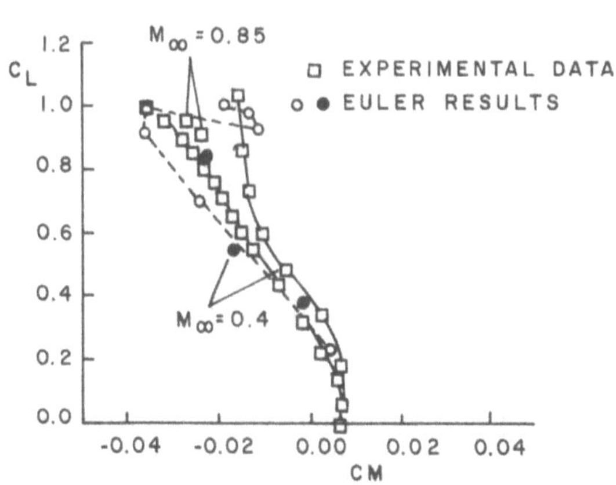

Fig. 8.17b. Variation of pitching moment coefficients as a function of angles of attack for the cropped delta wing with rounded leading-edges (Sirbaugh 1987)

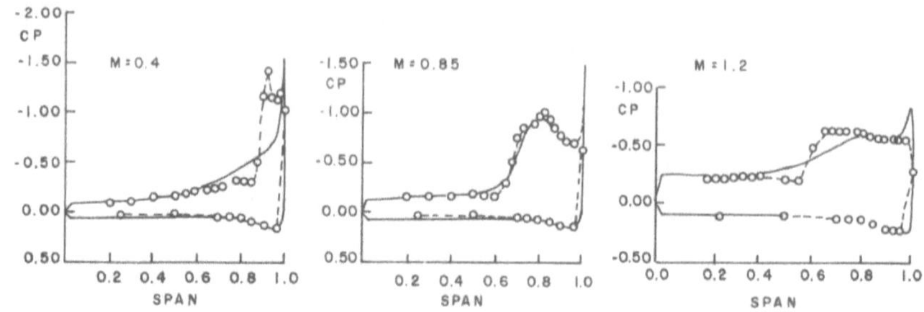

Fig. 8.18. Spanwise pressure distributions for the cropped-delta wing with rounded leading-edges at the 80% chord station and at $\alpha = 10^o$ and M=0.4, 0.85, 1.2 (Sirbaugh 1987)

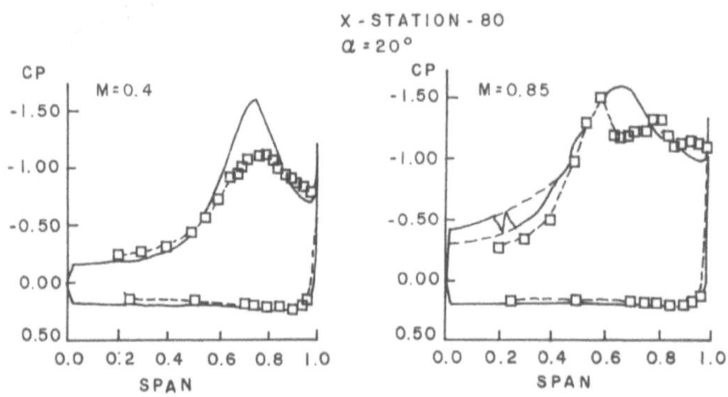

Fig. 8.19. Spanwise pressure distributions for the cropped-delta wing with rounded leading-edges at the 80% chord station and at $\alpha = 20^o$ and $M = 0.4$ and 0.85 (Sirbaugh 1987)

unsteady flow around the rolling delta wing. The steady state results illustrate very well the leading-edge vortex structure over the delta wing at increasing angles of attack. The unstructured grid on the wing is shown in Fig. 8.20. The grid, which was generated using an advancing front method, has a total of 4226 nodes and 8299 elements. The grid was designed to be fine on the leeward side of the wing and be more coarse on the windward side. The steady state solutions for the total pressure loss, calculated as the angle of attack is increased from 10° to 20° and 30°, is illustrated in Fig.

8.21. At the low angle of attack ($\alpha = 10^o$), the contours indicate that the flow separates from the leading-edge to form two small, widely spaced circular vortices. At 20°, the contours indicate larger circular vortices with two vertically oriented cross-flow shock waves on the outboard portion of the wing. At $\alpha = 30^o$, the contours indicate that the flow separating from the leading-edge produces two larger, more closely spaced vortices, as shown in Fig. 8.21. In this case, vertically oriented cross-flow shock waves are formed on the outboard portions of the wing. In addition, a weaker shock wave is formed on the top of each vortex and also between the two vortices.

8.2.6.1.1 Calculation of Vortex Structure Over Delta Wing. The process of the rolling-up of the shear layer, which is separated from the leading-edge of a highly swept delta wing, into a concentrated vortex above the wing surface is a most intriguing problem. The capability of the inviscid Euler solutions to capture this flow process was explained by the effect of the "numerical viscosity" which initiates this process in analogy to the initiation by the physical viscosity. The capability of the Euler solutions to generate the leading-edge rolled-up vortex is now well established as shown, for example, in Fig. 8.13 on the 55° delta wing and in Fig. 8.14 for the cropped-delta wing. Moreover, it was noted that these Euler solutions for the rolled-up concentrated vortices indicate a phenomenon which resembles the effect of "breakdown" of the leading-edge vortex as the vortex develops downstream over the wing, a phenomenon described by Sirbaugh (1987). This phenomenon can be seen in the sudden expansion of the rolling vortex lines (particle traces) as shown in Fig. 8.22. It is observed also in comparing the cross plane views of the velocity field and the vorticity contours, as in Fig. 8.23a which are calculated in the concentrated vortex zone with the vortex "break down" zone in Fig.8.23b. A similar significant change in the flow field, calculated from solutions of the Euler equations for flows over a delta wing, was also observed by Scherr and Das (1988) and interpreted as indicating the appearance of "breakdown". Here, the nature of the roll-up around the leading-edge at a chord station of 80% chord is seen to change drastically as the delta wing angle of attack is increased from 20° to 24°, calculated at a flow Mach number of 0.85. It can be seen in Fig. 8.24 that at $\alpha = 20^o$, the typical coiling of the rolled-up vortex is observed, while at $\alpha = 24^o$ at the same chord station of 80% chord, the presence of high radial velocities in the region of the vortex core is clearly observed. The presence of the radial velocities destroys the high surface velocities induced by the concentrated vortex core, so that the vortex induced low pressure is lost, and this initiates the loss in vortex lift associated with the appearance of the vortex breakdown.

It will be shown in Chapter 9 that similar effects of the vortex breakdown can also be obtained in the solutions of the Navier-Stokes equations. However, the ability of the Euler solutions to capture the vortex roll-up and follow its development up to its breakdown indicates that this is basically an inviscid flow effect.

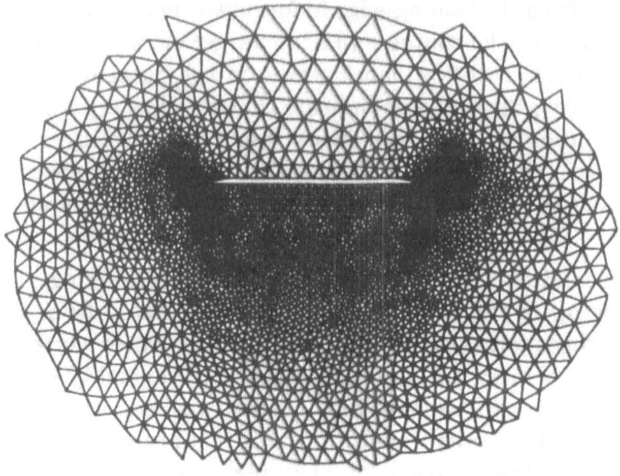

Fig. 8.20. Unstructured grid about a sharp-edge 75° swept delta wing (Lee and Batina 1991)

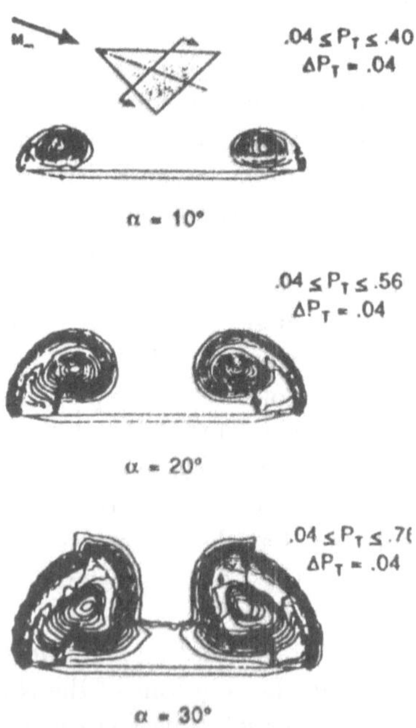

Fig. 8.21. Steady-state total pressure contours for a 75° swept delta wing at $M = 1.2$ (Lee and Batina 1991)

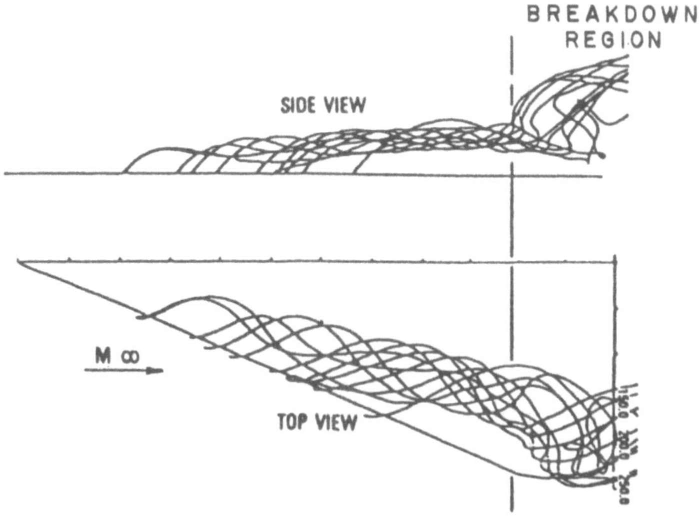

Fig. 8.22. Vortex trajectories (particle traces) over the Cropped-Delta wing at $M = 0.85$ and $\alpha = 21^o$ illustrating the vortex breakdown (Sirbaugh 1987)

8.2.6.2 Strake - Wing - Body Configuration. Solutions of the Euler equations on configurations more complex than that of the single delta type wing, discussed in the previous section, are feasible with the presently available vector supercomputers. Calculations on wing-canard configurations are presented by Eriksson and Rizzi (1984) and by Scherr and Das (1988). Some results of calculations on simple wing-body configurations are presented by Agarwal and Deese (1983) and by Moitra (1985). Interesting results were also obtained by Raj et al. (1986) on a strake-wing-body configuration, which will be discussed in this section as an example of Euler solutions over intermediately complex configurations before calculations are made on full aircraft and missile configurations.

The strake-wing-body configuration has a cropped-delta wing of AR=2.5 and taper ratio $\lambda = 0.2$ with a leading-edge angle of 50°, where the strake area is 28% of the wing area. The wing profile has biconvex airfoils of 6% thickness at the wing root, 4% thickness at the wing tip, and both the strake and the wing have sharp leading-edges. A C-H type grid is generated using 98,304 (96 × 32 × 32) cells, as shown in an oblique view in Fig. 8.25. Calculations were made for the configurations with strake on and strake off using the TEAM code.

Comparison of the computed lift coefficients with experimental data at a Mach number of 0.5 is presented in Fig. 8.26. There is a good agreement between the computed values with the experimental lift coefficients for the strake on case up to angles of attack of about 20°, and for the strake off

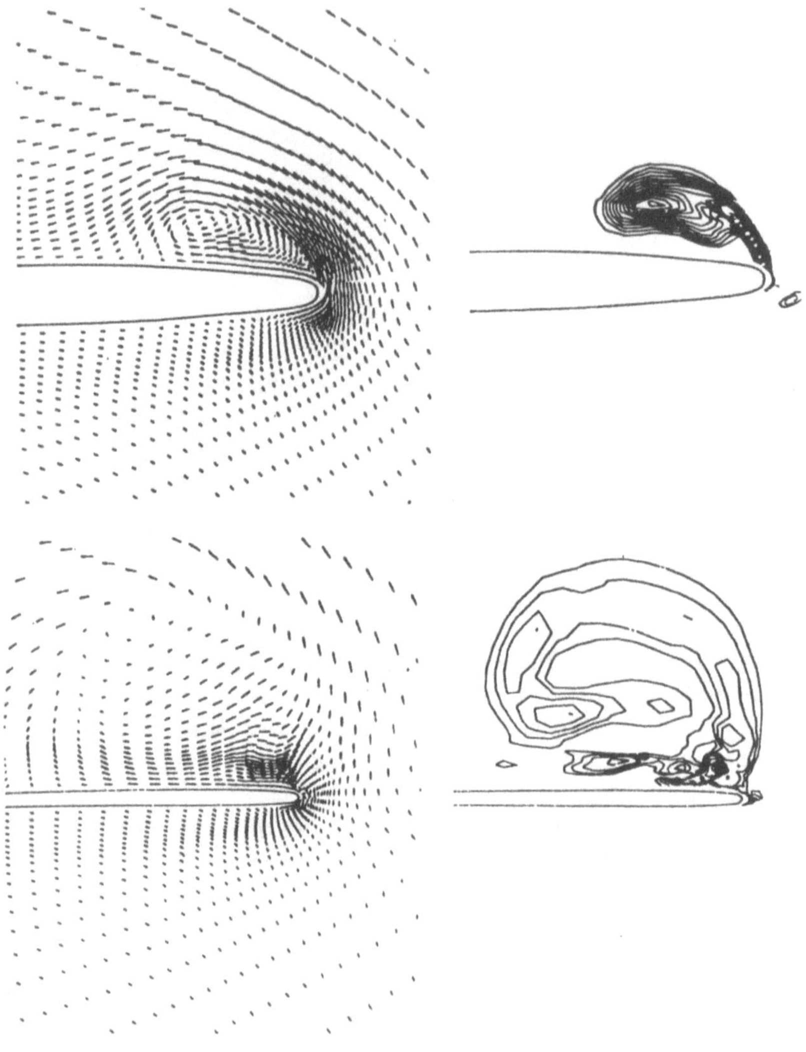

a. Plane in forward part – concentrated leading-edge vortex

b. Plane in aft part – spread of vorticity, vortex "breakdown"

Fig. 8.23. Cross plane velocity field and vorticity contours over the cropped-delta wing at $M = 0.85$ and $\alpha = 21^\circ$ (Sirbaugh 1987)

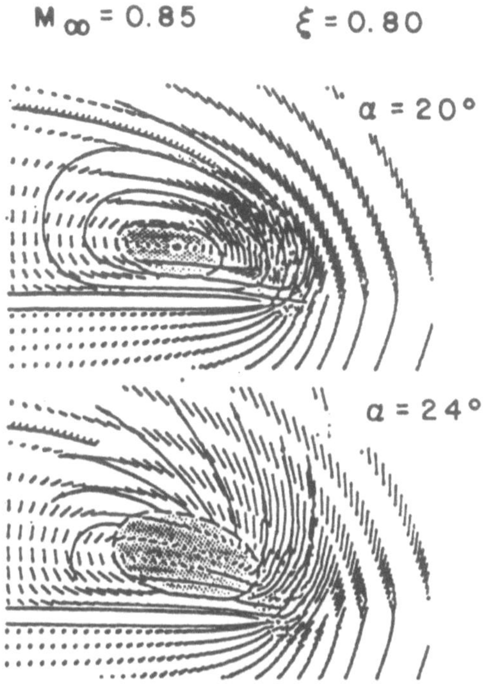

Fig. 8.24. Rolled-up flow in the 80% chord cross plane of a delta wing calculated from the Euler equations at $M = 0.85$ and at $\alpha = 20^o$ and $\alpha = 24^o$ (Scherr and Das 1988)

case up to angles of attack of about 15°. At higher angles of attack, the computations did not converge, as noted by the asterisk in Fig. 8.26, on the TEAM points at $\alpha = 30^o$ and $\alpha = 18^o$ for the strake on and strake off cases, respectively. Examining the flow patterns at these conditions, Figs. 8.27 and 8.28 offer a very interesting explanation, namely, that the velocity field in the cross planes over the configurations in these cases show outward flow directions rather than the concentrated rolled-up pattern of the leading-edge vortex. This can be interpreted as a "vortex burst", as noted by Raj et al. (1986), or as an indication of vortex breakdown as these conditions coincide with the beginning of loss of lift in the experimental data. The velocity field at two cross-planes on the wing at strake off and strake on cases at an angle of attack of 18° is shown in Fig. 8.27. The outflow velocities can be clearly seen for the strake off case, in contrast to the vortex flow in the strake on case. This is an excellent example of the stabilizing effect of the strake at high angles of attack. This phenomena is well known from experimental observation; however, in this case, it is obtained numerically from the Euler solution.

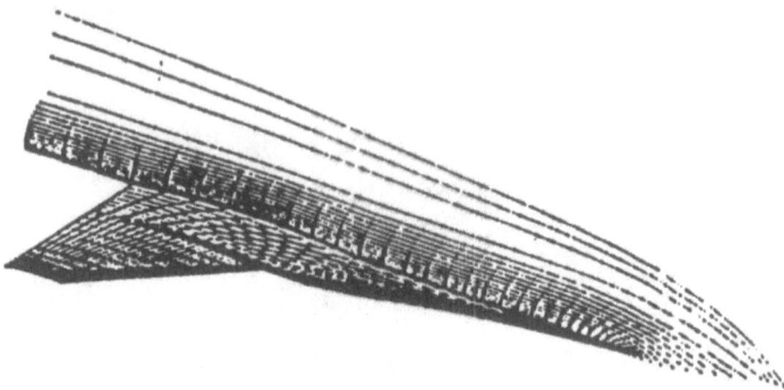

Fig. 8.25. Oblique view of the C-H grid about the Strake-Wing-Body configuration (Raj et al. 1986)

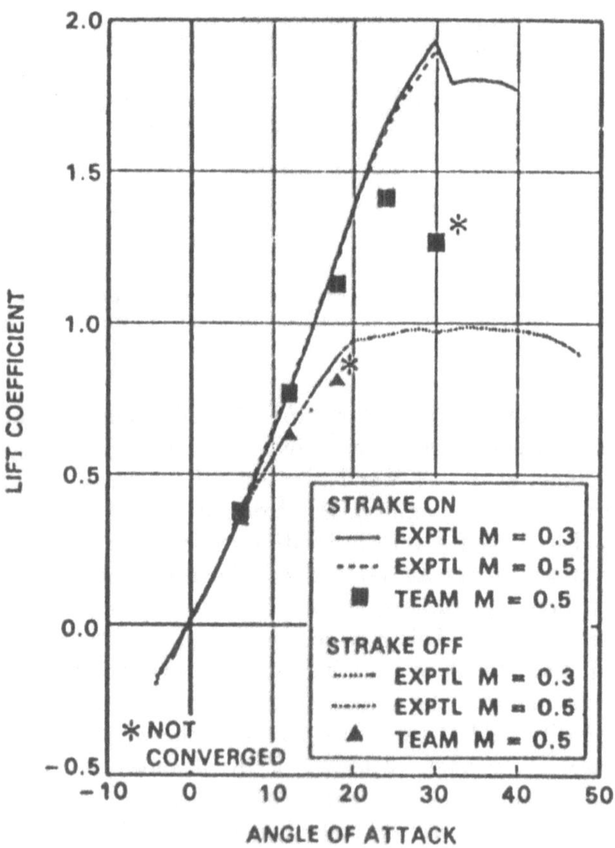

Fig. 8.26. Comparison of the experimental and computed lift coefficients for the strake on and strake off configurations (Raj et al. 1986)

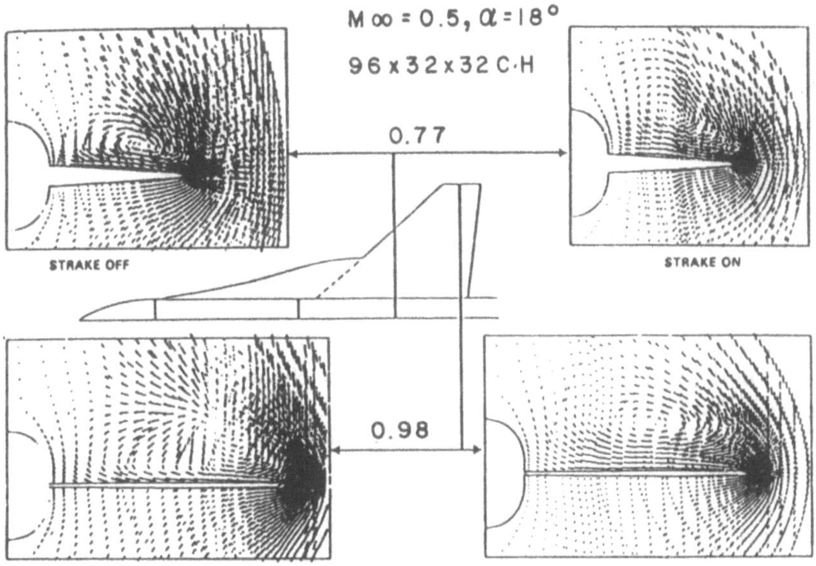

Fig. 8.27. Velocity field in the cross-plane over the wing with Strake on and Strake off configurations at $\alpha = 18^o$ (Raj et al. 1986)

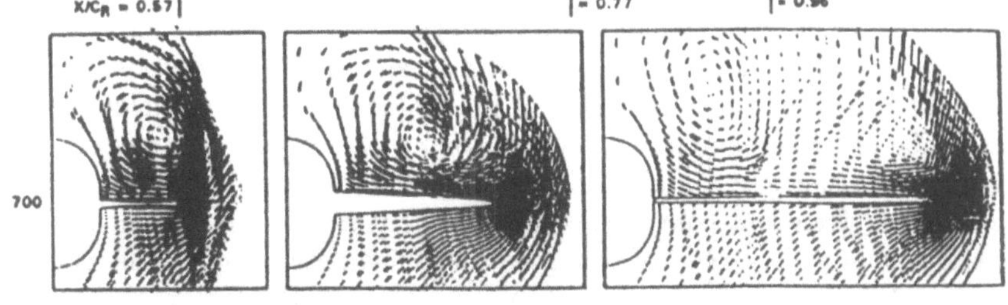

Fig. 8.28. Velocity field in the cross-plane over strake-wing-body configuration at $\alpha = 30^o$(Raj et al. 1986)

8.2.6.3 Euler Solutions Over Aircraft and Missile Configurations. The presently available supercomputers make it possible to solve the Euler equations with realistic boundary conditions of complete aircraft type configurations in reasonable running time. Of course, methods for the solutions of the Navier-Stokes equations are available, but considering the much larger computing effort required and the much longer running times, the Euler solutions are becoming an essential tool in the aerodynamic design of aircraft and missiles. It was shown that many of the important flow phenomena at high angles of attack are dominated by the inviscid vortical flow. Therefore, the solutions of the inviscid Euler equations offer a very good compromise giving relatively economical reliable computations fairly quickly as well as a reasonable representation of the flow field in comparison to the full Navier-Stokes solutions.

The various methods of solution of the Euler equations which have been developed are described in the first part of this Chapter. All of these numerical methods are based on a discretization of the flow equations. The discretization of the physical space about the configuration is then required. There are grid generation methods now available which are capable of achieving extensive meshes about very complex configurations. However, these can be very complex, requiring considerable experience on the part of the user, and are generally so time consuming as to make them uneconomical for preliminary design use.

There are a number of examples of solutions of the Euler equations for flows over aircraft and missile configurations. Some of these calculations will be discussed here in order to indicate the capabilities of these methods in obtaining useful results for the aerodynamic characteristics at high angles of attack.

An efficient method for computing the flows about aircraft configurations is presented by Volpe and Jameson (1988). In this method, a single-block H-O mesh is used to represent the space of a large variety of configurations including fairly complex ones. With this mesh, a fast numerical scheme is designed to take advantage of the ordered sequence of mesh cells without the encumbrance of internal mesh boundaries. The method uses the finite-volume algorithm of Jameson for the integration of the time dependent Euler equations to steady state via a Runge-Kutta scheme. Acceleration to steady state is achieved by the use of a local time-step. Additional acceleration is provided by smoothing of the residuals and by enthalpy damping. The ordered regular mesh system then makes it possible to imbed the numerical scheme in a multigrid sequence of meshes to provide further acceleration. Excellent convergence rates are claimed for this method. Flow fields over a wide variety of configurations have been computed in as little as 5 to 10 minutes of computing time on a Cray-XMP class machine.

The grid generated around a generic fighter configuration is shown in Figs. 8.5 to 8.7. The appropriate boundary conditions are applied on the configuration surfaces, across the wake, and on the outer boundaries of the mesh. For a symmetric configuration with zero sideslip, consideration of symmetry enables the calculation on only one half of the flow field by applying the symmetry boundary conditions on the plane of symmetry.

An example of calculation results of the isomach contours on the surface and on a mid-fuselage cross-plane on a fighter configuration is shown in Fig. 8.29. At a flow Mach number of 0.80 and at a 20° angle of attack, a shock can be discerned as a clustering of the Mach contour lines in the mid-fuselage cross-sectional plane, and the region of supersonic flow can be seen to extend as a "bubble" in the flow field over the wing and fuselage.

Fig. 8.29. Isomach contours on upper surface and in the cross-plane of a fighter aircraft (Volpe and Jameson 1988)

An example illustrating the capability of the Euler solution to capture the vortex structure at reasonably high angles of attack at subsonic speeds is presented in Fig. 8.30 The flow structure over a chine-forebody fighter configuration is calculated at $M = 0.2$ and at an angle of attack of 25°. The interactions between the vortices from the chine-forebody and those of the wing leading-edges are clearly seen in the different views presented in this figure. This vortex structure is in good qualitative agreement with experimental results.

The solutions of the Euler equation on more complex configurations, for example those involving configurations with multiple bodies such as underwing pylon/nacelle, require complicated grid generation systems. A grid generation system suitable for complex aircraft configurations was developed at the ARA in England and was presented recently by Bocci (1990). This is a multiblock system for the generation of grids about three-dimensional aircraft configurations which is now being extended to deal with more complex elements such as external bodies. The extension program includes the possibility of developing a more flexible hybrid system containing regions of unstructured grid embedded within the overall structured grids.

The multiblock system is based on the division of the flow field into non-overlapping blocks, arranged so that body-conforming grids are defined by the node-to-node interface of the grids at the block faces. The grid is generated by a set of elliptic partial differential equations which map a curvilinear coordinate system of points in each block in real space to uniformly spaced points in the three coordinate directions of a cube in computational space. On each face of the cube, the appropriate boundary conditions are imposed. Having defined a suitable grid, the solution of the Euler equations is obtained using the finite-volume cell center algorithm of Jameson et al. (1981).

An illustration of surface grid for an advanced military aircraft configuration with modeling of the intake and exhaust flows is shown in Fig. 8.4. For this configuration, the field grid is defined by 468 blocks with about 300,000 grid cells. In certain configurations with extra geometric components, such as underwing store or engine nacelle, a hybrid approach seems to be advantageous. In such cases, the structured grid is deleted in the block(s) where these components are located, and unstructured grids are generated over the surfaces of these components and within the block(s). An example of a surface grid defined by this hybrid system for a wing-body-pylon-nacelle configuration is shown in Fig. 8.31. The optimal approach for the hybrid system and the integration of these unstructured blocks into the computation scheme are still under development.

An illustration of a solution of the Euler equations by the finite-element method of Stoufflet et al. (1987) for the hypersonic flow over the Hermes type configuration at $M = 10$ and $\alpha = 40^o$ is presented by Viviand et al. (1987) and shown here in Fig. 8.32.

An example of the Euler solution for flows over a typical missile configuration is presented by Lijewski (1987). The missile configuration on which the surface grid is plotted is shown in Fig. 8.33. The field grid is obtained by the elliptic grid generation method extended to multiblock capability, and the Euler equations are solved by an explicit second order upwind scheme by Warming and Beam (1976). The Euler equations are written in strong conservation law form, and the computation scheme uses a finite volume formulation. Local time-stepping is used to accelerate convergence for steady state problems. Using characteristic variable boundary conditions,

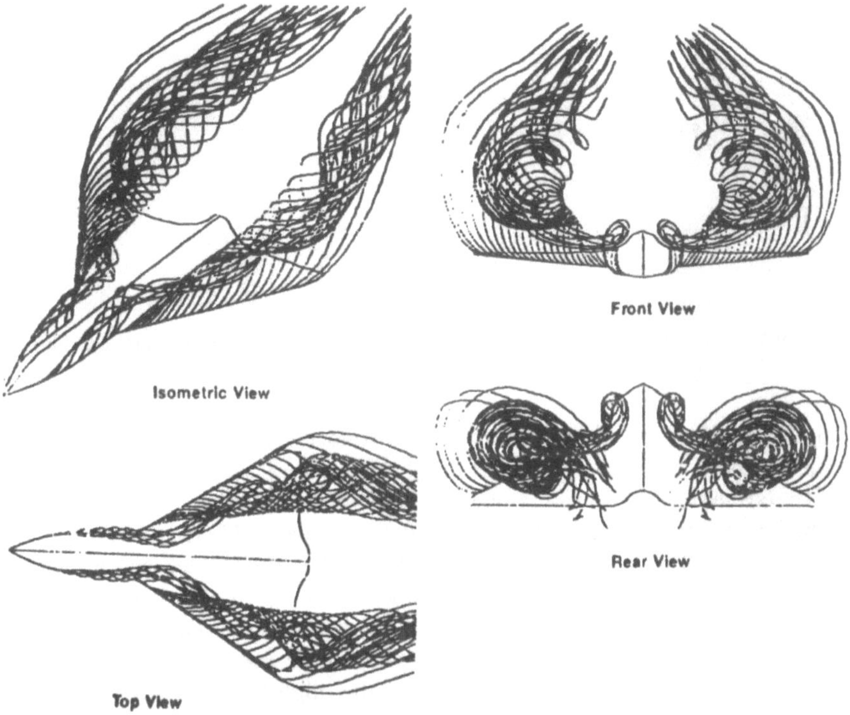

Fig. 8.30. Computed streamlines on Chine-Forebody configuration at flow $M = 0.2$ and $\alpha = 25°$ (Volpe and Jameson 1988)

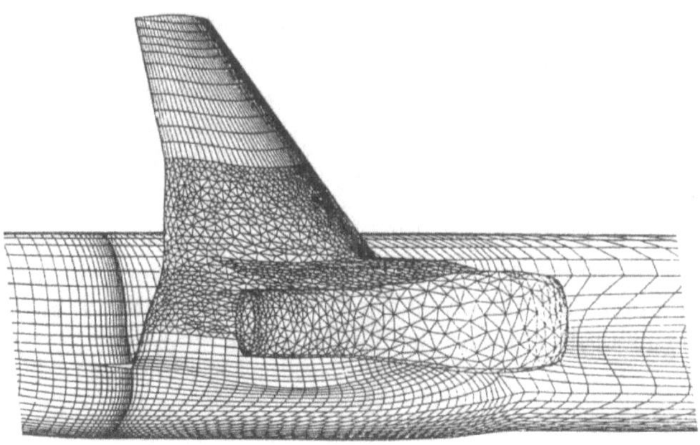

Fig. 8.31. Hybrid Surface Grid for Wing-Body-Pylon-Nacelle (Bocci 1990)

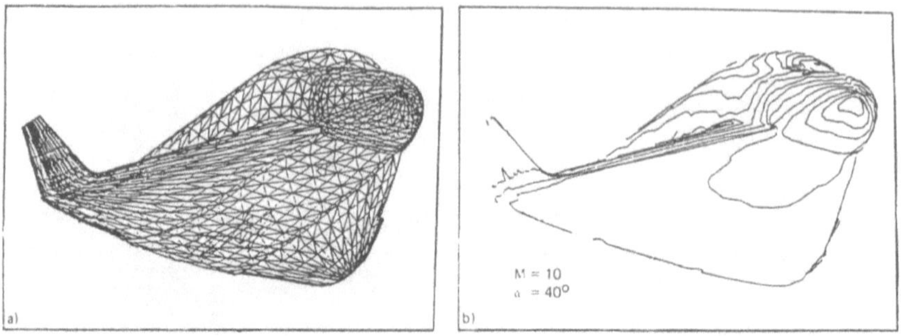

Fig. 8.32. Surface grid and isomach lines on Hermes type space plane calculated by Finite Element method at $M = 10$ and $\alpha = 40^o$ (Viviand et al. 1987)

the Euler equations are solved on the multiblock grid system for the missile configuration at $M = 0.95$ and 1.05 at angles of attack up to 12°. Results of the pressure distribution along the body surface, calculated by the Euler solutions for various grid densities and compared with experimental data, are shown in Fig. 8.34. The local pressure, calculated by all the three grid densities used, give approximately the same results and are in reasonable agreement with the measured values. The comparison of the calculated normal force and pitching moment coefficients with the experimental results is presented in Fig. 8.35. The Euler results slightly overpredict the normal force and give a somewhat more negative pitching moment than the experimental values.

The capability of the Cartesian grid method to evaluate the aerodynamic characteristics of a fighter aircraft configuration is presented by Epstein et al. (1989). The aircraft is a delta wing with a close coupled canard, shown with the surface grid in Fig. 8.36.

The calculated velocity field above the wing is shown in Fig. 8.37.

The calculated velocity vector field at three cross-sections on the aircraft configuration are presented in Fig. 8.38. This figure includes the view of the planform with the grids of several levels.

Comparison of the calculated lift and drag coefficients with wind tunnel measurements are presented in Fig. 8.39. Very good agreement is found at angles of attack of 8° and 16° but the calculated results are somewhat lower than the experimental values at the higher angles of attack.

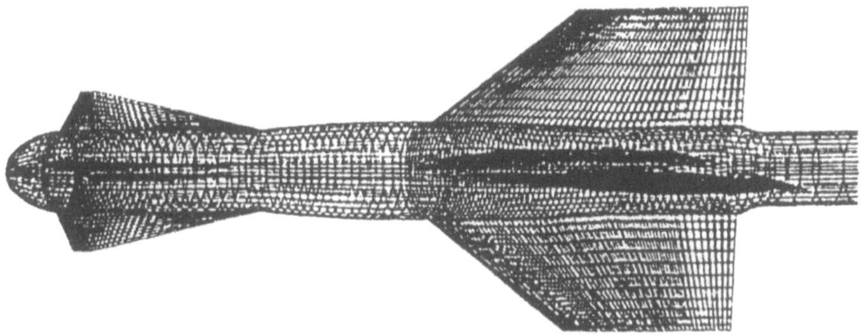

Fig. 8.33. Surface grid on the missile configuration (Lijewski 1987)

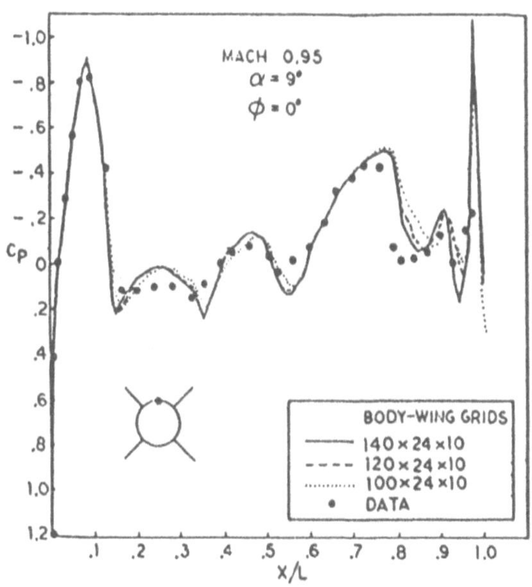

Fig. 8.34. Comparison of the calculated and experimental pressure distributions on the missile body surface at $M = 0.95$ and $\alpha = 9^o$ for various grid densities (Lijewski 1987)

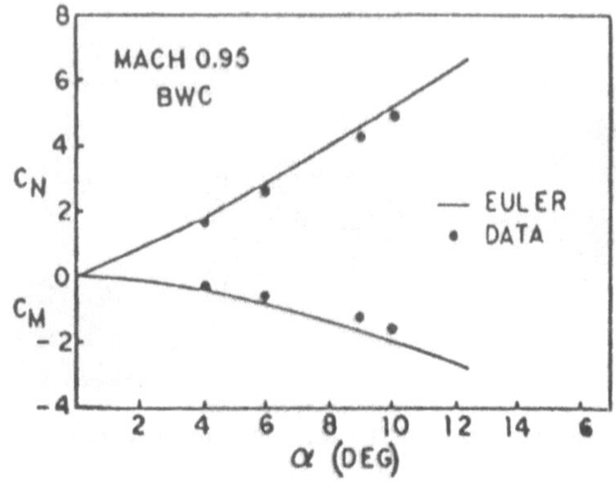

Fig. 8.35. Comparison of the calculated and measured normal force and pitching moment coefficients on missile configuration at $M = 0.95$ (Lijewski 1987)

Fig. 8.36. Delta wing-canard fighter configuration (Epstein et al. 1989)

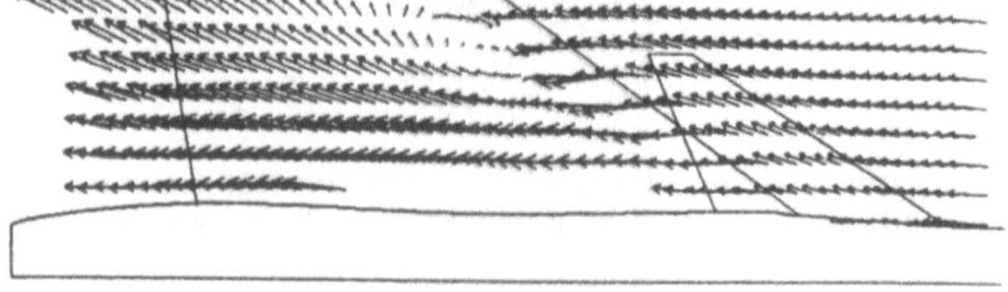

Fig. 8.37. Top view of the velocity vector field above the fighter canard-wing configuration at $M = 0.9$ and $\alpha = 24°$ (Epstein et al. 1989)

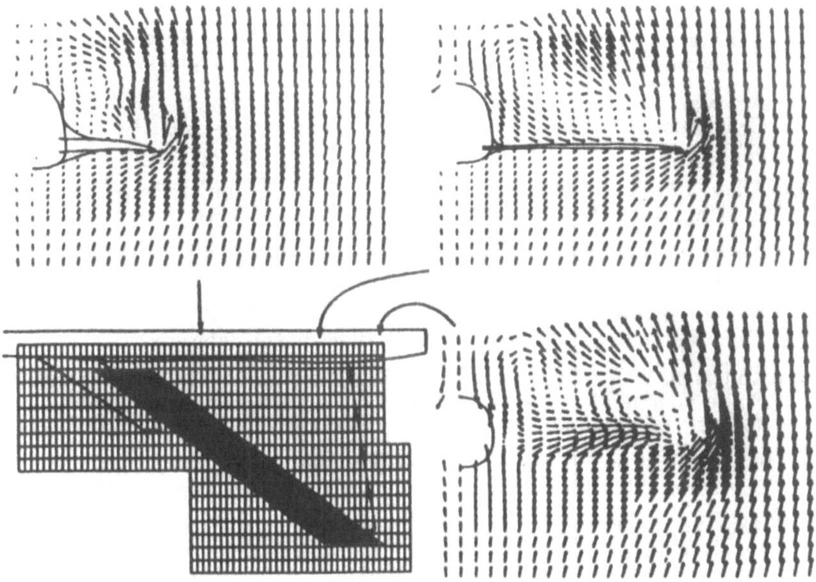

Fig. 8.38. Velocity vector field at three cross-sections on the fighter aircraft wing-canard configuration at $M = 0.9$ and $\alpha = 24^o$ (Epstein et al. 1989)

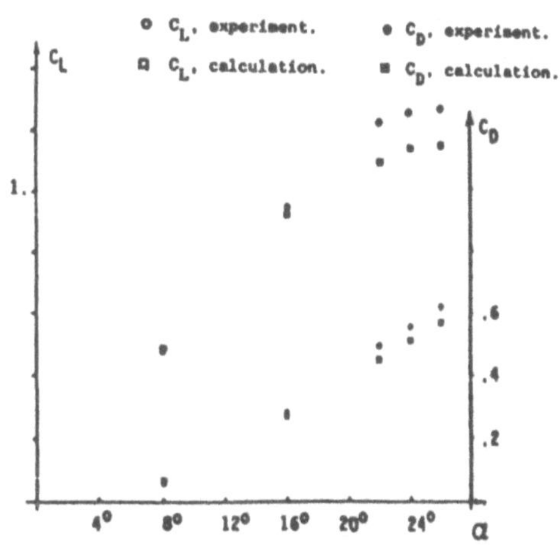

Fig. 8.39. Variation of the lift and drag coefficients as a function of angle of attack for the fighter aircraft configuration at $M = 0.9$ (Epstein et al. 1989)

References

1. Agarwal, S., Vermeland, R.E., Verhoff, A. and Lowery, R.B. (1988), "Euler Transonic Solutions over Finite Wings", AIAA Paper 88 0009.

2. Agarwal, R.K., Deese, J.E., Johnson, J. and Steinhoff, J. (1989), "Euler Calculations of Flow Over a Complete Fighter Aircraft", AIAA Paper 89-2221.

3. Allen, J.M. and Townsend, J.C. (1985), "Application of the SWINT Code to Wing/Body/Tail Geometries", AIAA Paper 85-1811.

4. Allwright, S.E. (1989), "Multiblock Topology Specification and Grid Generation for Complete Aircraft Configurations", AGARD CP-464, Paper 11.

5. Anderson, D.A., Tannehill, J.C. and Pletcher, R.H. (1984), "*Computational Fluid Mechanics and Heat Transfer*", Hemisphere Publishing Co.

6. Anderson, W.K., Thomas, J.L. and Van Leer, B. (1985), "A Comparison of Finite Volume Flux Vector Splitting for the Euler Equations", AIAA Paper 85-0122, January 1985. Also in AIAA Journal, Vol. 24, No. 9, pp. 1453–1460.

7. Anderson, W.K., Thomas, J.L. and Whitfield, D.L. (1986), "Multigrid Acceleration of the Flux Split Euler Equations", AIAA Paper 86-0105.

8. Angrand, F., Billey, V., Dervieux, A., Periaux, J., Pouletty, C. and Stoufflet, B. (1985), "2-D and 3-D Euler Flow Calculations with a Second-Order Accurate Galerkin Finite Element Method", AIAA Paper 85-1706.

9. Arlinger, B. (1984), "Computation of Supersonic Flow about Complex Configurations", ICAS Paper 84-1.3.2., pp. 266–274.

10. Arlinger, B. and Winzell, B. (1990), "Hypersonic Flow Fields Around HERMES Computed by Euler Codes", ICAS Paper 90-6.7R in the Proceedings of the 17th Congress, pp. 2183–2191.

11. Ashill, P.R., Fulker, J.L., Simmons, M.J. and Betts, C.J. (1990), "Flow Features of Highly-Swept Wings at Subsonic and Supersonic Speeds", ICAS Paper 90-3.9.1, in the Proceedings of the 17th Congress, pp. 1678–1689.

12. Baker, T.J. (1989), "Unstructured Mesh Generation by a Generalised Delaunay Algorithm", AGARD CP - 464, Paper 20.

13. Batina, J.T. (1990a), "Three-Dimensional Flux-Split Euler Schemes Involving Unstructured Dynamic Meshes", AIAA Paper 90-1649.

14. Batina, J.T. (1990b), "Development of Unstructured Grid Methods for Steady and Unsteady Aerodynamic Analysis", ICAS Paper 90 6.9.4. in the Proceedings of the 17th Congress, pp. 1833–1843.

15. Baxendale, A.J. (1990), "Application of a Multiblock CFD System to Obtaining Flowfield Predictions About Wing Body Pylon Store Configurations", ICAS Paper 90-6.6.2. in the Proceedings of the 17th Congress, pp. 1173–1183.

16. Beam, R.W. and Warming, R.F. (1976), "An Implicit Finite Difference Algorithm for Hyperbolic System in Conservation Form", Journal Comp. Phys., Vol. 23, pp. 87–110.

17. Belk, D.M., (1987), "Time-Accurate Euler Equations Solutions on Dynamic Blocked Grids", Proceedings of AIAA 8th Computational Fluid Dynamics Conference, Honolulu, HI, pp. 292–302.

18. Benek, J.A., Buning, P.G. and Steger, J.L. (1985), "A 3-D Chimera Grid Embedding Technique", AIAA Paper 85-1523.

19. Benek, J.A., Donegan, T.L. and Suhs, N.E. (1987), "Extended Chimera Grid Embedding Scheme with Applications to Viscous Flows", Proceedings of the 8th Computational Fluid Dynamics Conference, Honolulu, HI, pp. 283–291.

20. Berger, M. and Jameson, A. (1985), "Automatic Adaptive Grid Refinement for the Euler Equations", AIAA Journal, Vol. 23, pp. 561–568.

21. Berglind, T. (1990), "Flow Simulation Around a Realistic Fighter Airplane Configuration", ICAS Paper 90-6.4.1. in the Proceedings of the 17th Congress, pp. 745–754.

22. Bocci, A.J. (1990), "Recent Developments in CFD at ARA", ICAS Paper 90-6.8.3 in the Proceedings of the 17th Congress, pp. 1569–1585.

23. Botta, N. and Pandolfi, M. (1987), "Upwind Formulations for the Euler Equations in Steady Supersonic Flows", AIAA Paper 87-1150.

24. Brandt, A. (1977), "Multi-Level Adaptive Solutions to Boundary Value Problems", Math. Comp., Vol. 31, pp. 333–390.

25. Cannizzaro, F.E., Von Lavante, E. and Melson, N.D. (1988), "Calculations of Three-dimensional Flows Using the Isenthalpic Euler Equations with Implicit Flux-Vector Splitting", AIAA Paper 88-2516.

26. Cannizzaro, F.E., Elmiligui, A., Melson, N.D. and Von Lavante, E. (1990), "A Multiblock Multigrid Method for the Solution of the Three-Dimensional Euler Equations", AIAA Paper 90-0105.

27. Caughey, D.A. and Jamesom, A. (1980), "Progress in Finite-Volume Calculations for Wing-Fuselage Combinations", AIAA Journal, Vol. 18, No. 11, pp. 1281–1288.

28. Caughey, D.A. (1987), "A Diagonal Implicit Multigrid Algorithm for the Euler Equations", AIAA Paper 87-453.

29. Causon, D.M. and Ford, P.J. (1984), "Numerical Solutions of the Euler Equations Governing Axisymmetric and Three-Dimensional Transonic Flow", ICAS Paper 84-1.3.3., pp. 275–286.

30. Chakravarthy, S.R. and Szema, K.Y. (1985), "An Euler Solver for Three-Dimensional Supersonic Flows with Subsonic Pockets", AIAA Paper 85-1703.

31. Chakravarthy, S.R. (1986), "The Versatility and Reliability of Euler Solvers Based on High Accuracy TVD Formulations", AIAA Paper 86-0424.

32. Chakavarthy, S.R. and Ota, D. (1986), "Numerical Issues in Computing Inviscid Supersonic Flow Over Conical Delta Wings", AIAA Paper 86-0440.

33. Chima, R.V. and Johnson, G.M. (1983), "Efficient Solution of the Euler and Navier-Stokes Equations with a Vectorized Multiple Grid Algorithm", AIAA Paper 83-1893, Proceedings of the 6th Computational Fluid Dynamics Conference, pp. 72–89.

34. Doe, R.H., Brown, T.W. and Pagano, A. (1986), "The Development of Practical Euler Methods for Aerodynamic Design", ICAS Paper 86-1.4.2.

35. Drikakis, D. and Tsangaris, S. (1990), "Improved Mesh Sequencing Method for the Accelerated Solution of the Compressible Euler and Navier-Stokes Equations", ICAS Paper 90-6.10.1. in the Proceedings of the 17th Congress, pp. 1999–2011.

36. Eberle, A. (1985), "3D Euler Calculations Using Characteristic Flux Extrapolation", AIAA Paper 85-0119.

37. Eberle, A., Schmatz, M.A. and Schafer, O. (1986), "High Order Solution of the Euler Equations by Characteristic Flux Averaging", ICAS Paper 86-1.3.1.

38. Epstein, B., Luntz, A.L. and Nachshon, A. (1989a), "Multigrid Transonic Computations About Arbitrary Aircraft Configurations", Journal of Aircraft, Vol. 26, No. 8, pp. 751–759.

39. Epstein, B., Luntz, A.L. and Nachshon, A. (1989b), "Multigrid Euler Solver About Arbitrary Aircraft Configurations with Cartesian Grids and Local Refinements", AIAA Paper 89-1960.

40. Epstein, B., Luntz, A.L. and Nachshon, A. (1990), "Euler Solver for Arbitrary Configurations: Further Applications", Proceedings of the 31st Israeli Conference on Aeronautics and Astronautics, pp. 205–216.

41. Eriksson, L.E. (1982), "Generation of Boundary Conforming Grids around Wing-Body Configurations Using Transfinite Interpolation", AIAA Journal, Vol. 20, pp. 1313–1320.

42. Eriksson, L.E. and Rizzi, A. (1984), "Computation of Vortex Flow Around a Canard/Delta Combination", Journal of Aircraft, Vol. 21, No. 11, pp. 858–865.

43. Fedorenco, R.P. (1964), "The Speed of Convergence of One Iterative Process", USSR Comp. Math. and Math. Phys., Vol. 4, pp. 227–235.

44. Fujii, K. and Schiff, L.B. (1987), "Numerical Simulation of Vortical Flows over a Strake-Delta Wing", AIAA Paper 87-1229.

45. Fujii, K. and Obayashi, S. (1989), "High Resolution Upwind Scheme for Vortical Flow Computations", Journal of Aircraft, Vol. 26, No. 12, pp. 1123–1129.

46. Gaffney, R.L., Hassan, H.A. and Salas, M.D. (1987), "Euler Calculations for Wings using Cartesian Grids", AIAA Paper 87-0356.

47. Goodsell, A.M., Madson. M.D. and Melton, J.E. (1989), "TranAir and Euler Computations of a Generic Fighter Including Comparisons with Experimental data", AIAA Paper 89-0263.

48. Goodsell, A.M., Melton, J.E. and Madson, M.D. (1990), "A Transonic/Supersonic CFD Analysis of a Generic Fighter", ICAS Paper 90-6.4.2. in the Proceedings of the 17th Congress, pp. 755–769.

49. Gumbert, C., Lohner, R., Parikh, P. and Pirzadeh, S. (1989), "A Package for Unstructured Grid Generation and Finite Element Flow Solvers", AIAA Paper 89-2175 CP.

50. Hessenius, K.A. and Rai, M.M. (1986), "Three-Dimensional, Conservative, Euler Computations using Patched Grid Systems and Explicit Methods", AIAA Paper 86-1081.

51. Hitzel, S.M. and Schmidt, W. (1984), "Slender Wings with Leading-Edge Vortex Separation; a Challenge for Panel Methods and Euler Solvers", AIAA Journal of Aircraft, Vol. 21, No. 10, pp. 751–759.

52. Hoeijmakers, H.W.M., Jacobs, J.M.J.W. and Van der Berg, J.I. (1990), "Numerical Simulation of Vortical Flow over a Delta Wing at Subsonic and Transonic Speeds", ICAS Paper 90-3.3.3. in the Proceedings of the 17th Congress, pp. 486–499.

53. Jacobs, J.M.J.W., Kassies, A., Boerstoel, J.W. and Buijsen, F. (1988), "Numerical Interactive Grid Generation for 3-D Flow Calculations", in *Numerical Grid Generation in Computational Fluid Mechanics 1988,* Pineridge Press, pp. 925–943.

54. Jameson, A. and Turkel, E. (1981), "Implicit Schemes and LU Decompositions", Math. Comp., Vol. 37, pp. 385–389.

55. Jameson, A., Schmidt, W. and Turkel, E. (1981), "Numerical Solutions of the Euler Equations by Finite Volume Methods Using Runge-Kutta Time-Stepping Schemes", AIAA Paper 81-1259.

56. Jameson, A. (1983), "Solution of the Euler Equations by a Multigrid Method", Applied Math. and Computation, Vol. 13, pp. 327–356.

57. Jameson, A. and Baker, T.J. (1983), "Solution of the Euler Equations for Complex Configurations", AIAA Paper 83-1929CP, in the Proceedings of the AIAA Computational Fluid Dynamics Conference.

58. Jameson, A. (1985a), "A Nonoscillatory Shock Capturing Scheme using Flux Limited Dissipation", *Lectures in Applied Mathematics*, Vol. 22, Part 1, *Large Scale Computations in Fluid Mechanics*, ed. by Enquist, B.E., Osher, S. and Sommerville, R.C.J., A.M.S., pp. 345–370.

59. Jameson, A. (1985b), "Transonic Flow Calculations for Aircraft", *Lecture Notes in Mathematics*, Vol. 1127, *Numerical Methods in Fluid Dynamics*, ed. by Brezzi, F., Springer Verlag, pp. 156–242.

60. Jameson, A. and Yoon, S. (1985), "Multigrid Solution of the Euler Equations Using Implicit Schemes", AIAA Paper 85-0293, also AIAA Journal, Vol. 24, No. 11, November 1986, pp. 1737–1743.

61. Jameson, A. and Yoon, S. (1986), "LU Implicit Schemes with Multiple Grids for the Euler Equations", AIAA Paper 86-0105.

62. Jameson, A., Baker, T.J. and Weatherill, N.P. (1986), "Calculation of Inviscid Transonic Flow over a Complete Aircraft", AIAA Paper 86-0103.

63. Jameson, A. and Marviplis, D.J. (1986), "Finite Volume Solution of the Two-Dimensional Euler Equations on a Regular Triangular Mesh", AIAA Journal, Vol. 24, No. 4, pp. 611–618.

64. Jameson, A. (1987), "Successes and Challenges in Computational Aerodynamics", AIAA Paper 87-1184.

65. Jameson, A. and Baker, T.J. (1987), "Improvements to the Aircraft Euler Method", AIAA Paper 87-0452.

66. Lee, E.M. and Batina, J.T. (1991), "Conical Euler Methodology for Unsteady Vortical Flows About Rolling Delta Wings", AIAA paper 91-0.730.

67. Lee, K.D. and Brandt, S.A. (1988),"Modeling of Vortex Dominated Flowfields in the Euler Formulation", ICAS Paper 88-5.9.3. in the Proceedings of the 16th ICAS Congress, pp. 1437–1450.

68. Lijewski, L.E. (1986), "Transonic Flow Solutions on a Blunt, Finned Body of Revolution Using Euler Equations", AIAA Paper 86-1082.

69. Lijewski. L.E. (1987), "Transonic Flow Solutions on a Blunt, Body-Wing-Canard Configuration Using the Euler Equations", AIAA Paper 87-2273.

70. Lohner, R., Morgan, K. and Peraire, J. (1986), "Improved Adaptive Refinement Strategies for the Finite Element Aerodynamic Configurations", AIAA Paper 86-0499.

71. Lohner, R. (1987), "The Efficient Simulation of Strongly Unsteady Flows by the Finite Element Method", AIAA Paper 87-0555.

72. Lohner, R., Morgan, K., Peraire, J. and Vahdati, M. (1987), "Finite Element Flux-Corrected Transport (FEM-FCT) for the Euler and Navier-Stokes Equations", Int. J. Num. Meth. Fluids, No. 7, pp. 1093–1109.

73. Longo, J.M.A. (1989), "The Role of Numerical Dissipation on the Computational Euler Equations Solutions for Vortical Flows", AIAA Paper 89-2232.

74. MacCormack, R.W. and Paullay, A.J. (1972), "Computational Efficiency Achieved by Time Splitting of Finite Difference Operators", AIAA Paper 72-154.

75. Manie, F., Neron, M. and Schmitt, V. (1984), "Experimental and Computational Investigation of the Vortex Flow over Swept Wing", ICAS Paper 84-2.8.1., pp. 881–895.

76. Martinez, A., Chae, Y.S. and Thompson, J.F. (1987), "Applications of Numerical Grid Generation to Advanced Weapon Airframe Configurations", AIAA Paper 87-2294 CP.

77. Mavriplis, D.J. (1988), "Multigrid Solution of the Two-Dimensional Euler Equations on Unstructured Triangular Meshes", AIAA Journal, Vol. 26, No. 7, pp. 824–831.

78. Mavriplis, D.J. (1990), "Accurate Multigrid Solution of the Euler Equations on Unstructured and Adaptive Meshes", AIAA Journal, Vol. 28, No. 2, pp. 213–221.

79. McGrory, W.D., Walters, R.W. and Lohner, R. (1990), "A Three-Dimensional Space Marching Algorithm for the Solution of the Euler Equations on Unstructured Grids", AIAA Paper 90-0014.

80. McMillin, S.N., Thomas, J.L. and Murman, E. (1987), "Euler and Navier-Stokes Solutions for the Leeside Flow Over Delta Wings at Supersonic Speeds", AIAA Paper 87-2270.

81. Moretti, G. (1987), "An Efficient Euler Solver, with Many Applications", AIAA Paper 87-0352.

82. Morgan, K., Praire, J., Thareja, R.R. and Stewart, J.R. (1987), "An Adaptive Finite Element Scheme for the Euler and Navier-Stokes Equations", AIAA Paper 87-1172.

83. Murman, E., Powell, K.G., Miller, D.S. and Wood. R.M. (1986), "Comparison of Computations and Experimental Data for Leading-Edge Vortices – Effects of Yaw and Vortex Flaps", AIAA Paper 86-0439.

84. Murman, E.M. and Rizzi, A. (1986), "Applications of Euler Equations to Sharp-Edge Delta Wings with Leading Edge Vortices", AGARD CP 412, Paper 15.

85. Newsome, R.W. and Kandil, O.A. (1987), "Vortical Flow Aerodynamics – Physical Aspects and Numerical Simulation", AIAA Paper 87-0205.

86. Ni, R.H. (1982), "A Multiple-Grid Scheme for Solving the Euler Equations", AIAA Journal, Vol. 20, No. 11, pp. 1565–1571.

87. O'Neil, F.J., Barnett, R.M. and Louis, C.M. (1989), "Numerical Simulation of Leading-Edge Vortex Breakdown Using an Euler Code", AIAA Paper 89-2189.

88. Peraire, J., Vahdati, M., Morgan, K. and Zienkiewicz, O.C. (1987), "Adaptive Remeshing for Compressible Flows Computations" Journal of Comp. Physics, Vol 72, No. 2, pp. 449–466.

89. Peraire, J., Piero, J., Formaggia, L., Morgan, K. and Zienkiewicz, O.C. (1988), "Finite Element Euler Computations in Three Dimensions", AIAA Paper 87-0032.

90. Powell, K.G., Murman, E.M., Perez, E. and Baron, J. (1985), "Total Pressure Loss in Vortical Solutions of the Conical Euler Equations", AIAA Paper 85-1701.

91. Priolo, F.J. and Wardlaw, A.B. (1987), "A Comparison of Inviscid Computational Methods for Supersonic Tactical Missiles", AIAA Paper 87-0113.

92. Pulliam, T.H. and Steger, J.L. (1980), "Implicit Finite Difference Simulations of Three-Dimensional Compressible Flow", AIAA Journal, Vol. 18, pp. 159–167.

93. Pulliam, T.H. and Steger, J.L. (1985), "Recent Improvements in Efficiency, Accuracy and Convergence for Implicit Approximate Factorization Algorithms", AIAA Paper 85-0360.

94. Rai. M.M. (1985a), "A Relaxation Approach to Patched-Grid Calculations with the Euler Equations", AIAA Paper 85-0295.

95. Rai, M.M. (1985b), "An Implicit, Conservative, Zonal-Boundary Scheme for Euler Equation Calculations", AIAA Paper 85-0488.

96. Rai, M.M. (1986), "A Relaxation Approach to Patched Grid Calculations with the Euler Equations", Journal Comp. Physics, Vol. 66, pp. 99–131.

97. Raj, P. (1984), "A Multigrid Method for Transonic Wing Analysis and Design", Journal of Aircraft, Vol. 21, No. 2, pp. 143–150.

98. Raj, P. (1984), "Computational Simulation of Free-Vortex Flows using an Euler Code", ICAS Paper 84-1.3.1.

99. Raj, P. and Sikora, J.S. (1984), "Free-Vortex Flows; Recent Encounters with an Euler Code", AIAA Paper 84-0135.

100. Raj, P., Sikora, J.S. and Keen, J.M. (1986), "Free-Vortex Flow Simulation using a Three-Dimensional Euler Aerodynamic Method", ICAS Paper 86-1.5.2., pp. 604-617. Also Journal of Aircraft, Vol. 25, No. 2, 1988, pp. 128–134.

101. Raj, P. (1988), "An Euler Code for Nonlinear Aerodynamic Analysis: Assessment of Capabilities", SAE SP-757, pp. 215–230.

102. Raj, P., Keen, J.M. and Singer, S.W. (1988), "Applications of an Euler Aerodynamic Method to Free-Vortex flow Simulation", AIAA Paper 88-2517 CP.

103. Raj, P., Oiling, C.R. and Singer, S.W. (1990), "Applications of Multizone Euler/Navier-Stokes Aerodynamic Methods to Aircraft Configurations", ICAS Paper 90-6.4.4. in the Proceedings of the 17th Congress, pp. 782–794.

104. Reddy, K.C. and Jackocks, J.L. (1987), "A Locally Implicit Scheme for the Euler Equations", AIAA Paper 87-1144.

105. Rieger, H. and Jameson, A. (1988), "Solution of Steady Three-Dimensional Compressible Euler and Navier-Stokes Equations by an Implicit LU Scheme", AIAA Paper 88-0619.

106. Rizzetta, D.P. and Shang, J.S. (1986), "Numerical Simulation of Leading-Edge Vortex Flows", AIAA Journal, Vol. 24, No. 2, pp. 237–245.

107. Rizzi, A. (1982), "Damped Euler Equation Method to Compute Transonic Flow Around Wing-Body Combinations", AIAA Journal, Vol. 20, No. 10, pp. 1321–1328.

108. Rizzi, A. and Eriksson, L.E. (1984), "Computation of Flow around Wings Based on the Euler Equations", Journal of Fluid Mechanics, Vol. 148, pp. 45–71.

109. Rizzi, A. and Purcell, C.J. (1985), "Numerical Experiment with Inviscid Vortex-Stretched Flow Around a Cranked-Delta Wing: Subsonic Speed", AIAA Paper 85-4080.

110. Rizzi, A. and Purcell, C.J. (1986), "Disordered Vortex Flow Computed Around a Cranked-Delta Wing at Subsonic Speed and High Incidence", ICAS Paper 86-1.4.1., pp. 457–470.

111. Rizzi, A. and Purcell, C.J. (1987), "Comparisons of Euler and Navier-Stokes Solutions for Vortex Flow over Delta Wing", AIAA 87-2347.

112. Scherr, S. and Das, A. (1988), "Basic Analysis of the Flow Fields of Slender Delta Wings Using the Euler Equations", ICAS Paper 88-5.9.2 in the Proceedings of the 16th Congress, pp. 1428–1436.

113. Schmidt, W., Jameson, A. and Whitfield, D.L. (1981), "Finite Volume Solution for the Euler Equations for Transonic Flow over Airfoils and Wings Including Viscous Effects", AIAA Paper 81-1265.

114. Schonfeld, T. and Wilde, P. (1990), "Methods to Enhance the Accuracy of Finite Volume Schemes", ICAS Paper 90-6.10.4. in the Proceedings of the 17th Congress, pp. 2035–2038.

115. Sengupta, S., et al. (Ed.), (1988), "*Numerical Grid Generation in Computational Fluid Mechanics*", Pineridge Press.

116. Shapiro, R.A. and Murman, E. (1987), "Cartesian Grid Finite Element Solutions to the Euler Equations", AIAA Paper 87-0559.

117. Sirbaugh, J.R. (1987), "Euler Analysis of the AFWAL 65° Delta Wing", AIAA Paper 87-2272, pp. 34–53.

118. Smith, R.E. (1986), "Algebraic Grid Generation About Wing-Fuselage Bodies", ICAS Paper 86-1.4.3., pp. 471–481.

119. Steger, J.L. and Chausses, D.S. (1980), "Generation of Body-Fitted Coordinates using Hyperbolic Partial Differential Equations", SIAM Journal, Scientific and Statistical Computing, Vol. 1, pp. 431–437.

120. Steger, J.L. and Warming, R.F. (1981), "Flux Vector Splitting of the Inviscid Gas Dynamics Equations with Applications to Finite Difference Methods", Journal Comp. Phys., Vol. 40, pp. 263–293.

121. Stoufflet, B. (1985), "Implicit Finite Element Methods for the Euler Equations", in *Numerical Methods for the Euler Equations of Fluid Dynamics*, Ed. Angrand, F. et al. SIAM, pp. 409–434.

122. Stoufflet, B., Periaux, J., Fezoui, F. and Dervieux, A. (1987), "Numerical Simulation of 3-D Hypersonic Euler Flows Around Space Vehicles using Adaptive Finite Elements", AIAA Paper 87-0560.

123. Swanson, R.C. and Turkel, E. (1987), "Artificial Dissipation and Central Difference Schemes for the Euler and the Navier-Stokes Equations", AIAA Paper 87-1107 CP, pp. 55–69.

124. Szema, K.Y., Chakravarthy, S.R., Riba, W.T., Byerly, J. and Dresser, H.S. (1987), "Multi-Zone Euler Marching Technique for Flow over Single and Multi-Body Configurations", AIAA Paper 87-0592.

125. Takakura, Y. (1987), "On the Recent Difference Schemes for the Three-Dimensional Euler Equations", AIAA Paper 87-1151.

126. Thomas, J.L., Van Leer, B. and Walters, R.W. (1985), "Implicit Flux-Split Schemes for the Euler Equations", AIAA Paper 85-1680.

127. Thompson, J.F. and Warsi, Z.U.A. (1983), "Three-Dimensional Grid Generation from Elliptic Systems", AIAA Paper 83-1905.

128. Thompson, J.F. (1987), "A Composite Grid Generation Code for General Three-Dimensional Regions", AIAA Paper 87-0275.

129. Turnock, S.R. (1990), "Parallel Implementation of an Explicit Finite Volume Euler Solver on an Array of Transputers", ICAS Paper 90-6.8.2. in the Proceedings of the 17th Congress, pp. 1557–1568.

130. Usab, W.J. and Murman, E.M. (1983), "Embedded Mesh Solution of the Euler Equations Using a Multiple Grid Method", AIAA Paper 83-1946 CP.

131. Van der Berg, J.I. and Boerstoel, J.W. (1990), "Development and Validation of a Characteristic Boundary Condition for a Cell-Centered Euler Method", ICAS Paper 90-6.9.3 in the Proceedings of the 17th Congress, pp. 1826–1832.

132. Van Leer, B. (1982), "Flux-Vector Splitting for the Euler Equations", ICASE Report 82-30, 1982, also in the Proceedings of the 8th International Conference on Numerical Methods in Fluid Dynamics, L. Krause (ed.), Springer Verlag, 1982, pp. 507–512.

133. Van Leer, B., Thomas, J.L., Roe, P.L. and Newsome, R.W. (1987), "A Comparison of Numerical Flux Formulas for the Euler and Navier-Stokes Equations", AIAA Paper 87-1104.

134. Van Leer, B., Tai, C.H. and Powel, K.G. (1989), "Design of Optimally Smoothing Multi-Stage Schemes for the Euler Equations", AIAA Paper 89-1933.

135. Verhoff, A. and O'Neil, P.J. (1984), "A Natural Formulation for Numerical Solution of the Euler Equations", AIAA Paper 84-0163.

136. Verhoff, A. and O'Neil, P.J. (1987), "Accurate Efficient Prediction Method for Supersonic Hypersonic Inviscid Flow", AIAA Paper 87-1165.

137. Verhoff, A. and Stookesberry, D. (1990), "Prediction of Inviscid Supersonic/Hypersonic Aircraft Flowfields", ICAS Paper 90-6.7.1. in the Proceedings of the 17th Congress, pp. 1394–1404.

138. Viviand, H., Lecomte, C. and Morice, Ph. (1987), "Computational Fluid Dynamics in France", AIAA Paper 87-1131.

139. Volpe, G., Siclari, M.J. and Jameson, A. (1987a), "A New Multigrid Euler Method For Fighter-Type Configurations", AIAA Paper 87-1160, pp. 627–646.

140. Volpe, G., Siclari, M.J. and Jameson, A. (1987b), "Computation of Aircraft flow Fields by a Multigrid Euler Method", AIAA Paper 87-2268, pp. 1–13.

141. Volpe, G. and Jameson, A. (1988), "An Efficient Method for Computing Transonic and Supersonic Flows about Aircrafts", ICAS Paper 88.4.8.3, pp. 1224–1236.

142. Von Lavante, E. and Haertl, A. (1985), "Numerical Solutions of Euler Equations Using Simplified Flux Vector Splitting", AIAA Paper 85-1333.

143. Walters, R.W. and Dwoyer, D.L. (1985), "An Efficient Iteration Strategy for the Solution of the Euler Equations", AIAA Paper 85-1529.

144. Walters, R.W., Reu, T., McGrory, W.D. and Thomas, J.L. (1988), "A Longitudinally-Patched Grid Approach with Applications to High Speed Flows", AIAA Paper 88-0715.

145. Wardlaw, A.B. Jr. and Davies, S.F. (1989), "Euler Solutions for Delta Wings", AIAA Paper 89-3398.

146. Warming, R.F. and Beam, R.M. (1976), "Upwind Second-Order Difference Schemes and Applications in Aerodynamic Flows", AIAA Journal, Vol. 14, No. 9, pp. 1241–1249.

147. Weatherhill, N.P. and Forsey, C.R. (1984), "Grid Generation and Flow Calculations for Complex Aircraft Geometries Using a Multi Block Scheme", AIAA Paper 84-1665.

148. Weatherhill, N.P. and Forsey, C.R. (1985), "Grid Generation and Flow Calculations for Aircraft Geometries", Journal of Aircraft, Vol. 22, No. 10, pp. 855–860.

149. Weatherill, N.P. (1988), "On the Combination of Structured-Unstructured Meshes", in *Numerical Grid Generation in Computational Fluid Mechanics '88,* S. Sengupta et al. (eds.), Pineridge Press.

150. Whitaker, D.L., Grossman, B. and Lohner, R. (1989), "Two-Dimensional Euler Computations on a Triangular Mesh Using an Upwind Finite-Volume Scheme", AIAA Paper 89-0470.

151. Whitfield, D.L. and Janus, J.M. (1984), "Three-Dimensional Unsteady Euler Equations Solution Using Flux Vector Splitting", AIAA Paper 84-1552.

152. Yoon, S. and Jameson, A. (1987), "An LU-SSOR Scheme for the Euler and Navier-Stokes Equations", AIAA Paper 87-0600.

153. Zienkiewicz, O.C. and Morgan, K. (1983), "*Finite Elements and Approximations*", John Wiley and Sons.

Chapter 9

Solutions of the Navier-Stokes Equations for Flows over Configurations at High Angles of Attack

The Navier-Stokes equations represent the conservation of mass, momentum and energy in a flow of a compressible, viscous and heat conducting fluid. They are known to simulate correctly the compressibility and viscous stresses effects. As the Mach number increases from subsonic to transonic, supersonic and hypersonic speeds, the Navier-Stokes equations are able to capture the shock waves in their proper positions and strengths as given by the Rankin-Hugoniot relations. The Navier-Stokes equations are capable as well to calculate the effects of viscosity on the flow fields, including the effects of three-dimensional separations. It is expected that the Navier-Stokes formulation is capable of capturing the vortical structures, such as the rolled-up vortices and the three-dimensional separating vortex sheets, generated in the flow over configurations at high angles of attack. It was shown in Chapter 8 that the Euler equations solutions are able to capture vortical structures due to the effects of "numerical viscosity" which is introduced by the numerical differencing schemes, the grid schemes and the artificial viscosity. However, since the "numerical viscosity" in the Euler solution is different from the physical viscosity of the flow, the Euler solutions must be used for the calculation of vortical flows with serious reservations. It is hoped that introducing the correct physical viscous laws into the Navier-Stokes equations may enable correct simulation of the viscous effects. This is the case for laminar flows. But the problem of correct formulation of the turbulent viscous shear is still unresolved, and approximate relations must be use for the turbulent shear. Therefore, the uncertainty concerning the turbulent model detracts from the expectations of the ultimate accuracy that can be achieved with the present state of knowledge on transitional and turbulent flows. Taking this problem into account, we can still expect important results from solutions of the Navier-Stokes equations which justify the efforts to obtain such solutions.

The unsteady, compressible Navier-Stokes equations are a mixed set of hyperbolic-parabolic nonlinear partial differential equations, while the unsteady, incompressible Navier-Stokes equations are a mixed set of elliptic-parabolic nonlinear partial differential equations. Therefore, different numerical methods must be used for the solutions of these equations in the incompressible or compressible flow regimes. Furthermore, while the unsteady compressible Navier-Stokes equations are a mixed set of hyperbolic-parabolic equations in the case of steady compressible flow, the corresponding Navier-Stokes equations are a mixed set of elliptic-parabolic equations. Therefore, it is advantageous to use the unsteady compressible flow equations even when steady flow solutions are sought. In this case, the nonsteady equations are integrated in time until a steady state solution is achieved. Therefore, in order to calculate a three-dimensional flow field, it is necessary to solve the Navier-Stokes equations in a four-dimensional space (x, y, z and t). In all of these cases, the solution of the complete Navier-Stokes equations requires a very large amount of computer time and memory capacity. Fortunately, there are many viscous flow problems where approximations which reduce the set of equations can be justified. Such approximations enable the reduction of the complex set of equations so that they fall between the complete Navier-Stokes equations and the boundary layer equations in terms of complexity. These reduced equations are generally of the form of either the thin-layer approximation for the unsteady time marching integration or the Parabolized Navier-Stokes method for the solutions of the approximate steady state equations.

We begin by presenting the formulation of the full Navier-Stokes equations in Cartesian coordinates and in generalized curvilinear coordinates and then the thin-layer and parabolized equations approximations. The methods of solutions of these equations, used for flows at high angles of attack, are then presented. The significant results of such calculations on various three-dimensional configurations at moderate and high angles of attack are reviewed and discussed.

9.1 Formulation of the Navier-Stokes Equations

9.1.1 Navier-Stokes Equations in Cartesian Coordinates

The Navier-Stokes equations in Cartesian coordinates, neglecting body forces and external heat addition, can be written in the following form:

$$\frac{\partial \mathbf{Q}}{\partial t} + \frac{\partial \mathbf{F}}{\partial x} + \frac{\partial \mathbf{G}}{\partial y} + \frac{\partial \mathbf{H}}{\partial z} = 0 \tag{9.1}$$

where $\mathbf{Q}$, $\mathbf{F}$, $\mathbf{G}$ and $\mathbf{H}$ are vectors given by

$$\mathbf{Q} = \begin{bmatrix} \rho \\ \rho u \\ \rho v \\ \rho w \\ E_t \end{bmatrix} \qquad \mathbf{F} = \begin{bmatrix} \rho u \\ \rho u_u^2 + p - \tau_{xx} \\ \rho u v - \tau_{xy} \\ \rho u w - \tau_{xz} \\ (E_t + p)u - u\tau_{xx} - v\tau_{xy} - w\tau_{xz} + q_x \end{bmatrix}$$

$$\mathbf{G} = \begin{bmatrix} \rho v \\ \rho uv - \tau_{xy} \\ \rho v^2 + p - \tau_{yy} \\ \rho vw - \tau_{yz} \\ (E_t + p)v - u\tau_{xy} - v\tau_{yy} - w\tau_{yz} + q_y \end{bmatrix}$$

$$\mathbf{H} = \begin{bmatrix} \rho w \\ \rho uw - \tau_{xz} \\ \rho vw - \tau_{yz} \\ \rho w^2 + p - \tau_{zz} \\ (E_t + p)w - u\tau_{xz} - v\tau_{yz} - w\tau_{zz} + q_z \end{bmatrix}$$

and the components of the shear stress tensor and the heat flux vector are

$$\tau_{xx} = \frac{2}{3}\mu\left(2\frac{\partial u}{\partial x} - \frac{\partial v}{\partial y} - \frac{\partial w}{\partial z}\right), \qquad \tau_{yy} = \frac{2}{3}\mu\left(2\frac{\partial v}{\partial y} - \frac{\partial u}{\partial x} - \frac{\partial w}{\partial z}\right),$$

$$\tau_{zz} = \frac{2}{3}\mu\left(2\frac{\partial w}{\partial z} - \frac{\partial u}{\partial x} - \frac{\partial v}{\partial x}\right), \qquad \tau_{xy} = \mu\left(\frac{\partial u}{\partial y} + \frac{\partial v}{\partial x}\right) = \tau_{yx},$$

$$\tau_{xz} = \mu\left(\frac{\partial w}{\partial x} + \frac{\partial u}{\partial z}\right) = \tau_{zx}, \qquad \tau_{yz} = \mu\left(\frac{\partial v}{\partial z} + \frac{\partial w}{\partial y}\right) = \tau_{zy},$$

$$q_x = -k\frac{\partial T}{\partial x}, \qquad q_y = -k\frac{\partial T}{\partial y}, \qquad q_z = -k\frac{\partial T}{\partial z}$$

The pressure is related to the conservative flow variables by the equation of state, given by

$$p = (\gamma - 1)[e - \frac{1}{2}\rho(u^2 + v^2 + w^2)] \tag{9.2}$$

9.1.2 Nondimensional Navier-Stokes Equations in Generalized Curvilinear Coordinates

Retaining the Cartesian velocity components as dependent variables, the three-dimensional, unsteady Navier-Stokes equations can be transformed to the generalized curvilinear coordinates ξ, η, ζ, τ shown in Fig. 9.1, in the following formulation

$$\partial_\tau \mathbf{Q} + \partial_\xi(\mathbf{F} - \mathbf{F}_v) + \partial_\eta(\mathbf{G} - \mathbf{G}_v) + \partial_\zeta(\mathbf{H} - \mathbf{H}_v) = 0 \tag{9.3}$$

where

$$\mathbf{Q} = J^{-1}\begin{bmatrix} \rho \\ \rho u \\ \rho v \\ \rho w \\ e \end{bmatrix} \qquad \mathbf{F} = J^{-1}\begin{bmatrix} \rho U \\ \rho uU + \xi_x p \\ \rho vU + \xi_y p \\ \rho wU + \xi_z p \\ (e + p)U + \xi_t p \end{bmatrix}$$

$$
\mathbf{G} = J^{-1} \begin{bmatrix} \rho V \\ \rho u V + \eta_x p \\ \rho v V + \eta_y p \\ \rho w V + \eta_z p \\ (e+p)V + \eta_t p \end{bmatrix} \qquad \mathbf{H} = J^{-1} \begin{bmatrix} \rho W \\ \rho u W + \zeta_x p \\ \rho v W + \zeta_y p \\ \rho w W + \zeta_z p \\ (e+p)W + \zeta_t p \end{bmatrix}
$$

and U, V, W are contravariant velocities written without metric normalization

$$
\begin{aligned}
U &= \xi_t + \xi_x u + \xi_y v + \xi_z w \\
V &= \eta_t + \eta_x u + \eta_y v + \eta_z w \\
W &= \zeta_t + \zeta_x u + \zeta_y v + \zeta_z w
\end{aligned}
$$

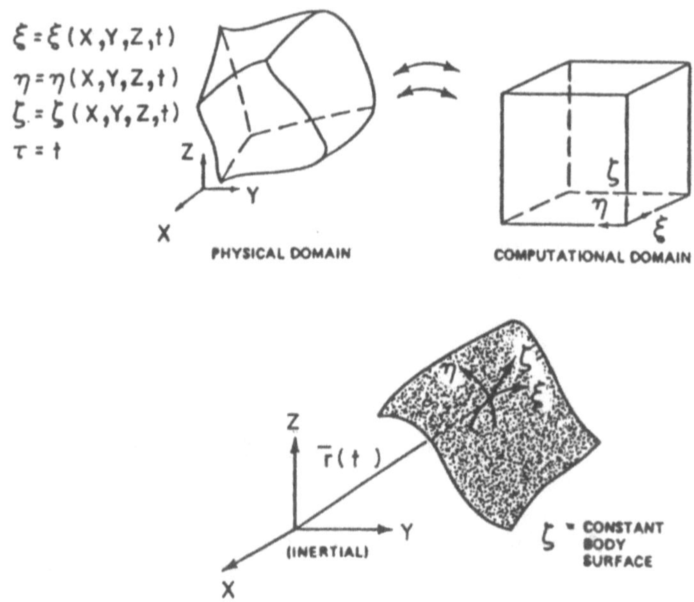

Fig. 9.1. Transformation and body coordinates

The viscous flux terms are

$$
\mathbf{F}_v = \frac{1}{J} \begin{bmatrix} 0 \\ \xi_x \tau_{xx} + \xi_y \tau_{xy} + \xi_z \tau_{xz} \\ \xi_x \tau_{yx} + \xi_y \tau_{yy} + \xi_z \tau_{yz} \\ \xi_x \tau_{zx} + \xi_y \tau_{zy} + \xi_z \tau_{zz} \\ \xi_x \beta_x + \xi_y \beta_y + \xi_z \beta_z \end{bmatrix} \qquad \mathbf{G}_v = \frac{1}{J} \begin{bmatrix} 0 \\ \eta_x \tau_{xx} + \eta_y \tau_{xy} + \eta_z \tau_{xz} \\ \eta_x \tau_{yx} + \eta_y \tau_{yy} + \eta_z \tau_{yz} \\ \eta_x \tau_{zx} + \eta_y \tau_{zy} + \eta_z \tau_{xz} \\ \eta_x \beta_x + \eta_y \beta_y + \eta_z \beta_z \end{bmatrix}
$$

$$\mathbf{H}_v = \frac{1}{J}\begin{bmatrix} 0 \\ \zeta_x\tau_{xx} + \zeta_y\tau_{xy} + \zeta_z\tau_{xz} \\ \zeta_x\tau_{yx} + \zeta_y\tau_{yy} + \zeta_z\tau_{yz} \\ \zeta_x\tau_{zx} + \zeta_y\tau_{zy} + \zeta_z\tau_{xz} \\ \zeta_x\beta_x + \zeta_y\beta_y + \zeta_z\beta_z \end{bmatrix}$$

and

$$\tau_{xx} = \lambda(u_x + v_y + w_z) + 2\mu u_x$$
$$\tau_{xy} = \tau_{yx} = \mu(u_y + v_x)$$
$$\tau_{xz} = \tau_{zx} = \mu(u_z + w_x)$$
$$\tau_{yy} = \lambda(u_x + v_y + w_z) + 2\mu v_y$$
$$\tau_{yz} = \tau_{zy} = \mu(v_z + w_y)$$
$$\tau_{zz} = \lambda(u_x + v_y + w_z) + 2\mu w_z$$
$$\beta_x = \gamma\kappa \mathrm{Pr}^{-1}\partial_x e_I + u\tau_{xx} + v\tau_{xy} + w\tau_{xz}$$
$$\beta_y = \gamma\kappa \mathrm{Pr}^{-1}\partial_y e_I + u\tau_{yx} + v\tau_{yy} + w\tau_{yz}$$
$$\beta_z = \gamma\kappa \mathrm{Pr}^{-1}\partial_z e_I + u\tau_{zx} + v\tau_{yz} + w\tau_{zz}$$

where $e_I = e\rho^{-1} - \frac{1}{2}(u^2 + v^2 + w^2)$ and the Cartesian derivatives $\partial_x, \partial_y, \partial_z$ are expressed in terms of the general coordinates, ξ, η, ζ, via the chain-rule relations $u_x = \xi_x u_\xi + \eta_x u_\eta + \zeta_x u_\zeta$.

The Cartesian velocity components u, v, w are nondimensionalized with respect to a_∞, while the density ρ is nondimensionalized with ρ_∞ and e with $\rho_\infty a_\infty^2$. The pressure is related to the conservative flow variables by the equation of state, as written in Eq. 9.2.

The metric terms are obtained from the chain-rule relations

$$\text{(9.3a)} \quad \begin{cases} \xi_x = J(y_\eta z_\zeta - y_\zeta z_\eta), & \eta_x = J(z_\xi y_\zeta - y_\xi z_\zeta), \\ \xi_y = J(z_\eta x_\zeta - x_\eta z_\zeta), & \eta_y = J(x_\xi z_\zeta - x_\zeta z_\xi), \\ \xi_z = J(x_\eta y_\xi - y_\eta x_\xi), & \eta_z = J(y_\xi x_\zeta - x_\xi y_\zeta), \end{cases}$$

$$\text{(9.3b)} \quad \begin{cases} \zeta_x = J(y_\xi z_\eta - z_\xi y_\eta), & \xi_t = -x_\tau\xi_x - y_\tau\xi_y - z_\tau\xi_z, \\ \zeta_y = J(x_\eta z_\xi - x_\xi z_\eta), & \eta_t = -x_\tau\eta_x - y_\tau\eta_y - z_\tau\eta_z, \\ \zeta_z = J(x_\xi y_\eta - y_\xi x_\eta), & \zeta_t = -x_\tau\zeta_x - y_\tau\zeta_y - z_\tau\zeta_z, \end{cases}$$

with

$$\text{(9.3c)} \quad J^{-1} = x_\xi y_\eta z_\zeta + x_\zeta y_\xi z_\eta + x_\eta y_\zeta z_\xi - x_\xi y_\zeta z_\eta - x_\eta y_\xi z_\zeta - x_\zeta y_\eta z_\xi$$

9.1.3 Thin-Layer Approximation

In high Reynolds number flows, the dominating viscous effects are usually confined to regions adjacent to the solid boundaries of the configuration. Furthermore, the numerical solution of the Navier-Stokes equations require a large number of grid points in the viscous dominated regions. Therefore, noting that at high Reynolds numbers the viscous dominated layers are usually thin, it is possible to formulate the "thin-layer" approximation for the Navier-Stokes equations. In the thin-layer formulation, the grid lines can be clustered near the surface of the configuration, enabling better resolution and improved accuracy in the numerical solutions.

In the development of the thin-layer approximation, it is assumed that the derivatives normal to the surface are much larger than those along the surface. In this case, it is allowable to neglect all derivatives along the body surface, in the ξ and η directions, and to retain only the derivatives normal to the surface in the ζ direction. In this thin-layer formulation, the Navier-Stokes equations, Eq. 9.3, can be written as

$$\partial_\tau \mathbf{Q} + \partial_\xi \mathbf{F} + \partial_\eta \mathbf{G} + \partial_\zeta \mathbf{H} = Re^{-1}\partial_\zeta \mathbf{S} \tag{9.4}$$

where

$$\mathbf{S} = J^{-1}\begin{bmatrix} 0 \\ \mu m_1 u_\zeta + (\mu/3)m_2\zeta_x \\ \mu m_1 v_\zeta + (\mu/3)m_2\zeta_y \\ \mu m_1 w_\zeta + (\mu/3)m_2\zeta_z \\ \mu m_1 m_3 + (\mu/3)m_2(\zeta_x u + \zeta_y v + \zeta_z w) \end{bmatrix}$$

and

$$m_1 = \zeta_x^2 + \zeta_y^2 + \zeta_z^2, \qquad m_2 = \zeta_x u_\zeta + \zeta_y v_\zeta + \zeta_z w_\zeta,$$
$$m_3 = (u^2 + v^2 + w^2)/2 + \mathrm{Pr}^{-1}(\gamma - 1)^{-1}(a^2)_\zeta$$

The pressure is again expressed by the equation of state as written in Eq. 9.2, and the metric terms are defined by the relations presented in Eqs. 9.3a, 9.3b and 9.3c.

The boundary conditions, valid for steady or nonsteady viscous flows, require that the tangency conditions along the surface $\zeta(x, y, z, t)$ be constant as well as the no-slip condition be fulfilled; therefore, $W = 0$ (tangency condition), and $U = V = 0$ (no slip condition) on the surface. The velocities u, v and w are then obtained from the following relation

(9.5)

$$\begin{bmatrix} u \\ v \\ w \end{bmatrix} = \frac{1}{J}\begin{bmatrix} (\eta_y\zeta_z - \eta_z\zeta_y) & -(\xi_y\zeta_z - \xi_z\zeta_y) & (\xi_y\eta_z - \xi_z\eta_y) \\ -(\eta_x\zeta_z - \eta_z\zeta_x) & (\xi_x\zeta_z - \xi_z\zeta_x) & -(\xi_x\eta_z - \xi_z\eta_x) \\ (\eta_x\zeta_y - \eta_y\zeta_x) & -(\xi_x\zeta_y - \xi_y\zeta_x) & (\xi_x\eta_y - \xi_y\eta_x) \end{bmatrix}\begin{bmatrix} U - \xi_t \\ V - \eta_t \\ W - \zeta_t \end{bmatrix}$$

The pressure along the body surface is obtained from a normal momentum relation given by (where n is the normal direction to the body surface)

$$(9.6)\quad \begin{cases} p_n(\zeta_x^2+\zeta_y^2+\zeta_z^2)^{1/2} = (\xi_x\zeta_x+\xi_y\zeta_y+\xi_z\zeta_z)p_\xi + \\ +(\eta_x\zeta_x+\eta_y\zeta_y+\eta_z\zeta_z)p_\eta+(\zeta_x^2+\zeta_y^2+\zeta_z^2)p_\zeta = \\ = \rho[\partial_\tau\zeta_t+u\partial_\tau\zeta_x+v\partial_\tau\zeta_y+w\partial_\tau\zeta_z] \end{cases}$$

9.1.4 The Parabolized Navier-Stokes Equations

The steady Navier-Stokes equations are parabolized with respect to the streamwise coordinate direction by assuming that the viscous derivatives in the streamwise direction are negligible in comparison with those in the crossflow directions. This assumption is generally considered valid for high Reynolds number flows. For supersonic and hypersonic flows, this approximation results in parabolic-hyperbolic type equations which can be solved by stepping procedures. In certain respects, the Parabolized Navier-Stokes approximation is equivalent to a steady state formulation of the thin-layer approximation. The equations for the Parabolized Navier-Stokes approximation can be written in a generalized coordinate system as

$$(9.7)\qquad \frac{\partial \mathbf{F}}{\partial \xi}+\frac{\partial \mathbf{G}}{\partial \eta}+\frac{\partial \mathbf{H}}{\partial \zeta}=0$$

where

$$\begin{aligned} \mathbf{F} &= J^{-1}(\xi_x\mathbf{F}_i+\xi_y\mathbf{G}_i+\xi_z\mathbf{H}_i) \\ \mathbf{G} &= J^{-1}[\eta_x(\mathbf{F}_i-\mathbf{F}_v')+\eta_y(\mathbf{G}_i-\mathbf{G}_v')+\eta_z(\mathbf{H}_i-\mathbf{H}_v'] \\ \mathbf{H} &= J^{-1}[\zeta_x(\mathbf{F}_i-\mathbf{F}_v')+\zeta_y(\mathbf{G}_i-\mathbf{G}_v')+\zeta_z(\mathbf{H}_i-\mathbf{H}_v')] \end{aligned}$$

where the subscript i indicates the inviscid vector terms and the subscript v indicates the viscous vector terms of the Navier-Stokes equations, as presented in Section 9.1.2, while the prime indicates that terms containing partial derivatives with respect to ξ have been omitted. Hence, the shear stress and heat flux terms are reduced to

$$\tau_{xx}' = \frac{2}{3}\mu[2(\eta_x u_\eta+\zeta_x u_\zeta)-(\eta_y v_\eta+\zeta_y v_\zeta)-(\eta_z w_\eta+\zeta_z w_\zeta)]$$

$$\tau_{yy}' = \frac{2}{3}\mu[2(\eta_y v_\eta+\zeta_y v_\zeta)-(\eta_x u_\eta+\zeta_x u_\zeta)-(\eta_z w_\eta+\zeta_z w_\zeta)]$$

$$\tau_{zz}' = \frac{2}{3}\mu[2(\eta_z w_\eta+\zeta_z w_\zeta)-(\eta_x u_\eta+\zeta_x u_\zeta)-(\eta_y v_\eta+\zeta_y v_\zeta)]$$

$$\begin{aligned} \tau_{xy}' &= \mu(\eta_y u_\eta+\zeta_y u_\zeta+\eta_x v_\eta+\zeta_x v_\zeta) \\ \tau_{xz}' &= \mu(\eta_z u_\eta+\zeta_z u_\zeta+\eta_x w_\eta+\zeta_x w_\zeta) \\ \tau_{yz}' &= \mu(\eta_z v_\zeta+\zeta_z v_\zeta+\eta_y w_\eta+\zeta_y w_\zeta) \\ q_x' &= -k(\eta_x T_\eta+\zeta_x T_\zeta) \\ q_y' &= -k(\eta_y T_\eta+\zeta_y T_\xi) \\ q_z' &= -k(\eta_z T_\eta+\zeta_z T_\zeta) \end{aligned}$$

9.1.5 The Reynolds-Averaged Equations and Turbulence Models

The Reynolds-averaged equations are used for calculations of turbulent flows instead of the Navier-Stokes equations with the Newtonian viscous terms, which are appropriate for laminar flows. Using the Boussinesq approximation for the turbulent eddy viscosity, μ_T, the Reynolds-averaged equations can be obtained from the Navier-Stokes equations by using the average values for all physical variables and replacing the viscosity coefficient μ with $(\mu + \mu_T)$ and the coefficient of thermal conductivity k with $(k + k_T)$, where k_T is the turbulent thermal conductivity. The turbulent Prandtl number is defined by $\mathrm{Pr}_T = Cp\mu_T/k_T$.

At present, the main efforts in the computations of turbulent flows are directed towards the solution of the time-averaged Navier-Stokes equations (the Reynolds-averaged equations). As a result of the time averaging of the equations of motion, terms of "apparent" stress gradients and heat-flux quantities associated with the turbulent motion are obtained. These turbulent stress and heat flux quantities must be related to the mean flow variables through assumed turbulent models. A comprehensive introduction to the various turbulent models can be found in the book by Anderson et al. (1984). In the case of viscous separated flows, the algebraic turbulence models have been used extensively. The algebraic models are rather simple to use and have proved to be extendible to handle turbulent flows with massive crossflow separation. For attached turbulent flows, the two-layer zero equation model developed by Baldwin and Lomax (1978) has proven to be very successful. In the Baldwin-Lomax model, the Reynolds number in the thin-layer equation, Eq. 9.4, is written as

$$\mathrm{Re}^{-1} = \mathrm{Re}_{\mathrm{lam}}^{-1} + \mathrm{Re}_{\mathrm{turb}}^{-1} \tag{9.8}$$

and

$$\mathrm{Re}_{\mathrm{turb}}^{-1} = \begin{cases} \left(\mathrm{Re}_{\mathrm{turb}}^{-1}\right)_{\mathrm{inner}} & y \le y_{\mathrm{crossover}} \\ \left(\mathrm{Re}_{\mathrm{turb}}^{-1}\right)_{\mathrm{outer}} & y > y_{\mathrm{crossover}} \end{cases} \tag{9.9}$$

where y is the local distance normal to the body surface and $y_{\mathrm{crossover}}$ is the smallest value of y for which $\mathrm{Re}_{\mathrm{turb}}^{-1}\mid_{\mathrm{inner}} = \mathrm{Re}_{\mathrm{turb}}^{-1}\mid_{\mathrm{outer}}$.

In the inner region

$$\mathrm{Re}_{\mathrm{turb}}^{-1}\mid_{\mathrm{inner}} = l^2 \mid \omega \mid \tag{9.10}$$

where

$$l = ky\{1 - \exp[-(y^+/A^+)]\} \tag{9.11}$$

and

$$\mid \omega \mid = [(u_y - v_x)^2 + (v_z - w_y)^2 + (w_x - u_z)^2]^{1/2}; \tag{9.12}$$

the law-of-the-wall coordinate is

$$y^+ = y(\mathrm{Re}_{\mathrm{lam}} \partial u/\partial y \mid_{\mathrm{wall}})^{1/2} \tag{9.13}$$

In the outer region

$$(\mathrm{Re}_{\mathrm{turb}})^{-1}_{\mathrm{outer}} = K C_{cp} F_{\mathrm{wake}} F_{\mathrm{Kleb}} \tag{9.14}$$

where

$$F_{\mathrm{wake}} = \min[C_{wk}(\Delta\tilde{\bar{V}})^2 y_{\max}/F_{\max}; y_{\max} F_{\max}] \tag{9.15}$$

and

$$\Delta\tilde{\bar{V}} = (u^2+v^2+w^2)^{1/2}_{\max} - (u^2+v^2+w^2)^{1/2}_{\min},$$

and

$$F_{\max} = \max[F(y)], \qquad y_{\max} = y(F_{\max})$$

where

$$F(y) = \mid \omega \mid y[1 - e^{-(y^+/A^+)}] \tag{9.16}$$

The function F_{Kleb} is the Klebanoff intermittency factor

$$F_{\mathrm{Kleb}}(y) = \left[1 + 5.5\left(\frac{C_{\mathrm{Kleb}}y}{y_{\max}}\right)^6\right]^{-1} \tag{9.17}$$

The constants in Equations 9.11 to 9.17 are given by Baldwin and Lomax (1978) to be $A^+ = 26$; $C_{CP} = 1.6$; $C_{\mathrm{Kleb}} = 0.3$; $C_{WK} = 0.25$; $k = 0.4$ and $K = 0.168$.

The Baldwin-Lomax turbulence model is applicable for thin viscous layers obtained in boundary layers of attached flows or thin free shear layers. This turbulence model must be modified for proper use in regions of massive separation, such as those obtained on the leeward side of an elongated body at high angles of attack. Suitable modifications to the Baldwin-Lomax turbulent model are presented by Degani and Schiff (1986). The application, of the modification to transitional flows is presented by Hartwich and Hall (1990). Hartwich and Hall propose a procedure to determine the extent of the massive separation zone and the use of the Degani-Schiff modifications in the algebraic turbulence model to handle flow with massive transitional and fully turbulent crossflow separated regions in the leeward part of elongated bodies.

Consider the separated flow in the crossflow plane, shown in Fig. 9.2a. For the transitional flow case, the attached windward boundary layer is assumed to be laminar, and transition to turbulence is assumed to occur as the shear layer is separated into the massive crossflow separation zone, as illustrated in Fig. 9.2b. The flow in the massive separation zone is assumed to be fully turbulent. The procedure suggested by Hartwich and Hall (1990) for the determination of the extent of the massive separated zone for this

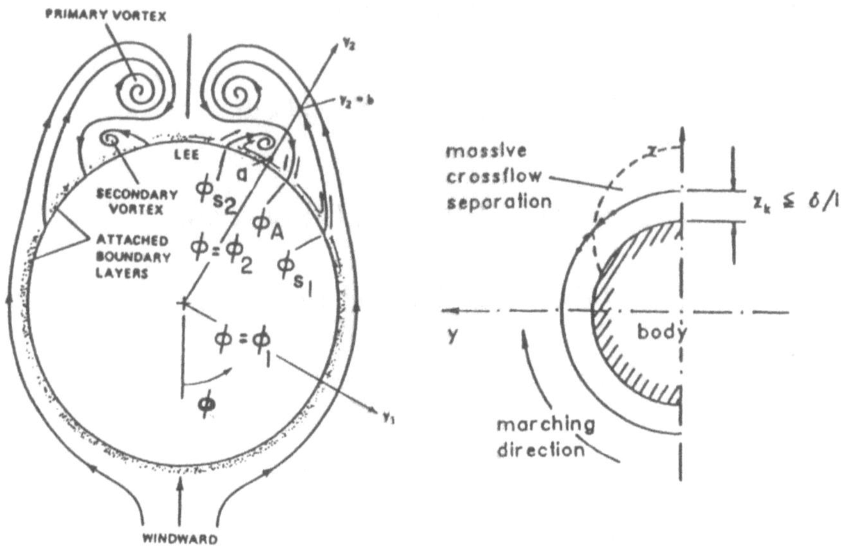

a. Flow structure in the crossflow plane (Degani and Schiff 1986)

b. Crossflow separation model (Hartwich and Hall 1990)

Fig. 9.2. Flow structure with massive separation region

transitional crossflow pattern is as follows: the solution in each crossplane is monitored by marching from the windward-side to the leeward-side meridian along a circumferential grid line a few step sizes off the body surface, as illustrated in Fig. 9.2b. The distance between the circumferential grid line and the body surface must be larger than the local laminar boundary layer thickness, $y_k > \delta/l$, where the boundary layer thickness can be estimated by the relation $\delta/l = 5.4\mathrm{Re}_l^{-1/2}$ and $l = \pi d/2$. The onset of massive crossflow separation is detected by the first occurrence of a negative product of two consecutive crossflow velocities. For fully turbulent flows, the Degani-Schiff model, described in the following section, can be used.

In the fully developed turbulent separated flow case, the Baldwin-Lomax turbulent model underestimates the scale of the turbulent layer. The evaluation of the proper turbulent shear layer thickness is included in a modification to the Baldwin-Lomax model by Degani and Schiff (1986). The need for this modification in the massive separation zone can be understood by examining the variation of the length scale $y_{\max}$ and the corresponding function $F(y)$ in the windward attached boundary layer and by comparing with the corresponding variations in the leeward-side separated zone, as illustrated in Fig. 9.2a, in the ϕ_1 and the ϕ_2 azimuthal directions, respectively. The variation of the $F(y)$ function (Eq. 9.16) in the windward boundary layer (the ϕ_1 direction) is shown in Fig 9.3a, with the corresponding variation of $F(y)$ in the separated zone (the ϕ_2 direction) in Fig. 9.3b. Based on experimental observations, Degani and Schiff (1986) assume that the flow in the massive separated zone, in the leeward-side of a body at high angles of attack, consists of a series of attached boundary layers underlying the leeward-side vortex structure. This overlying vortex structure is assumed to be essentially inviscid and governed primarily by the convection of vorticity generated within the boundary layer at the body surface. The underlying boundary layer at the body surface is expected to be turbulent, so that the application of the eddy-viscosity model, which depends only on the local velocity profiles, is justified within this attached turbulent layer. As indicated in Fig. 9.3a, on the windward-side along ray ϕ_1, the attached boundary layer gives rise to a profile $F(y)$ which has only one well defined peak. Thus, the determination of $F_{\max}(\phi_1), y_{\max}(\phi_1)$, and $F_{\text{wake}}(\phi_1)$ for the attached boundary layer on the windward-side is straightforward. On the leeward-side, as indicated in Fig. 9.3b, in addition to the local peak in $F(y)$ in the underlying attached boundary layer at $y_2 = a$, the overlying vortex structure causes a larger peak in $F(y)$ at $y_2 = b$. Without the modification, the computer code would select the maximum value of $F(y_2)$ at $y_2 = b$, with the corresponding value of $F_{\text{wake}}(\phi_2)$ which will determine a very high value for the outer layer eddy-viscosity, $(\mu_T)_{\text{outer}}$. This value of the eddy-viscosity is at least one order of magnitude (and may be as much as two orders of magnitude) larger than the value of the eddy-viscosity, $(\mu_T)_{\text{outer}}$, resulting from the evaluation of $F_{\text{wake}}(\phi_2)$ from the peak at $y_2 = a$ (Fig. 9.3b). Therefore, without modification, the very large eddy-viscosity, computed in the massive separated zone in the leeward-side, will cause major distortions of the computed flow field.

To eliminate these difficulties, it is proposed by Degani and Schiff (1986) to modify the procedure for the calculation of the turbulent model in regions of massive separation as follows: at each axial station, the code searches radially outward along successive rays, sweeping from the windward to the leeward plane of symmetry. Along each of these rays, the code determines the first peak in $F(y)$, by defining $F_{\max} = F(y')$ when $F(y) = 0.9F(y')$ for $y > y'$, and the search is cut off after this peak is found, thus determining a "cutoff" distance. For most rays in the crossflow separation region, the two peaks in $F(y)$ are easily distinguished. However, these peaks may come very close or even merge in the vicinity of the primary and secondary separation points. It is then necessary to use a specified cutoff distance determined from a previous ray. In this way, using proper numerical criteria for the determination of the cutoff distances along the rays in the massive separated region results in the correct value of $F_{\max}$

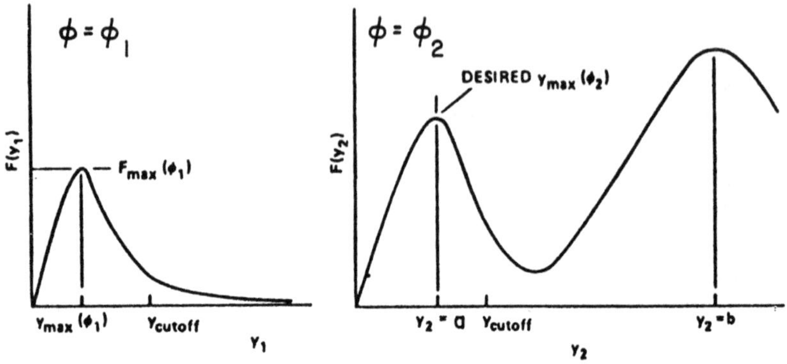

a. Windward-side, $\phi = \phi_1$ b. Leeward-side, $\phi = \phi_2$

Fig. 9.3. Behavior of $F(y)$ on rays on a body at high angles of attack (Degani and Schiff 1986)

also in the vicinity of the primary and secondary separation points. Thus, the Degani and Schiff (1986) modification to the Baldwin-Lomax turbulent model enables the evaluation of reasonable eddy-viscosity coefficients in regions of massive separation, even in the vicinity of the crossflow separation points.

9.2 Numerical Methods for Solutions of the Navier-Stokes Equations

The Navier-Stokes equations are a set of nonlinear partial differential equations which are "very difficult" (an understatement) to solve in their complete form. In particular, this is true since the compressible Navier-Stokes equations are a mixed set of elliptic-parabolic equations for steady flow and a mixed set of hyperbolic-parabolic equations for unsteady flow. It is, therefore, advantageous in many cases to obtain the steady state solution by solving the time-dependent compressible equations until a steady state solution is achieved. Thus, for a three-dimensional flow field, a four-dimensional (3 space and 1 time) problem must be solved. For incompressible flow, the unsteady Navier-Stokes equations are of a mixed set of elliptic-parabolic equations. Therefore, different numerical methods must be used to solve the Navier-Stokes equations in the compressible and incompressible flow regimes. An introduction to the various numerical techniques which may be used can be found in various textbooks, particularly that by Anderson et al. (1984) which is cited as a comprehensive reference.

9.2.1 Methods of Solution of the Compressible Navier-Stokes Equations by a Time-Dependent Method

Most successful solutions of the compressible Navier-Stokes equations are obtained by solving the unsteady compressible equations by marching in time until convergence to the steady state solution is achieved. This procedure is denoted as the *time-dependent method.* Both explicit and implicit finite-difference methods have been used with the time-dependent approach. Usually, methods which are second order accurate in space and which are first or second order accurate in time are used. If only steady state solutions are desired, it is generally sufficient to use a less accurate scheme in time, since it is then possible to obtain the steady state solutions with fewer time steps.

An explicit finite difference scheme which is second order accurate in both space and time is the explicit predictor-corrector method of MacCormack (1969). It can be used both for time-dependent solutions and for convergent solutions to steady state. The original method was modified by MacCormack (1971), by incorporating time splitting into the scheme and was further refined by MacCormack and Baldwin (1975). The explicit MacCormack algorithm is applicable to solutions of steady and nonsteady viscous-compressible flows of low to moderate Reynolds numbers. At high Reynolds numbers, the mesh size must be highly refined in order to resolve accurately the thin shear layers. This leads to small steps and subsequently long computer time when an explicit scheme, such as the explicit MacCormack method is used.

An implicit two-stage method was developed by MacCormack (1981). The first stage uses the original MacCormack scheme while the second stage uses an implicit scheme in order to eliminate any stability restrictions.

A finite difference implicit method utilizing the ADI (Alternating Direction Implicit) scheme was developed by Beam and Warming (1978). This implicit method is expected to be useful for moderate to high Reynolds number flows.

The details of the algorithms of these methods, the explicit MacCormack, the implicit Beam and Warming and the implicit MacCormack methods, can be found in the book by Anderson et al. (1984) and will not be presented here.

9.3 Method of Solution of the Thin-Layer Equations

The number of grid points required for accurate solution of the Navier-Stokes equations are excessive even for the presently available large supercomputers. At high Reynolds numbers, it is possible to limit the domain of calculations to thin layers into which enough grid points can be introduced so as to resolve the viscous terms there. In the thin layer equations, the streamwise terms are neglected and only the normal viscous terms are retained. Furthermore, the normal momentum equation is not simplified, and the convection terms are not one-sided differenced for marching. Thus, the thin-layer equations are numerically well behaved also at separation and require no special treatment there. The results of the thin-layer solutions for the supersonic flow over a ramp compares very favorably with results of the full Navier-Stokes solutions, including the effects of separation as shown by Degani and Steger (1983). The thin-layer equations can be expected to simulate well the viscous flow and enable reasonable numerical results with presently available computer resources.

The thin-layer equations in general curvilinear coordinates are presented in Section 9.1.3. In differencing these equations, it is often advantageous to difference about a known base solution, in this case using the free stream reference denoted by the subscript ∞, as follows

$$\text{(9.18)} \quad \begin{cases} \delta_\tau(\mathbf{Q}-\mathbf{Q}_\infty)+\delta_\xi(\mathbf{F}-\mathbf{F}_\infty)+\delta_\eta(\mathbf{G}-\mathbf{G}_\infty)+ \\ +(\mathbf{H}-\mathbf{H}_\infty)-\mathrm{Re}^{-1}\delta_\zeta(\mathbf{S}-\mathbf{S}_\infty)= \\ =[-\partial_\tau\mathbf{Q}-\partial_\xi\mathbf{F}-\partial_\eta\mathbf{G}-\partial_\zeta\mathbf{H}+\mathrm{Re}^{-1}\partial_\zeta\mathbf{S}]_\infty=0 \end{cases}$$

where δ indicates a general difference operator, ∂ is the differential operator, and since the free stream is used as the base solution, then the right-hand side of Eq. 9.18 is zero.

The time-marching thin-layer code, F3D, developed at the NASA - Ames Laboratory by Steger, Ying and Schiff (1986), is used for flow calculations on various configurations in many of their recent research programs, for example those by Ekaterinaris and Schiff (1990) and by Cummings et al. (1990). The algorithm uses flux-vector splitting and upwind spatial differencing for the convective terms in the streamwise direction. By using upwind differencing for the convective terms in the streamwise direction while retaining central differencing in the other directions, a two-factor implicit approximately factored algorithm is obtained which is unconditionally stable for a representative model wave equation. The algorithm is written for the thin-layer Navier-Stokes equation in the form

$$\text{(9.19)} \quad \begin{aligned} &[I+h\delta_\xi^b(\mathbf{A}^+)^n+h\delta_\zeta\mathbf{C}^n-h\mathrm{Re}^{-1}\bar{\delta}_\zeta J^{-1}\mathbf{M}^n J-D_{i|\zeta}] \\ &\times[I+h\delta_\xi^f(\mathbf{A}^-)^n+h\delta_\eta\mathbf{B}^n-D_{i|\eta}]\Delta\mathbf{Q}^n= \\ &-\Delta t\{\delta_\xi^b[(\mathbf{F}^+)^n-\mathbf{F}_\infty^n]+\delta_\xi^f[(\mathbf{F}^-)^n-\mathbf{F}_\infty^n]+\delta_\eta(\mathbf{G}^n-\mathbf{G}_\infty) \\ &+\delta_\zeta(\mathbf{H}^n-\mathbf{H}_\infty)+\mathrm{Re}^{-1}\bar{\delta}_\zeta(\mathbf{S}^n-\mathbf{S}_\infty)\}-D_e(\mathbf{Q}^n-\mathbf{Q}_\infty) \end{aligned}$$

where free stream base solution is used and where $h=\Delta t$ or $\Delta t/2$ for first or second-order time accuracy, respectively. Second order time accuracy is used when unsteady solutions are studied. In Eq. 9.19, δ is typically a three-point, second-order accurate central difference operator, $\bar{\delta}$ is a midpoint operator used with the viscous terms, and δ_ξ^b and δ_ξ^f are the backward and forward three-point difference operators, respectively. The flux $\mathbf{F}$ has been split into $\mathbf{F}^+$ and $\mathbf{F}^-$ according to its eigenvalues. The matrices $\mathbf{A}^\pm$, $\mathbf{B}$, $\mathbf{C}$ and $\mathbf{M}$ result from local linearization of the fluxes about the previous time level. J is the Jacobian of the coordinate transformation. The numerical dissipation terms, D_e and D_i, are used in the central space differencing directions η and ζ and are given as combinations of second and fourth differences in the form

$$
\begin{aligned}
D_{e|\eta} &=(\Delta t)J^{-1}\{\varepsilon_2\bar{\delta} \mid \mathbf{B} \mid \beta\bar{\delta} + \varepsilon_4\bar{\delta}\frac{\mid \mathbf{B} \mid}{1+\beta}\bar{\delta}^3\} \mid_\eta J, \\
D_{i|\eta} &=(\Delta t)J^{-1}\{\varepsilon_2\bar{\delta} \mid \mathbf{B} \mid \beta\bar{\delta} + \varepsilon_4\bar{\delta}\frac{\mid \mathbf{B} \mid}{1+\beta}\bar{\delta}^3\} \mid_\eta J
\end{aligned}
\tag{9.20}
$$

where

$$
\beta = \frac{\mid \bar{\delta}^2 p \mid}{\mid (1+\bar{\delta}^2)p \mid},
$$

$\mid \mathbf{B} \mid$ is the absolute value of the matrix $\mathbf{B}$, p is the nondimensional pressure and ε_2 is $O(1 + M_\infty^\gamma)$ while ε_4 is $O(0.01)$. In this form, the second order smoothing terms act to control numerical oscillations across shock waves, while the forth order smoothing is effective elsewhere. Accuracy can be improved by further scaling the fourth order smoothing terms by $\mathbf{Q}/\mathbf{Q}_\infty$.

9.4 Grid Topology, Boundary and Initial Conditions

The numerical solution of the Navier-Stokes equations require the generation of grids over the configuration surface as well as grids over the body in the flow field. These solvers can be very generally classified into three types according to the basic grid topology: 1. Cartesian, 2. unstructured and 3. multiblock structured. The Cartesian grid solvers lead to the simplest algorithms with the fewest operations and smallest memory per grid point, but there are serious difficulties in their application to complex curved surfaces. The unstructured and structured grids use body aligned grids, allowing accurate resolution of general curved surfaces. The unstructured grid solvers are the most general algorithms and their principal advantage is that grids over complex geometries can be generated in a relatively short time. However, these algorithms are also very costly in terms of operations per time step and memory per grid point due to non-sequential memory access and the explicit connections required between the cells of the mesh. The multiblock structured grids are more efficient in terms of operations per time step and of memory per grid point than the unstructured grids. However, more constraints are placed on the grid generation process in each block. The structured grid enables the application of efficient algorithms for treating viscous flows, since the grids can be highly stretched in the direction normal to the developing shear layers. In the future, it is very likely that hybrid grid systems will be used, for example, overlapping structured grids near the body linked to Cartesian grid in the far field used through an unstructured grid.

The methods for the generation of grids of various topological definitions were extensively discussed in reference to the solutions of the Euler equations in Chapter 8. Basically, the same grid topologies are also applicable to the numerical solution methods of the Navier-Stokes equations. Generally, complex configurations require multiblock structured grids similar to those presented for the Euler equations solution in Chapter 8. However, for the Navier-Stokes computations, the scale of the grid mesh in the viscous layers near surfaces must be much finer than those in the flow regions away from the body. Navier-Stokes calculations with embedded grids employing the Chimera scheme presented by Benek, Bunning and Steger (1985) can be used. The Chimera scheme enables solutions with several overlapping grids.

The Chimera scheme was applied to enable the determination of solutions for complex geometries (Buning et al. 1988) and for problems with multiple bodies in relative motions (Meakin and Suhs 1989), such as the store separation problem, where it is difficult to work with single-block structured grids. In such problems, it is necessary to construct individual grids over certain subdomains and these grids may overlap. In the application of the Chimera scheme, these grids are inputted into the Pegasus program, which searches for the overlapping regions and sets indices for the boundary points and interpolation points etc., as shown by Benek et al. (1985).

An adiabatic no-slip boundary condition is imposed at the body surface. The undisturbed free stream boundary conditions are maintained at the computational outer boundary. An implicit symmetry plane boundary condition is used at the circumferential edges of the fuselage grid, while at the downstream boundary a zero-axial-gradient extrapolation is applied.

The initial conditions are such that the flow field can be set initially to free-stream conditions throughout each grid and can be advanced sequentially in each component grid until a steady solution is obtained.

9.5 Solutions of the Navier-Stokes Equations for Flows over Three-Dimensional Configurations at High Angles of Attack

It is of interest to examine the capabilities of the available methods of solution of the Navier-Stokes equations under various formulations for the computation of the flow field at high angles of attack. Results of such computations were obtained for flows over elongated axisymmetric bodies for slender wings of delta and strake-delta planforms, and over complex aircraft configurations such as the F-18, F-15 and F-16.

9.5.1 Navier-Stokes Solutions for Flows over Elongated Axisymmetric Bodies at High Angles of Attack

The incompressible flows over elongated bodies at increasing angles of attack are dominated by the strong viscous-inviscid interactions. The calculations of such flows using various modifications of the boundary-layer approximations cannot correctly simulate the intricate flow field near and beyond the three-dimensional separation observed on the leeward side at the rear of these bodies. Given the present limitations on the capabilities of full Navier-Stokes solution, it is interesting to note that useful results for viscous flows over elongated bodies can be obtained by finding solutions to the parabolized and the thin-layer equations.

9.5.1.1 Flow on Prolate Spheroid and Ellipsoid in Subsonic Flow. The solution of the Parabolized Navier-Stokes equations for incompressible flows over a prolate spheroid at angles of attack up to 10° is presented by Rosenfeld et al. (1988). The steady, incompressible Parabolized Navier-Stokes (PNS) equations are obtained from the full steady Navier-Stokes equations by neglecting the streamwise diffusion terms in the momentum equations. These equations do not suffer from any singularity near separation because the pressure is retained as one of the dependent variables. The upstream boundary of the computed flow field starts a short distance downstream from the forward stagnation point of the prolate spheroid. In this region of the forward stagnation point, the boundary layer approximation can be justified and can provide the upstream boundary conditions for the PNS equations. The downstream boundary is placed ahead of the rear stagnation point in order to reduce the requirements for computer resources, due to the limitations of the computer used by the authors. The complete flow field at the rear of the body, including the regions of backward flow and strong viscous-inviscid interaction, can be calculated by this method providing that the downstream boundary is moved far enough into the wake. The finite difference equations are solved iteratively by marching in the streamwise direction. Grids of 15,000 to 20,000 points are used, and full convergence is obtained in 15-20 iterative sweeps, needing about 1-2 hrs of CPU time on an IBM 3081D computer. Although it is difficult to compare the results of these laminar calculations with experimental data, since at the Reynolds numbers of 10^6 the flow is probably turbulent, the pattern of the skin friction lines on the body surface is very interesting. The skin friction line pattern on the surface of the prolate spheroid at 10° angle of attack and Re of $8 \cdot 10^5$, as computed by Rosenfeld et al. (1988) is presented in Fig. 9.4. The calculated skin friction line pattern indicates that a long and swept convergence line is present, with skin friction lines merging into it from both sides. The precise origin of the convergence line cannot be defined with certainty and no singular points can be detected on the upstream part of the convergence line.

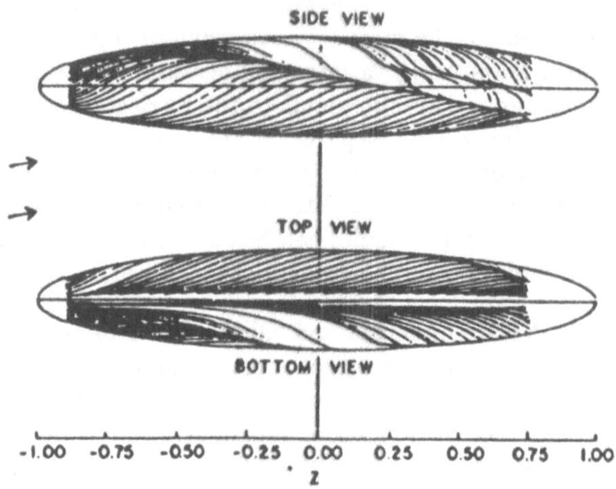

Fig. 9.4. Skin friction lines on the unwrapped 6:1 prolate spheroid at 10° angle of attack (Rosenfeld et. al 1988)

The solution of the thin layer Reynolds-averaged Navier-Stokes equations over a 6:1 ellipsoid placed at $\alpha = 10^o$ is presented by Pan and Pulliam (1986). The method of calculation and the numerical code are validated by comparing the computations with the experimental results of Meier et al. (1983). Computations are performed for both laminar flow over the complete model and for turbulent transition triggered at 1/4 body length from the upstream end since the flow Reynolds number is $1.6 \cdot 10^6$. The grid used is a $66 \times 66 \times 66$ grid with bilateral symmetry in η, as shown in Fig. 9.5. The first comparison is the agreement between the calculated surface shear stress vectors (a numerical simulation of the oil flow pattern) and the experimental oil flow pattern, as is presented in Fig. 9.6. The primary separation line computed for the laminar flow case seems to agree fairly well with the experimental separation line, certainly much better than for the turbulent case. Already at this moderate angle of attack of 10°, the calculated "particle traces", which are plotted in the cross-plane as shown in Fig. 9.7, indicate the formation of a symmetric pair of primary vortices. The primary vortex is generated by the separation at point S_1 (in Fig. 9.7) for both the laminar and turbulent computations. In the laminar case, the calculation results show an additional secondary separation at S_2 and a secondary vortex V_2. In the turbulent case, only the primary vortex is obtained. The calculated flow structure is in accord with that expected from the topological rules (discussed in Chapter 3) including the points of separation, S_1 and S_2, points of reattachment, R_1 and R_2, and the saddle point in the free stream, A_1.

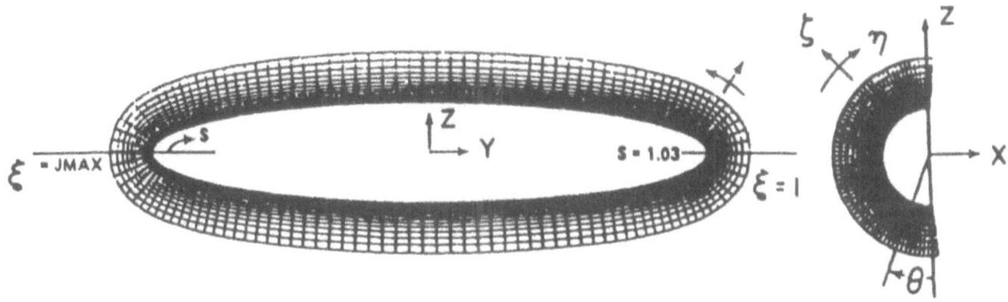

Fig. 9.5. Grid system for the ellipsoid (Pan and Pulliam 1986)

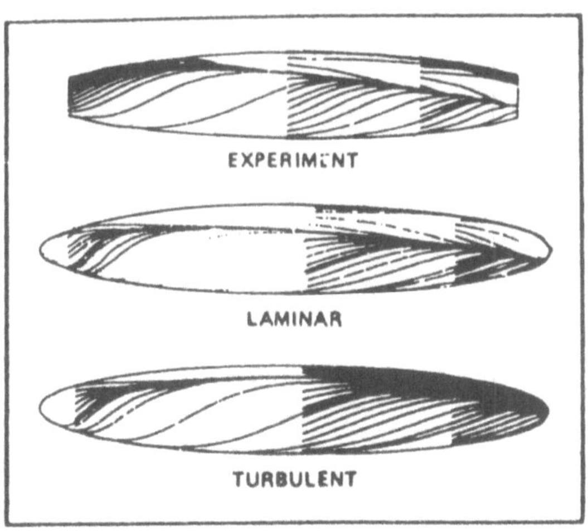

Fig. 9.6. Surface shear vector and oil flow pattern on the surface of Ellipsoid (Pan and Pulliam 1986)

The calculation of the flowfield about the prolate spheroid at higher angles of attack, up to 25° to 30°, using the F3D code is presented by Gee, Cummings and Schiff (1990). These calculations are performed at a high Reynolds number, $Re=7.2 \cdot 10^6$; therefore, a fully turbulent flow is expected. The effect of using various turbulent models is examined. The

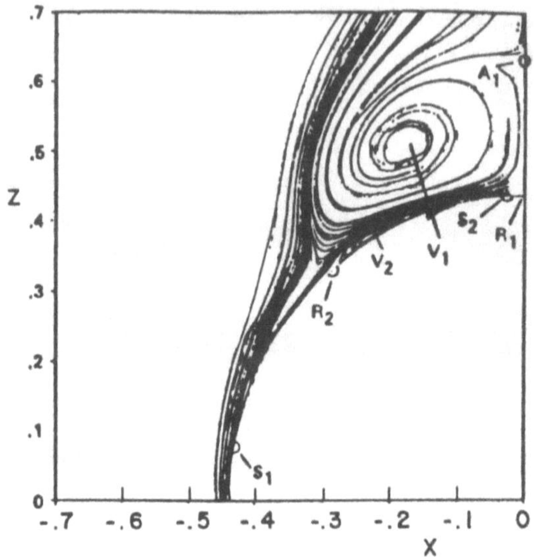

a. Laminar

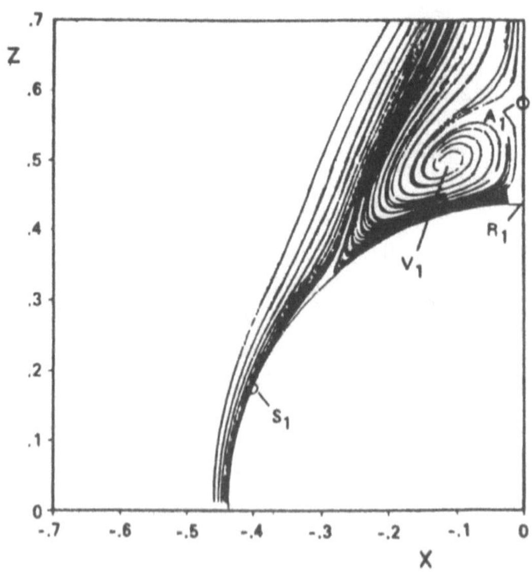

b. Turbulent

Fig. 9.7. Particle traces in the cross-plane above the ellipsoid (Pan and Pulliam 1986) comparison of the computed results with experimental data indicate that the Baldwin-Lomax model with the Degani-Schiff modification accords well, as seen in Fig. 9.8, and can be used for computations of flows with massive separations, as obtained at high angles of attack.

9.5.1.2 Flow on Pointed Axisymmetric Bodies. As the angle of attack of pointed axisymmetric bodies increases, the flow separates and the free shear layers first roll up into a pair of symmetric vortices. At increased angles of attack, the symmetric vortex pair may become asymmetric, inducing considerable lateral forces and moments. At higher angles of attack, the lateral force and moment decrease to nearly zero as the flow evolves into a pattern of multiple vortices which leave from alternate sides of the elongated body.

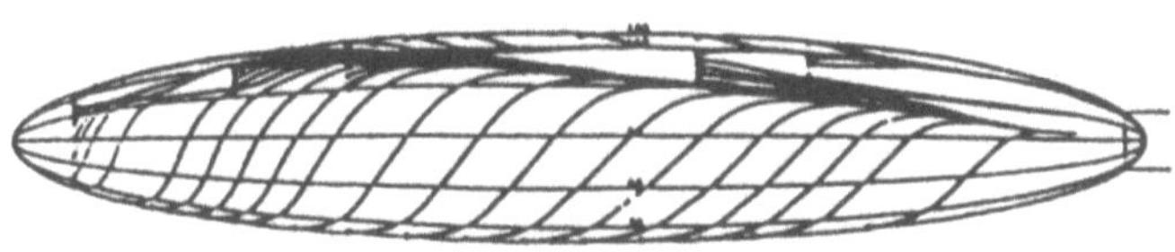

a. Experimental surface limiting streamline pattern at $M = 0.13$, $\alpha = 27^o$ and Re=$1.1 \cdot 10^6$ (turbulent flow)

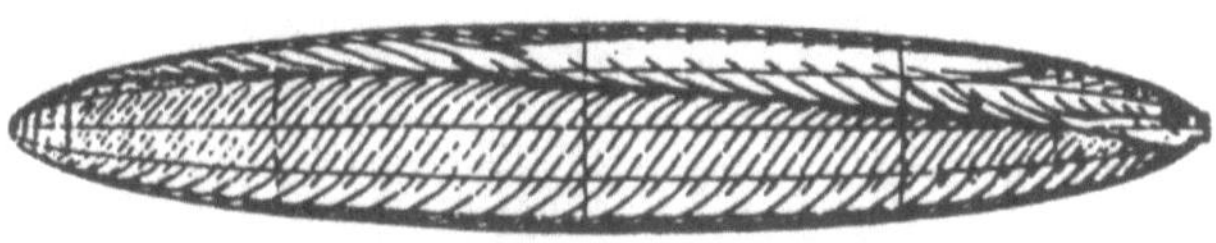

b. Calculated surface flow pattern at $M = 0.25$, $\alpha = 30^o$ and Re=$7.2 \cdot 10^6$ (Baldwin-Lomax with Degani-Schiff modified model)

Fig. 9.8. Comparison of experimental surface flow pattern with calculated results using turbulent model (Gee, Cummings and Schiff 1990)

The flow over pointed axisymmetric bodies can be computed by the solutions of the Navier-Stokes equations using the thin-layer or parabolized equations. Such calculations were performed on various ogive-cylinder models, and the results were compared with experimental data. The results of the Navier-Stokes computations indicate that as the angle of attack is increased, a pair of vortices is generated on the leeward side of the body. This pair of vortices remains symmetrical up to angles of attack of about 35°, as determined both experimentally and from the computed results. It is well known from experimental observations, that at higher angles of attack, the vortex pair becomes asymmetrical. Can the solution of Navier-

Stokes equations reproduce an asymmetric vortex structure at high angles of attack? This is a very intriguing question which we will address in the following discussions.

9.5.1.2.1 Subsonic Flow Solutions.

Symmetrical vortex flow. The capability of the Navier-Stokes solutions of flows over pointed bodies to simulate the symmetric separation case is demonstrated in a number of publications listed in the References. For example, the Navier-Stokes solutions for symmetrical vortical flows over a tangent-ogive cylinder are presented by Hartwich and Hall (1990). The C-O type grid used over the 3.5 caliber tangent-ogive cylinder is shown in Fig. 9.9.

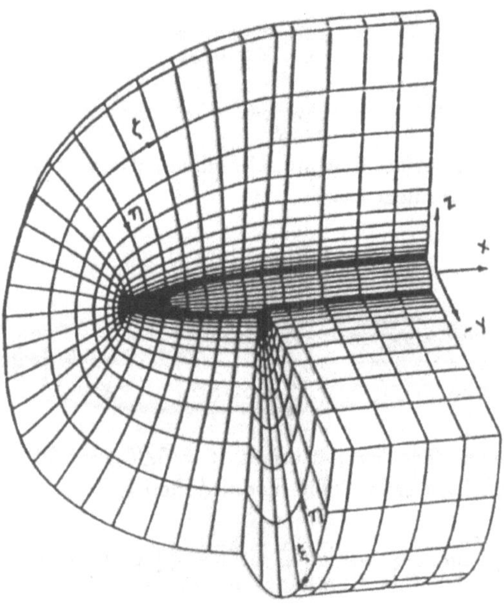

Fig. 9.9. View of C-O type grid around a 3.5 caliber tangent-ogive cylinder (Hartwich and Hall 1990)

The Navier-Stokes equations are solved in their thin-layer approximation for both laminar and turbulent flows. The calculated surface flow pattern on the tangent-ogive cylinder is compared with experimental results and shown in Fig. 9.10. The qualitative agreement between the calculated pattern with the experimental data is reasonable in the position of the primary separation line. Some more details can be observed in the particle traces over the body and in two cross planes in Fig. 9.11. The primary

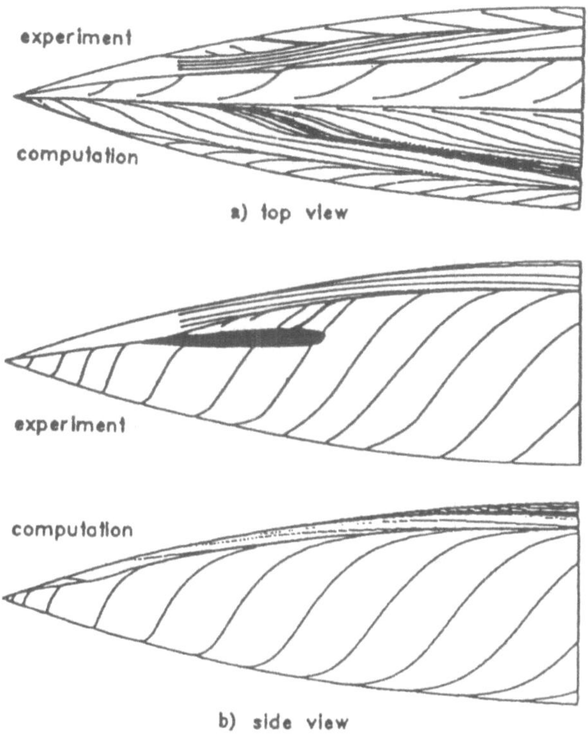

Fig. 9.10. Surface flow pattern on a tangent-ogive cylinder at $\alpha = 20^o$, $R_{eD} = 0.8 \cdot 10^6$ with $40 \times 91 \times 97$ grid (Hartwich and Hall 1990)

separation line can be seen beginning on the ogive and extending back over the cylindrical part. A secondary separation line and a reattachment line between the primary and secondary separation lines are also seen. This flow structure is shown clearly in Fig. 9.12, in which the helicity contours in the cross-section on the cylindrical portion are plotted.

The comparison of the computed and experimental circumferential pressure distributions can be seen in Fig. 9.13. The results of the computation using either the laminar, the Baldwin-Lomax turbulent model or Degani-Schiff modified turbulent model are indicated on Fig. 9.13. The use of the Degani-Schiff modified turbulent model provides better agreement with the experimental data on the cylindrical aft-portion of the body.

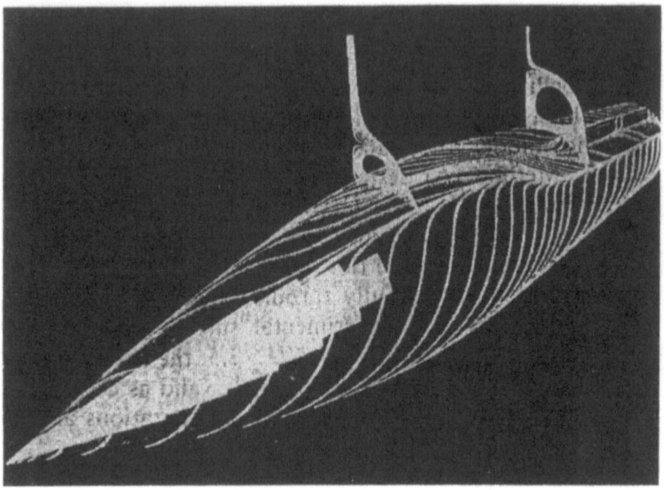

Fig. 9.11. Particle traces on surface and cross planes on tangent-ogive cylinder at $\alpha = 30^o$, $R_{eD} = 0.8 \cdot 10^6$

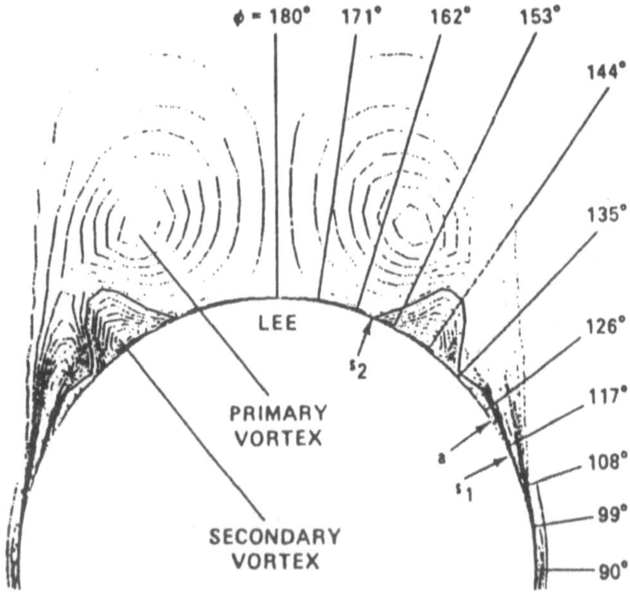

Fig. 9.12. Helicity contours in the cross-section at $x/D = 6$, $M = 0.2$, $\alpha = 20^o$ and $R_{eD} = 5.0 \cdot 10^6$ (Degani et al. 1990)

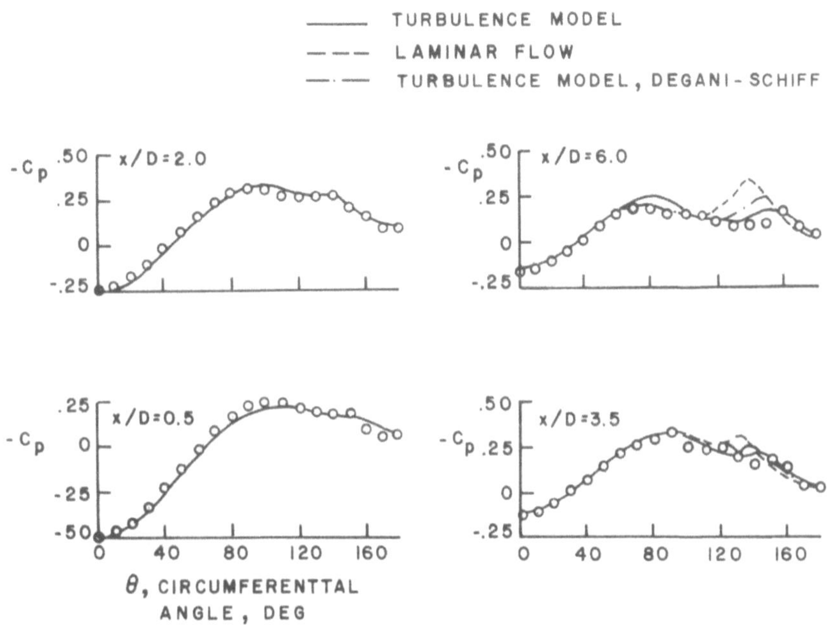

Fig. 9.13. Experimental and computed circumferential surface pressure distributions on tangent-ogive cylinder at $\alpha = 20°$, $R_{eD} = 0.2 \cdot 10^6$, 40 × 73 × 65 grid points (Hartwich an Hall 1990)

Asymmetric flow solution: The phenomenon of the appearance of asymmetric vortex structures at high angles of attack on axisymmetric pointed bodies, which is observed in experimental studies, has been attributed to effects of manufacturing imperfections in the body surface near its pointed tip. Such "imperfections" can also be induced in the computational scheme by adding a small bump on the body as done by Degani and Schiff (1989) and as studied by Degani (1990). These computational studies indicate that some disturbances must be introduced into the flow to trigger the asymmetry in the vortices and the appearance of lateral forces and moments. Such asymmetry in the vortices can be triggered by slight asymmetry in the freestream initial condition, and, as shown by Siclari and Marconi (1989), even very small disturbances such as machine round-off error may be sufficient to trigger the asymmetry of the vortical flow over the elongated pointed body. There are indications that the steady asymmetric flow solution is independent of the size of the initial triggering disturbance, although Degani (1990) shows that the size of the disturbance has some effect on the asymmetric flow solution.

The characteristics of the asymmetric vortical flow on a 3.5 caliber tangent-ogive forebody is studied by Deese et al. (1991). The thin layer approximation to the Reynolds averaged Navier-Stokes equations are solved by an explicit, fourth order, Runge-Kutta time-marching, finite-volume numerical scheme. The calculations are performed for $R_{eD} = (0.16\text{–}0.6) \cdot 10^6$, $M = 0.18$ and at angle of attack varying from $\alpha = 30^o$ to 70°. The results of the calculations at $\alpha = 30^o$ indicated a symmetrical vortex flow over the tangent-ogive body. By increasing the angle of attack to 50° without introducing asymmetric perturbation, the computation results actually indicated an asymmetric vortex structure on the leeward side of the ogive instead of the expected symmetric structure. This unexpected asymmetrical solution is attributed to numerical asymmetry hidden in the grid generation method and the cell volume calculation procedure used in this calculation. It is shown that the perturbation required to produce asymmetry is about five orders of magnitude smaller than the body diameter. This very small perturbation explains the great sensitivity of the symmetrical vortical flow, which is in agreement with the experimental results showing an asymmetric vortical flow for angles of attack in the range of 35° to 60°.

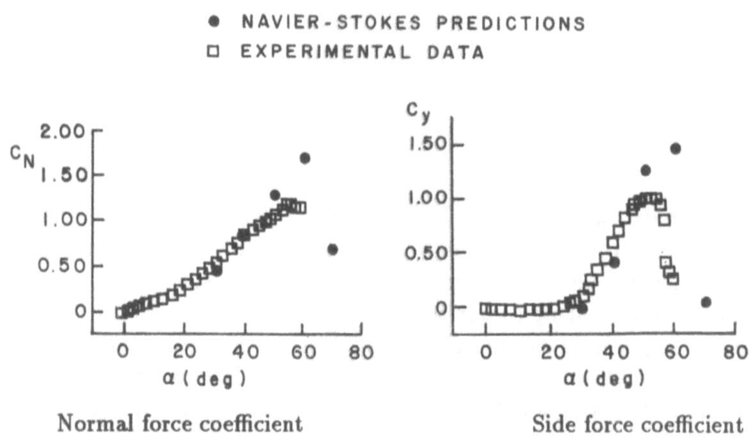

Fig. 9.14. Comparison of Navier-Stokes calculations with experimental data of the normal and side forces variation with angle of attack on tangent-ogive at $M = 0.18$, $R_{eD} = 0.16 \cdot 10^6$ (Deese et al. 1991)

The calculated normal and side force coefficients variation with increasing angle of attack on the 3.5 caliber tangent-ogive is presented in Fig. 9.14. These calculated results are also compared with experimental data in Fig. 9.14. Computationally, the start of the lift-off process for one of the two primary vortices on the leeward side and the development of significant side force occur for angles of attack greater than 30°. The experimental data show that the side force is already developing for an angle of attack of 25°.

The difference between the experimental and computed results may be explained by the different levels of perturbations causing the asymmetry in both cases. In any case, the calculated and experimental results agree well for both the normal and side forces for angles of attack from 30° to about 45°. At about this latter angle, the slopes of the experimental normal and side forces change. This change is due to the lift-off of the second primary vortex which starts at the rear end of the model and moves forward as the angle of attack increases. The experimental side force decreases from its maximum value to nearly zero as the angle of attack increases from 45° to 60°, corresponding to the lift-off point moving from the trailing edge up to the ogive apex. The Navier-Stokes calculations overestimate the peak normal force and the peak side force for angles of attack from 50° to 60° and indicate a sharp drop in these forces as the angle of attack increases to 70°. The difference between these calculated results and the experimental data may again be attributed to the differences in the disturbances triggering the asymmetric vortical flows.

9.5.1.2.2 Supersonic Flow Solutions. Symmetric flow solutions: Computation of the supersonic flow field based on solutions of the Parabolized Navier-Stokes equations on cones and ogive-cylinder bodies is presented by Degani and Schiff (1983). The numerical algorithm used to march the solution downstream is an implicit, noniterative, approximately factored finite-difference scheme. The marching-stepback procedure was utilized to generate the conical solutions over the cones and ogive-cylinder bodies. The Degani-Schiff modification to the Baldwin-Lomax turbulence model was introduced into these calculations.

The calculations on a 12.5^o half-angle cone are compared with the experimental data of Rainbird (1968) at $M_\infty = 1.8$ and free stream Reynolds number of $25 \cdot 10^6$ based on the model axial length. The calculated circumferential pressure distribution on the 12.5^o cone at an angle of attack of 22.75^o at $x/L = 0.85$, where $\mathrm{Re}_x = 21.25 \cdot 10^6$, is presented in Fig. 9.15a. A similar calculation is performed on a tangent-ogive cylinder model at $M = 3.0$ and angle of attack of 10.4^o with $\mathrm{Re}_x = 7.03 \cdot 10^6$. The calculated circumferential pressure distribution at the base plane is presented in Fig. 9.15b. The improved agreement between the calculation, using the modified turbulent model, with the experimental data in the leeward side pressure distribution is evident.

Further evidence as to the capabilities of the Parabolized Navier-Stokes solutions at high incidence is obtained by examining the computational and experimental results on the 7.5^o cone at an angle of attack of $\alpha = 22.6^o$ and $M = 2.94$, as presented by Degani and Schiff (1983). The circumferential pressure distribution is presented in Fig. 9.16a and good agreement is found including the positions of the primary and secondary separation points. The leeward ray radial pressure distribution above the 7.5^o cone is presented in Fig. 9.16b. The experimental pressure indicates a jump (due to an imbedded shock wave) at $h/R = 0.17$. The calculated distribution does not capture

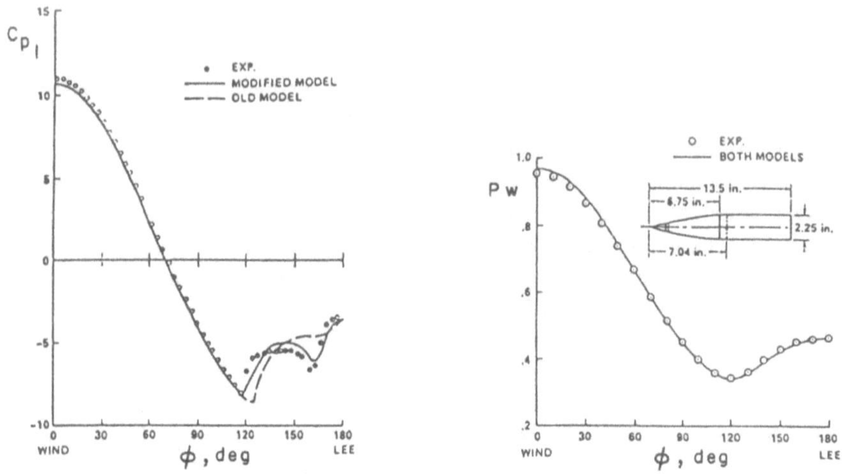

a. 12.5° cone at $M = 1.8$, $\alpha = 22.75°$, $Re_x = 21.25 \cdot 10^6$

b. Tangent-ogive cylinder at $M = 3.0$, $\alpha = 10.4°$, $Re_x = 7.03 \cdot 10^6$

Fig. 9.15. Circumferential surface pressure coefficient distributions (Degani and Schiff 1983)

this shock wave, probably because of insufficient grid mesh, so that the calculated pressure jump and the calculated value of the peak pressure are considerably smaller.

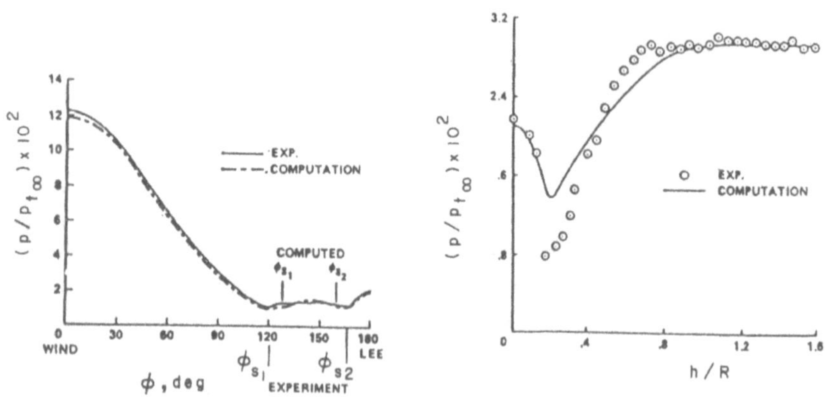

a. Circumferential surface pressure distribution

b. Leeward ray radial pressure distribution

Fig. 9.16. Pressure distributions on 7.5° cone at $M = 2.94$, $\alpha = 22.6°$ and $Re_x = 5.85 \cdot 10^6$ (Degani and Schiff 1983)

The solution of the Parabolized Navier-Stokes equations using an upwind algorithm to capture strong shock waves around a 10° cone in hypersonic flow, $M = 7.95$, is presented by Lawrence et al. (1987). The flow field geometries in the cross-plane are compared with experimental data at angles of attack of 12° and 24° in Fig. 9.17, and the corresponding pressure distributions in Fig. 9.18. These results illustrate the capability of the PNS solution to capture the hypersonic shock structure, including the internal shock waves, and to predict well the local pressure distribution.

Asymmetric flow solutions: The same mechanism that triggers the switch from the symmetric vortex pair to the asymmetric vortex structure at increasing angles of attack in the subsonic flow over pointed axisymmetric bodies, as discussed in the previous section, can also trigger the switch to the asymmetric vortex structure in supersonic flow over these bodies. However, it will be shown that at increasing Mach numbers, the flow tends to retain its symmetric structure. Solutions of the Navier-Stokes equations with some thin-layer approximations using a multigrid, central difference, finite volume scheme for flows over circular cones and over noncircular cross-section elongated bodies in supersonic flows are presented by Siclari and Marconi (1989) and by Siclari (1990). The computer code MIM3D (Multigrid Implicit Marching) uses a full plane grid and first computes a conical flow solution. Three-dimensional solutions are then obtained using a hybrid implicit marching method which computes the crossflow terms by the same finite volume method used for the conical flows and includes marching derivatives using a finite difference upwind approximation.

The development of the asymmetric structure from the initially symmetric flow is illustrated in the convergence behavior of the solution of the flow over an elliptic cone, as shown in Fig. 9.19. The solid line shows the variation of the maximum residual with multigrid cycles for computation started with symmetric initial freestream conditions. Initially the solution converges by almost 12 orders of magnitude, but then suddenly the residual error begins to increase to almost its original value. With further iterations, the residual error then decreases until the computation converges to the asymmetric flow structure, indicated on the figure by the crossflow streamline pattern at convergence. However, near the first minimum of the residual error, the crossflow streamline pattern at that point, as indicated in Fig. 9.19, is a symmetric pair of vortices. It is seen that after the residual error increases to almost its original value, this numerical error triggers the symmetrical flow and the computation converges to the asymmetric pattern. The same asymmetrical flow can also be obtained by a similar computation which is started with slightly asymmetric freestream initial conditions. In this case, the computation converges monotonically as indicated by the dashed line to the familiar asymmetric flow structure. The similarity between the convergence pattern of the slightly asymmetric freestream initial condition and that obtained after the residual error reached a maximum supports the hypothesis that in both cases the initial disturbance, either

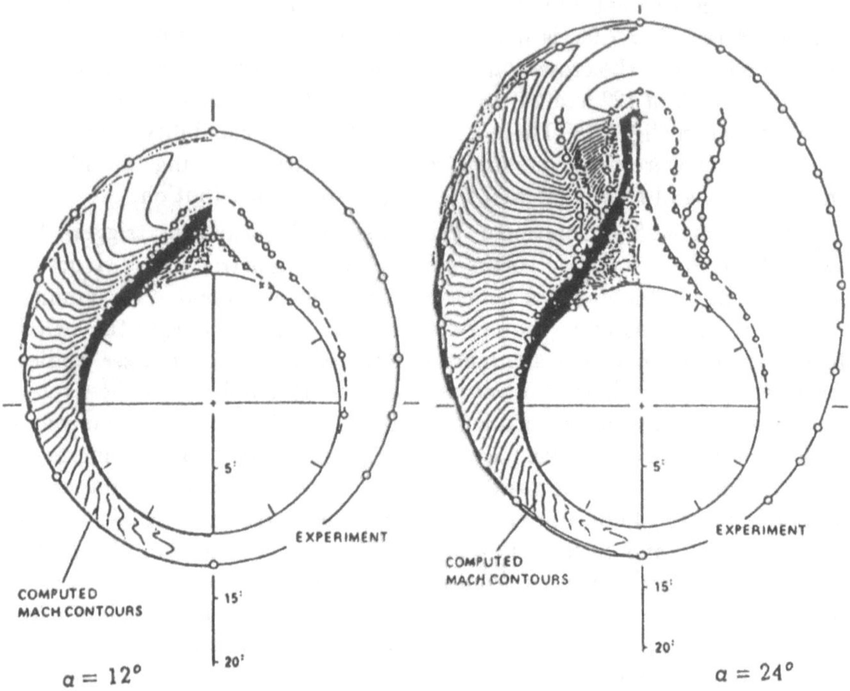

Fig. 9.17. Comparison of flowfield geometries on a 10° cone at $M = 7.95$ and $\mathrm{Re}_x = 0.4 \cdot 10^6$ (Lawrence et al. 1987)

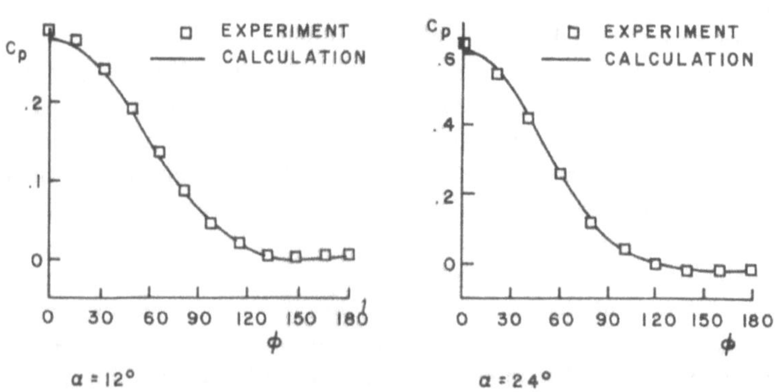

Fig. 9.18. Pressure distribution on a 10° cone at $M = 7.95$ and $\mathrm{Re}_x = 0.4 \cdot 10^6$ (Lawrence et al. 1987)

from the initial freestream asymmetry or from the residual error, acts to trigger the switch to the asymmetric vortex pattern.

The calculated streamline patterns for the asymmetric flow over a 5° circular cone at various free stream Mach numbers is shown in Fig. 9.20. As the Mach number increases from 1.4 to 3, the flow tends to become more symmetric, until at $M = 3$ the pattern returns to be fully symmetric. The calculations with 7° and 8° circular cones indicated that for the blunter cones, the Mach number for return to symmetry is lowered to 2.2 for the 7° cone and 1.8 for the 8° cone.

The separated vortical flow on bodies with various noncircular cross sections was also calculated by Siclari (1990). The results of the computation of the asymmetric flow around bodies with elliptic, biparabolic and biwedge cross-sections of different Aspect Ratios are presented in Fig. 9.21. It is evident that for all of these cross-sections, as the cross-section is flattened, the flow tends to become symmetrical. The calculated streamline patterns also indicate the secondary vortices which are generated. Of special interest are the cross-sections with sharp corners such as the biparabolic and biwedge models. It is clear that the asymmetric solutions are not unique to circular cones. Furthermore, the sharp corners which fix the position of separation on the body do not prevent the condition of asymmetric vortex structure. It seems that the Mach number, the major axis half angle and the ratio of major to minor axis or thickness ratio are the parameters which determine whether the asymmetrical flow occurs, while the cross-section shape has less of an effect.

9.5.1.3 Flow over Delta and Strake-Delta Wings.

9.5.1.3.1 Subsonic flow calculations. Symmetric flow calculations: An early calculation of the flow about a slender delta wing by the solution of the Navier-Stokes equations using the thin-layer approximation was presented by Fujii and Kutler (1984). The thin-layer Navier-Stokes equations were solved by an implicit noniterative approximate factorization scheme of Beam and Warming (1977). The computation was done for a delta wing of AR=1 (leading-edge sweep of 76°), with a maximum chord thickness of 1%, the symmetry condition was then imposed and the grid used $38 \times 51 \times 31$. The calculations were done for an angle of attack of 20.5^o and $Re=0.9 \cdot 10^6$, assuming laminar flow in the Navier-Stokes formulation. This calculation captured the primary leading-edge vortex as well as the secondary vortex and agreed reasonably well qualitatively with experimental results.

An improved solution of the incompressible Navier-Stokes equations for a flow over a delta wing was presented by Hartwich et al. (1988). The thin-layer approximationwas solved by an upwind scheme based on flux-difference splitting using the Baldwin-Lomax turbulent model with the Degani-Schiff modification in the massive separation regions. Central differencing of the viscous shear fluxes and a total variation diminishing (TVD) –

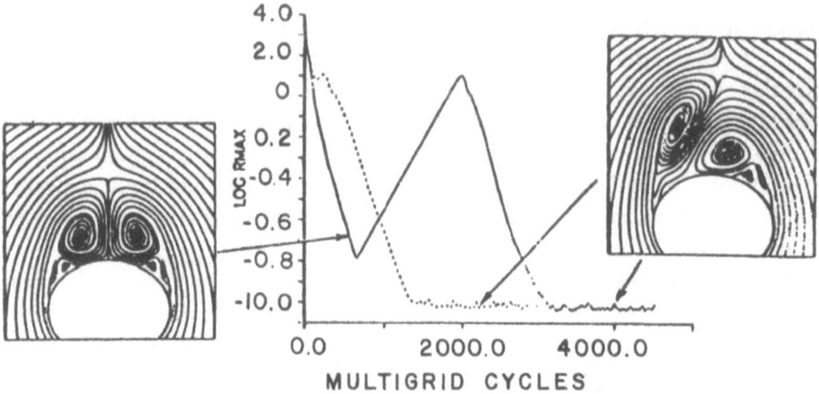

Fig. 9.19. Convergence behavior of an elliptic cone at $M = 1.8$ and $\alpha = 20^{o}$ (Siclari 1990)

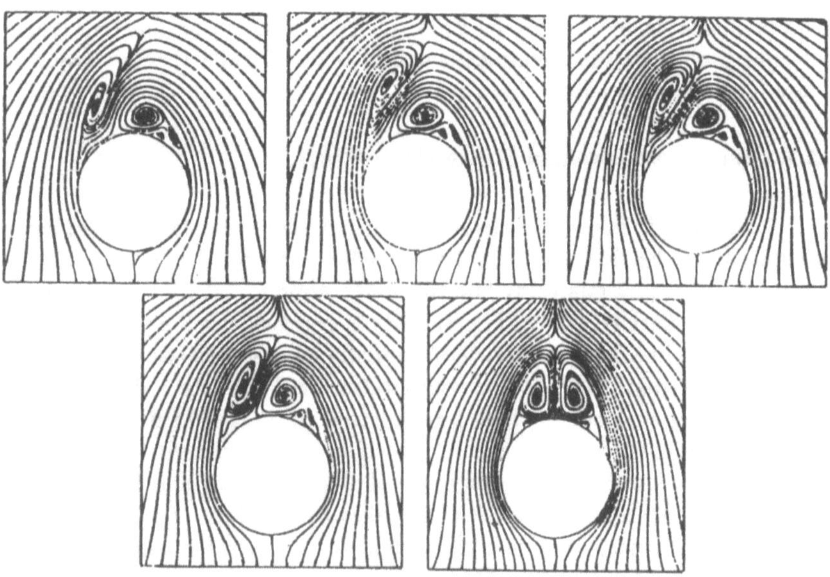

Fig. 9.20. Crossflow streamline pattern computed on a 5° circular cone at $\alpha = 20^{o}$ and Mach numbers 1.4 to 3 (Siclari 1990)

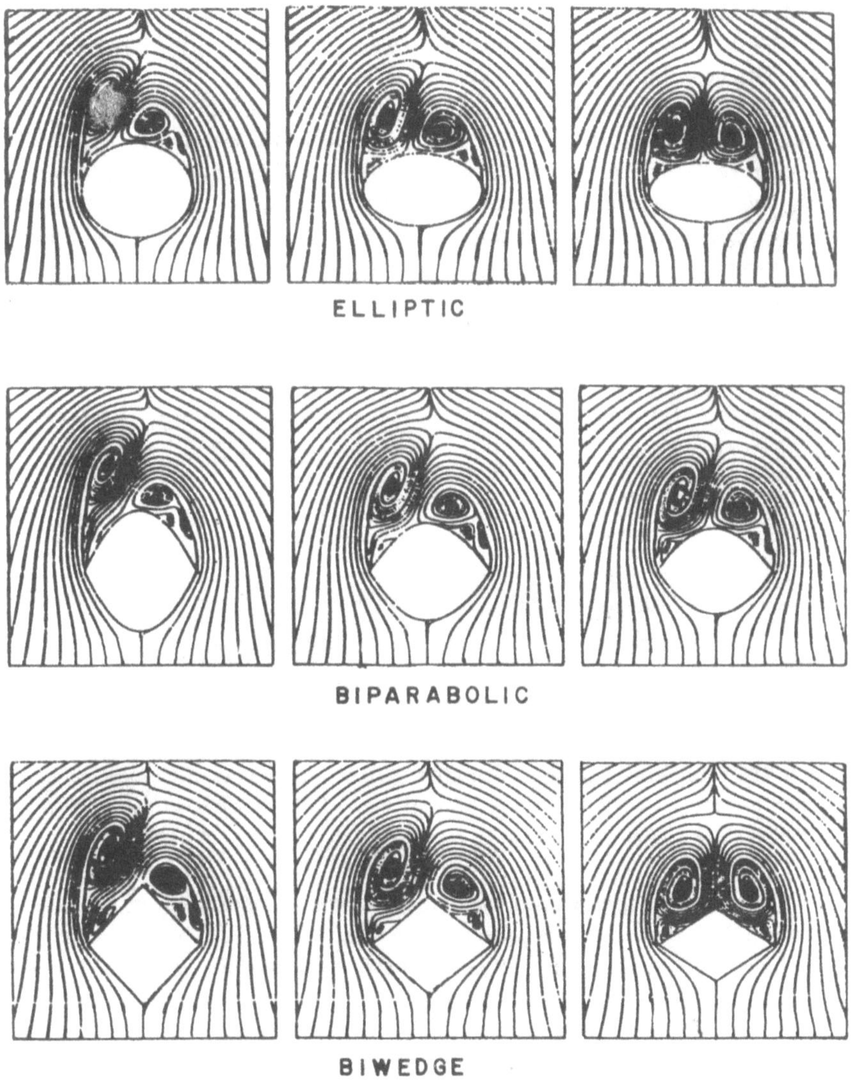

Fig. 9.21. Crossflow streamline patterns computed for three different cross sections with 5° half angles and varying Aspect Ratios at $M = 1.8$ and $\alpha = 20°$ (Siclari 1990)

like discretization of the inviscid fluxes yield second order spatial accuracy. The flow over a delta wing of AR=1 at an angle of attack of 20.5° and Re=$0.9 \cdot 10^6$ was calculated and the results compared with experimental data. The flow field was discretized using an H-C grid.

The comparison between the calculated surface streamlines and the surface oil flow pattern obtained experimentally is presented in Fig. 9.22. It is seen that the computed pattern simulates well the primary separation along the sharp leading-edge and its reattachment near the centerline with the flow directed outboard. Then, the induced secondary separation line is seen followed by the reattachment line very close to the leading-edge, which is in a good agreement with the experimental pattern shown in this figure.

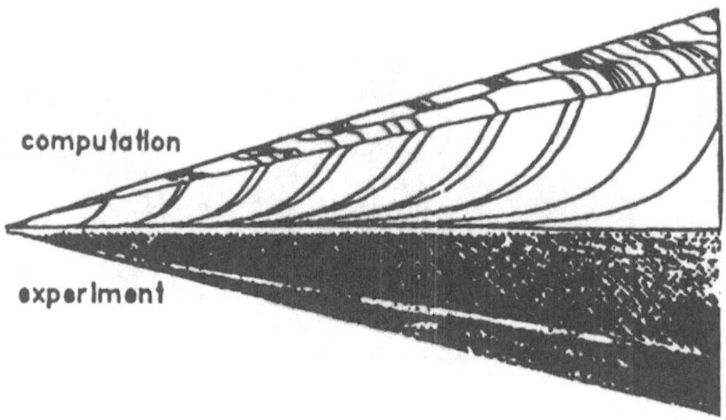

Fig. 9.22. Comparison of the calculated surface streamlines with experimental oil flow pattern on an AR=1 delta wing at $\alpha = 20.5°$, Re=$0.9 \cdot 10^6$ (Hartwich et al. 1988)

The spanwise pressure distributions at various chordwise x/c stations on this AR=1 delta wing is presented in Fig. 9.23. The flow is assumed to be transitional, so that the flow on the lower wing surface as well as on the forward part of the upper surface up to $x/c = 0.6$ is laminar while the flow on the upper surface from $x/c = 0.7$ to the trailing-edge is turbulent. This transitional model is used in the calculation of the pressure distribution, and reasonable good agreement is obtained with the experimental data, including the pressure variation due to the secondary separation.

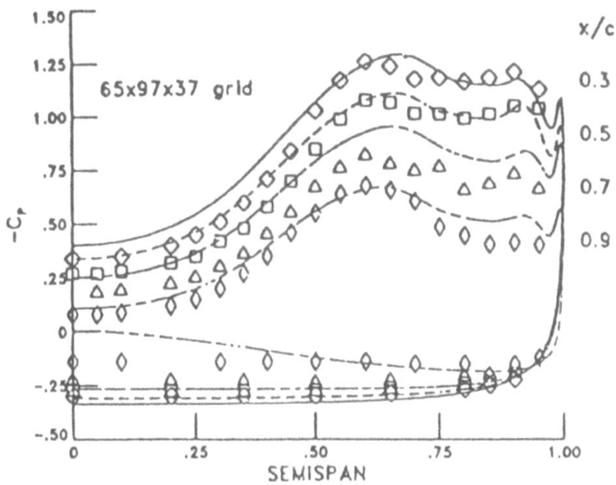

Fig. 9.23. Calculated spanwise pressure distributions (for transitional flow) and experimental data at various chordwise stations on the AR=1 delta wing at $\alpha = 20.5^o$, Re=$0.9 \cdot 10^6$ (Hartwich et al. 1988)

The shape of the core of the leading-edge vortex at a moderate angle of attack of 20.4^o, where no vortex breakdown effects are noticeable, is presented in Fig. 9.24.

The calculation of the lift coefficient variation with an angle of attack up to $\alpha = 40^o$ and the variation of the pitching moment coefficient is compared with experimental data in Fig. 9.25. Excellent agreement is obtained over the complete range of these angles of attack, including the diminishing rise of lift due to vortex breakdown onset at the trailing edge at $\alpha = 30^o$. At $\alpha = 40^o$, the vortex breakdown moves up towards the wing apex to x/c of about 0.4–0.45, causing considerable reduction in lift.

The ability of the presently available solutions of the Navier-Stokes equations to compute the flow field over a double delta (also called – strake-delta) wing is of particular interest because of the added complication of that flow field. Numerical solutions of the thin-layer approximation of the Navier-Stokes equations are used in the calculations of Fujii and Schiff (1989), Hartwich et al. (1988) and Ekaterinaris et al. (1991). The Baldwin-Lomax turbulent model with the Degani-Schiff modification is used in these calculations. The double delta wing geometries used are: 1. a 80° leading-edge sweep of the strake and 60° leading-edge sweep of the main wing with the leading-edge kink at $x/c = 0.5$ (used by Fujii and Schiff and by Hartwich et al.) and 2. a 76° leading-edge sweep of the strake and 40° sweep of the

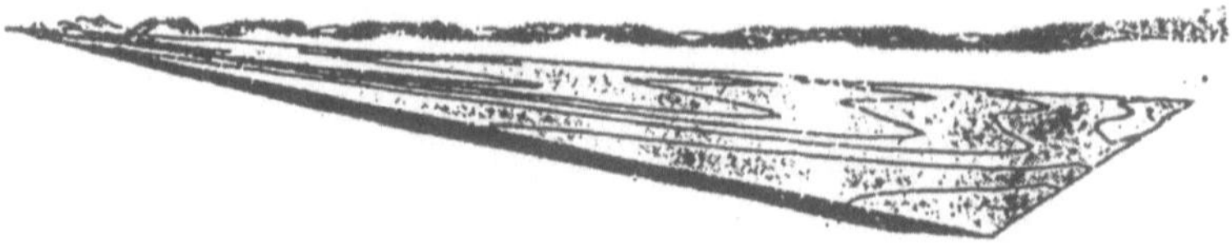

Fig. 9.24. Leading-edge vortex over an AR=1.07 delta wing, $M = 0.3$, Re=10^6 at $\alpha = 20.5^o$ (Ekaterinaris and Schiff 1990)

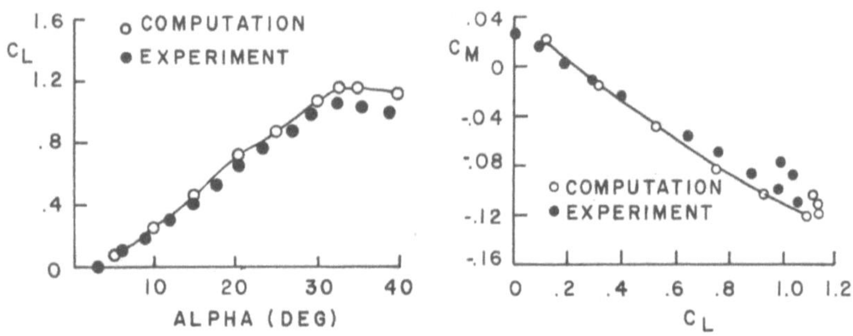

Fig. 9.25. Comparison of the calculated and experimental variation of the lift coefficient with angle of attack on AR=1 delta wing, Re=$0.9 \cdot 10^6$ (Hartwich et al. 1988)

main cropped delta wing (used by Ekaterinaris et al. 1991).

The primary inner and outer vortices on the $60^o/80^o$ double delta wing at a moderate angle of attack of $\alpha = 12^o$ and at a reasonably high angle of attack of $\alpha = 30^o$, as calculated by Fujii and Schiff (1989), are shown in Fig. 9.26. Good qualitative agreement is found between the calculated and experimental results.

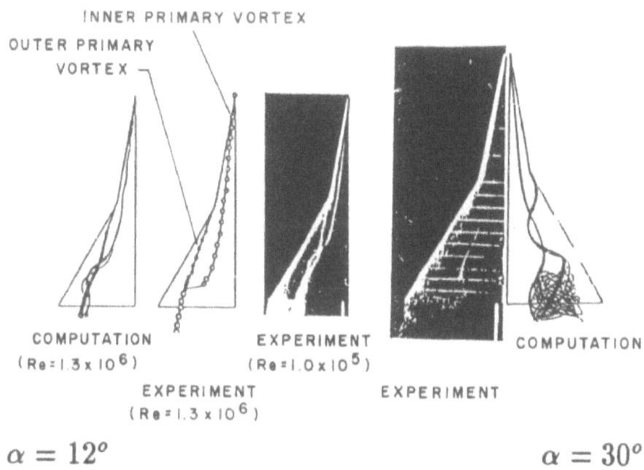

Fig. 9.26. Comparison of the vortex flow structure over the 80°/60° double delta wing at $M = 0.3$, $Re = 1.3 \cdot 10^6$ (Fujii and Schiff 1989)

Calculations on the 80°/60° double delta wing were also performed by Hartwich et al. (1988) using an upwind scheme of flux-difference splitting and applying some changes in the Degani-Schiff turbulence modification to the Baldwin-Lomax model. They used an H-C type grid with $65 \times 101 \times 81$ points. The calculation of Fujii and Schiff (1989), which was done on the same wing geometry, used an algorithm based on the LU-ADI factorization method and an H-O type grid with $119 \times 101 \times 71$ points. The lift coefficient on the double delta wing for angles of attack up to 40°, calculated by Hartwich et al. and by Fujii and Schiff, are compared with experimental data in Fig. 9.27. There is very good agreement between the Navier-Stokes solutions with the experimental data up to the maximum lift coefficient of about $\alpha \cong 25°$. At higher angles of attack, as the vortex breakdown point moves from the trailing-edge towards the wing apex, there is a good agreement between the calculation results of Hartwich et al. with the experimental data.

The vortical flowfield over the second configuration, 76°/40°, of a double delta wing (76° leading-edge sweep of the strake and 40° sweep of the leading-edge of the main wing) was calculated by Ekaterinaris et al. (1991). The thin layer Navier-Stokes equations are solved by using a partially flux-split numerical scheme and assuming the Baldwin-Lomax turbulent model modified by Degani and Schiff for the computation of separated vortical flow. Solutions are obtained using spherical grid topology with $82 \times 63 \times 64$ points; the double delta wing and the spherical grid are shown in Fig. 9.28. A global view of the calculated flow patterns on the leeward side at angles of attack of 10° and 19° is presented in Fig. 9.29. On the left side of Fig. 9.29, the helicity density contours in crossflow planes at several axial locations are shown. The strake and the wing vortex trajectories are denoted by lines through the centers of the vortices identified by the helicity den-

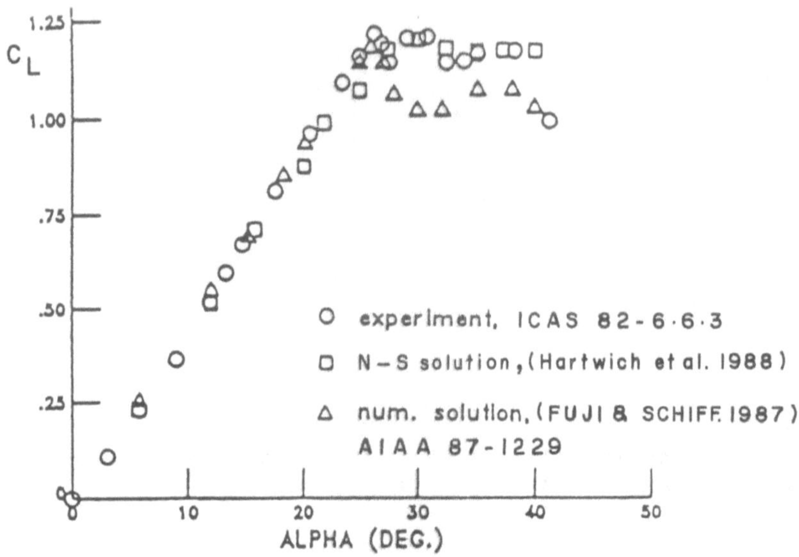

Fig. 9.27. Variation of the lift coefficient with angle of attack for the $80^o/60^o$ double delta wing in incompressible flow, Re=$1.3 \cdot 10^6$ (Hartwich 1988)

sity contours. On the right side, the computed surface flow pattern and the leading-edge vortices are shown. A schematic of the crossflow pattern at two representative locations, one on the strake and the other on the wing, is also sketched. The secondary separation is located approximately under the vortex center on the strake section. On the wing section, both the se-paration and the attachment skin-friction lines which originate at the strake have been moved further outward due to interaction with the vortical flowfield caused by the wing vortex. At $\alpha = 10^o$, the strake and wing vortices remain distinct over the wing surface beyond the trailing-edge, while at 19°, the strake and wing vortices are seen to merge near the trailing-edge.

Comparisons of the computed and experimental pressure distributions on the leeward side of the double delta wing are presented in Fig. 9.30 for angles of attack of 10°, 19° and 22.4^o. There is generally a good agreement between the measured and calculated pressures, although some of the suction peaks are underpredicted.

9.5.1.3.2 Supersonic flow calculations. Symmetrical flow case: It is recognized that the flow structure over a slender delta wing is determined by the components of the Mach number and the angle of attack, both measured normal to the leading-edge of the wing. The possible types of vortical flows which were observed experimentally can be classified in the map presented in Fig. 1.1 (in Chapter 1). In the case of supersonic free stream, the flow over the delta wing can be either of the subsonic vortical flow type for

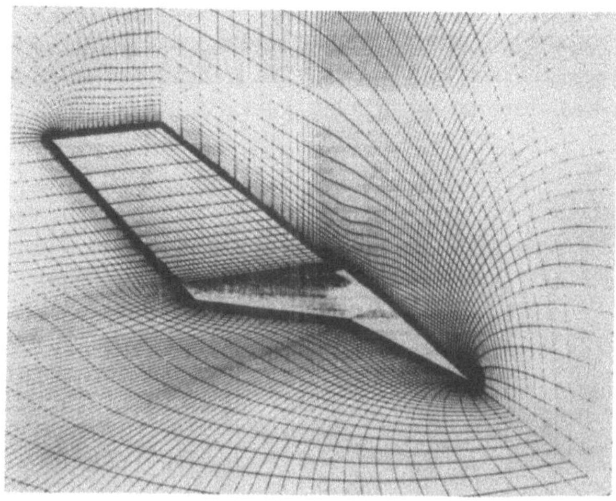

Fig. 9.28. Spherical grid over the double delta wing (Ekaterinaris et al. 1991)

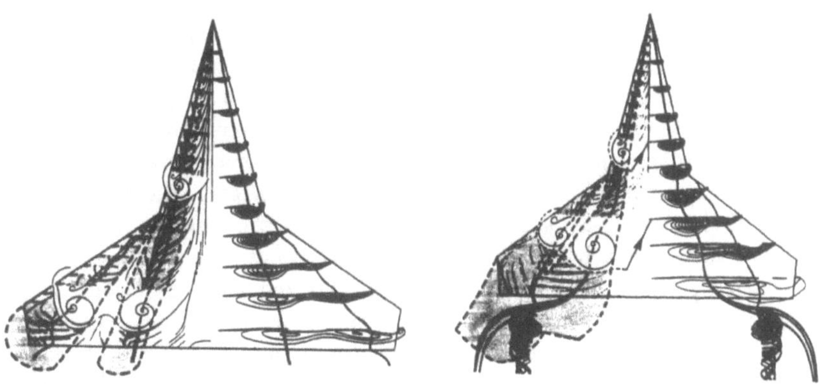

$\alpha = 10^o$ $\alpha = 19^o$

Fig. 9.29. Global view of the flow pattern on the leeward side of the double delta wing at $M = 0.22$, Re=$4 \cdot 10^6$ (Ekaterinaris et al. 1991) subsonic leading-edges or of the supersonic vortical flow (vortices with embedded shock waves) for the supersonic leading-edge geometries. At higher angles of attack, the vortical flow may be transformed to an asymmetric structure, as was observed in subsonic flows.

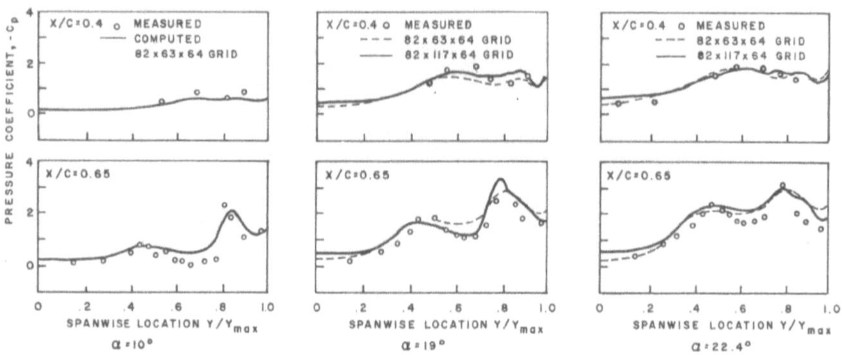

$\alpha = 10^o$ $\alpha = 19^o$ $\alpha = 22.4^o$

Fig. 9.30. Leeward side pressure distributions on the double delta wing at $M = 0.22$, Re$=4 \cdot 10^6$ (Ekaterinaris et al. 1991)

Numerical methods, including panel methods or Euler solutions, were used to calculate approximate values for the aerodynamic characteristics of these complicated flow structures. The application of the Navier-Stokes solutions for the calculations of the flow field over delta wings in supersonic flows is expected to produce more accurate results. Available solutions for the Navier-Stokes equations for delta wings both for supersonic and subsonic flows impose symmetric flow conditions. Therefore, there are no solutions of the Navier-Stokes equations at present showing the asymmetric modes of vortex separation on slender delta wings at high angles of attack.

Early solutions of the conical and parabolized approximations of the Navier-Stokes equations for supersonic flows over delta wings with sharp subsonic leading-edges were presented by Vigneron et al. (1978). The capability of the Navier-Stokes solutions to obtain the main flow features over the delta wing, including the effects of primary and secondary separations, has been demonstrated by Rizzetta and Shang (1986) using a grid with 20 x 64 x 40 points. Computations of the leeside flows over delta wings using the thin-layer and conical approximations of the Navier-Stokes equations were presented by Thomas and Newsome (1986). The calculations use an implicit algorithm employing upwind differencing for the pressure and convection terms and central differencing for the shear stresses and heat transfer terms. A comparison of the computed and experimental pressure distributions on a 75° swept sharp leading-edge delta wing at $M = 1.7$ and at angles of attack of 4°, 12° and 20° is shown in Fig. 9.31. The computations were made using the conical approximation with a grid of 151 points circumferentially and 75 points radially. The computed pressure distributions agree well with the experimental data including the correct indication of the effects of the primary and secondary vortices.

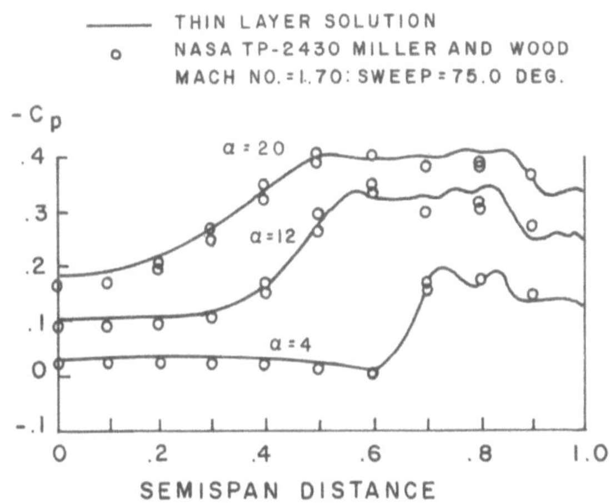

Fig. 9.31. Comparison of computed and experimental pressure coefficients on a 75° delta wing at $M = 1.7$, Re=$3.56 \cdot 10^6$ and at $\alpha = 4^o$, 12°, 20° (Thomas and Newsome 1986)

Ruffin and Murman (1988) extended the use of the conical approximation of the Navier-Stokes equations to the calculations of hypersonic flows over slender delta wings. It is observed experimentally that, although the viscous zone includes almost half of the region between the body and the leeside bow shock in the hypersonic flow over conical configurations, this flow is still dominated by the surrounding inviscid conical flow. Using this justification for the applicability of the conical approximation to the hypersonic flow over a slender delta wing, Ruffin and Murman (1988) calculated the hypersonic flow over a 75° swept sharp leading-edge delta wing at $M = 10.17$, $\text{Re}_x \cong 3.4 \cdot 10^5$ and at angles of attack up to 15°. The differencing scheme of flux vector splitting with MUSCL, developed by Van Leer and presented by Thomas, Van Leer and Walters (1985), is used. The results of the calculation with a grid of 65 × 105 at an angle of attack of 15° are presented in Figs. 9.32a to 9.32d. The schematic of the flow field structure determined from experimental data is illustrated in Fig. 9.32a. Figs. 9.32b, 9.32c and 9.32d present the cross-flow Mach number contours, the total pressure loss contours and the cross-flow streamlines, respectively. The calculated results simulate reasonably well the main feature of the experimental flow structure.

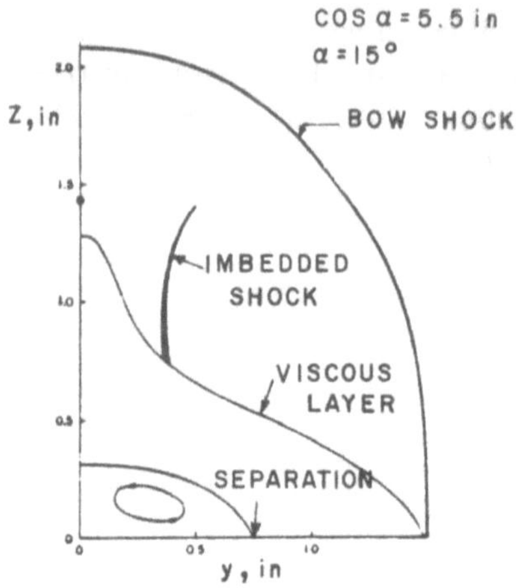

a. Experimental flow field

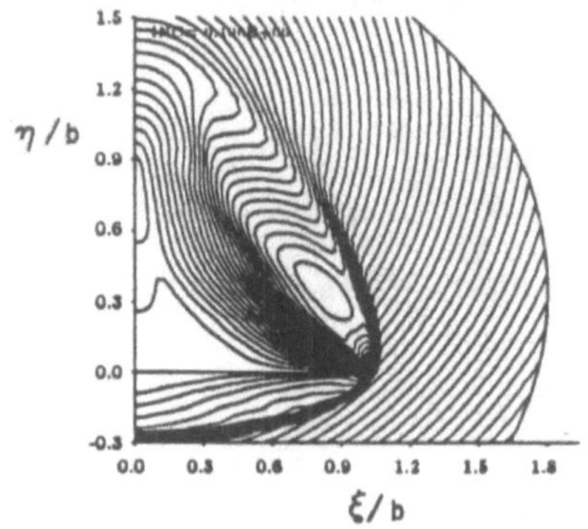

b. Cross-flow Mach number contour

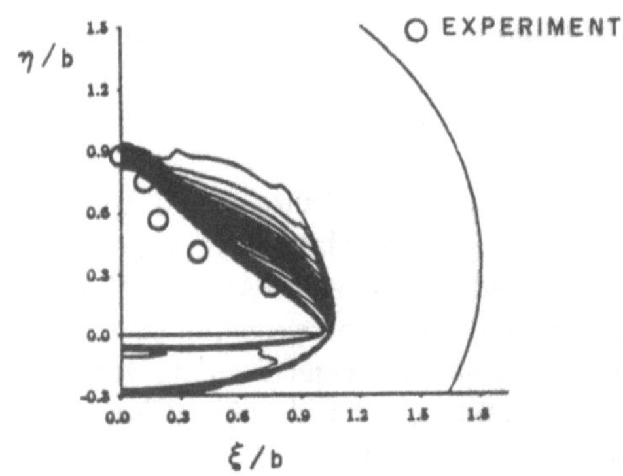

c. Total pressure for contours

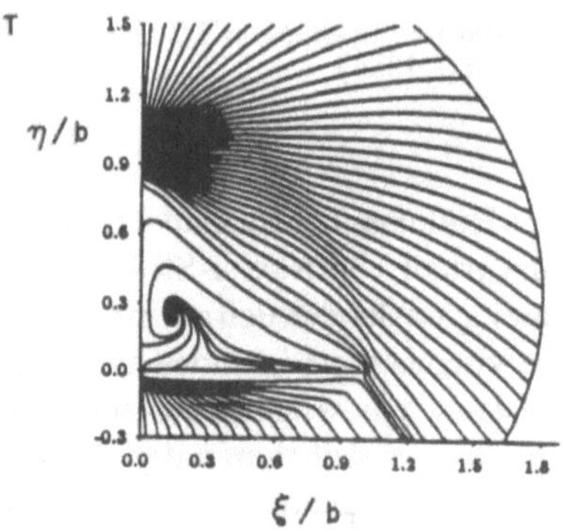

d. Cross-flow streamlines

Fig. 9.32. Calculation of the hypersonic flow over a 75° delta wing at $M = 10.17$, Re=$3.4 \cdot 10^5$ and $\alpha = 15°$ (Ruffin and Murman 1988)

9.5.1.4 Calculation of vortex breakdown on delta type wings. The verifications of the solutions of the Navier-Stokes equations proved that the present methods are able to simulate in a reasonably exact manner the flow field characteristics. Therefore, it is reasonable to expect that the detailed structure of the primary rolled-up vortices can be also obtained in these calculations. This capability was already demonstrated in the calculations of the vortex flow structures at moderate angles of attack, as seen in Figs. 9.24, 9.26, 9.29. It is known that that these leading-edge vortices are strengthened as the flow develops progressively away from the wing apex by continuous feeding of vorticity by the separating shear layers. As the angle of attack is increased, these vortices are strengthened until eventually a sudden change occurs in the nature of the vortex cores. The fluid elements entrained in the core are subject to an accelerating force, and the sudden change in the nature of the core involves the transition of the vortex core from a jet-like flow to a wake-like flow. This sudden change is known as vortex breakdown. The occurrence of breakdown was found to depend on a balance between the magnitude of the swirl and the external pressure gradient. Thus, the pressure gradient on the leeward side of the wing has a very strong effect. A pressure increase downstream drives the point of breakdown upstream towards the apex of the delta wing.

The phenomenological description of vortex breakdown was presented in Section 2.6, where examples of the bubble and the spiral types of vortex breakdown were shown. Since the vortex breakdown is in essence a dynamic, mainly inviscid effect, it was possible already to obtain an indication of this effect in the Euler solutions, as discussed in Chapter 8 (Section 8.2.6.1.1). Since the Euler equations cannot describe the flow structure in the core of the vortex, only an indication as to the occurrence of a sudden change (which can be denoted as "breakdown") is obtained in the Euler calculations, as described in Chapter 8. However, the Navier-Stokes solution includes the viscous-inviscid flow in the core and in the vortex flow surrounding the core, so it is reasonable to expect that the Navier-Stokes solutions will result in a more detailed description of the flow structure with the occurance of the vortex breakdown process. In order to capture the complex flow patterns in the breakdown region, as the vortex bursts at high angles of attack, a very high resolution of the grid is required. Small scale details of this flow, such as the flow structure at the tip of the bubble type breakdown can only be evaluated with high grid resolution. Simulation of vortex flow phenomena, including breakdown effects, requires fine grid resolution and needs large computer storage capability and long execution time. Therefore, various methods of grid adaptation or embedded grid approaches can be used to reduce the need for computer resources.

As a first example, the vortex breakdown on the AR=1 delta wing at $\alpha = 40^o$ is observed, which can be seen in the pattern of the trajectories of the vortices shed from the leading-edge, as shown in Fig. 9.33. The occurrence of vortex breakdown is seen in the shape of the vortex trajectories.

The vortex trajectories are first confined into a rolled-up core in the forward part of the wing; then, at about 0.4–0.45 of the chord, the core is suddenly expanded, destroying the core and resulting in breakdown into a bubble-shaped region of reversed flow.

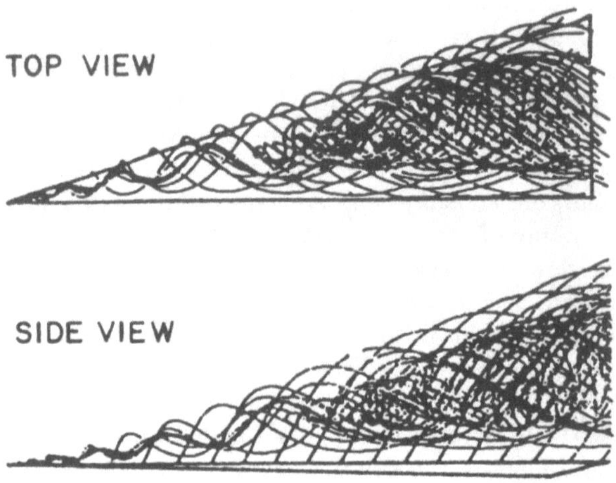

Fig. 9.33. Vortex trajectories over the AR=1 delta wing showing vortex breakdown at $\alpha = 40^o$, Re=$0.9 \cdot 10^6$ with $65 \times 101 \times 81$ grid points (Hartwich et al. 1988)

The vortex breakdown on a 75° swept delta wing (AR=1.07) was studied by Ekaterinaris and Schiff (1990). The vortex breakdown first appears on this wing at an angle of attack of 32°. The flow with the vortex breakdown was computed with a $57 \times 54 \times 70$ point cylindrical grid using the zonal method. The calculated shapes of the leading-edge vortex core and of the bubble type breakdown are shown in Fig. 9.34. The vortex core is shown by the streamlines starting from the wing apex, and the general shape of the vortex region around the core is shown by the stream surface originating at the leading-edge near the apex. The vortex breakdown is shown with the stream surface, which has a closed front end, indicating the tip of a bubble type breakdown; it then expands downstream of the burst point with the bubble remaining open in the wake region. At a higher angle of attack of 40°, the calculation is made using a spherical, single block grid of $112 \times 105 \times 70$ points. The vortex flow and breakdown region are shown in Fig. 9.35. The flow here is very similar to that obtained for $\alpha = 32^o$, except that the burst point is located further upstream towards the apex and the burst is larger.

Numerical experiments reported by Ekaterinaris and Schiff (1990) indicate that using a coarse grid of $33 \times 30 \times 35$ points for calculations of

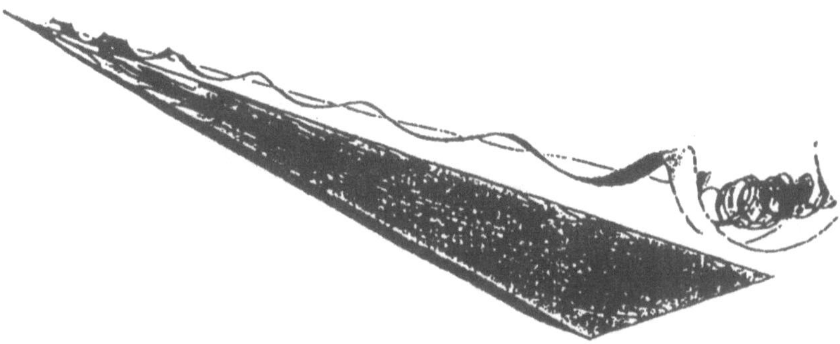

(a) Perspective view

(b) Detail of the vortex breakdown bubble

Fig. 9.34. Vortex core and vortex breakdown region on a 75° swept delta wing at $M = 0.3$, Re=10^6 and $\alpha = 32°$ with $112 \times 105 \times 70$ grid points (Ekaterinaris and Schiff 1990)

the flow field over the 75° delta wing at angles of attack above 35°, vortex breakdown was not observed. With somewhat better grid resolution, but still using a relatively coarse grid of $33 \times 54 \times 35$ points, the breakdown is observed. However, in the calculations with the coarse grid resolution, the position of the burst point is predicted to be further downstream and the extent of the vortex breakdown is much smaller than calculated with the fine grid resolution. This study of Ekaterinaris and Schiff also calculated the strong effect of the adverse axial pressure gradient on the vortex breakdown. It was shown that eliminating the effect of the adverse axial pressure gradient by excluding the wake region from the computations, vortex breakdown was eliminated from this calculations. An additional investigation involved the possible existence of unsteadiness in the region downstream of the tip of the bubble type breakdown. No such unsteadiness was observed for the bubble type vortex breakdown on the delta wing. However, there is evidence of self-excited unsteady flow behavior in the case of the spiral type breakdown observed on a double delta wing. The spiral type vortex breakdown observed on the 80°/60° double delta wing at an angle of attack of 35° is shown in Fig. 9.36.

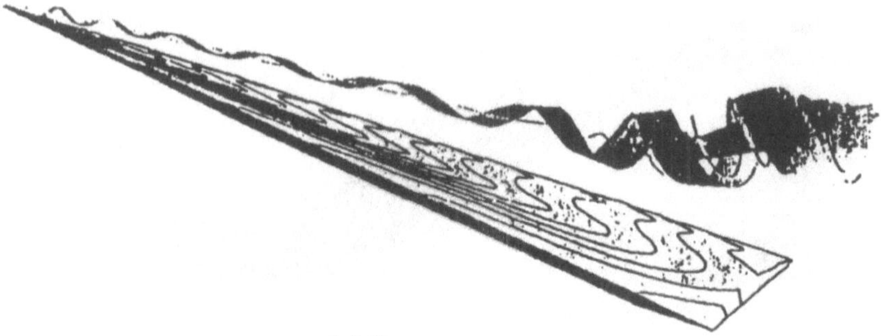

(a) Perspective view

(b) Detail of the vortex breakdown bubble

Fig. 9.35. Vortex core and vortex breakdown region on a 75° swept delta wing at $M = 0.3$, Re=10^6 and $\alpha = 40°$ with $112 \times 105 \times 70$ grid points (Ekaterinaris and Schiff 1990)

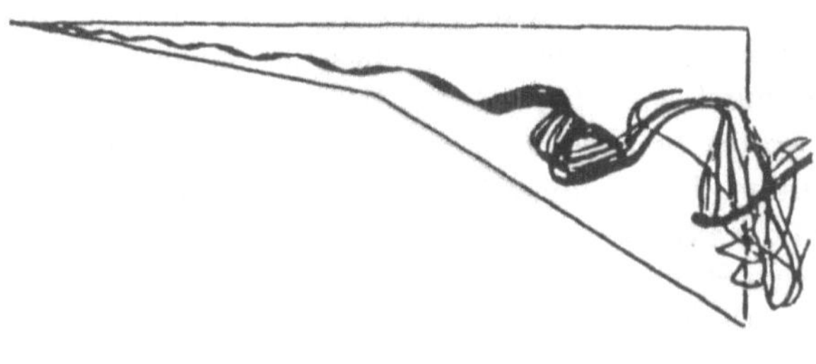

Fig. 9.36. Spiral type vortex breakdown on a double delta wing at $M = 0.3$, Re=$1.3 \cdot 10^6$ and $\alpha = 35°$ (Fujii and Schiff 1989)

9.5.1.5 Flow over Canard-Wing-Body Configuration. The calculations of the aerodynamic characteristics of the close-coupled wing-canard configuration at high angles of attack by the panel methods, NLVLM, and by the Euler solution were discussed in Chapters 7 and 8, respectively. It is most interesting now to observe the results of such calculations utilizing the Navier-Stokes solution. The thin-layer Navier-Stokes equations were solved by Tu (1991) for the flow over a coplanar close-coupled canard-wing-body configuration at transonic Mach number of 0.9 and at angles of attack of up to 12°. This calculation utilized the NASA Ames Research Center Transonic Navier-Stokes (TNS) code. The TNS code is a Reynolds-averaged thin-layer Navier-Stokes solver with structured zoning capability and its wing-fuselage version (TNSWF) has been successfully used to compute the flow about a F-16A wing-fuselage (Flores et al. 1988) and a complete F-16A aircraft (Flores and Chaderjian 1990). The Baldwin-Lomax turbulent model with the Degani-Schiff modification for the massive separation flow is used in this TNS code. The computations were performed on a Cray-YMP or a Cray-2 supercomputer and required 5 to 30 hours of CPU time, depending on the geometry, flow conditions and grid resolution.

The grid for the canard-wing-body configuration is generated as a single block H-O topology grid with $152 \times 32 \times 97$ points, shown in Fig. 9.37a, and the surface grid is presented in Fig. 9.37b.

The effect of the canard on the flow pattern on the wing is well illustrated in the surface flow pattern shown in Fig. 9.38. The rolled-up vortex of the canard influences the flow on the wing, affecting both the strengths and positions of the primary vortex separation and the reattachment and position of the secondary separation line (S_2), which can be seen by comparing the surface flow pattern for the canard-off case with the canard-on flow pattern. The interaction between these vortices is illustrated by the particle traces calculated on the canard-off and canard-on wing-body configuration at $\alpha = 12.38^o$ and is shown in Fig. 9.39. The wing leading-edge vortex on the canard-off configuration is seen to breakdown at the trailing edge of the wing, while with the canard-on, the interaction with the canard vortex stabilizes the wing vortex, delaying breakdown well beyond the trailing edge of the wing.

The pressure distributions are calculated at various spanwise stations on the wing using a refined grid, and the results are compared with experimental data. Results of such calculations for the canard-off and canard-on configurations at $M = 0.9$, Re$=1.52 \cdot 10^6$ and at $\alpha \cong 12^o$ are shown in Fig. 9.40. Good agreement is obtained between the calculated and experimental results for both the canard-off and on conditions, except near the wing tip in the canard-off configuration where the calculated results fail to predict the high suction peak of the primary vortex.

The calculated variation of the aerodynamic lift and drag forces and the pitching moment coefficients as a function of angle of attack are compared with experimental data in Fig. 9.41. There is a very good agreement between

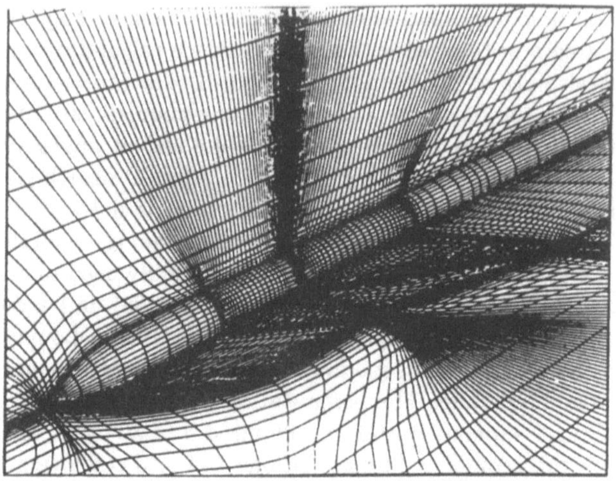

Fig. 9.37a. Canard-wing-body flowfield grid topology (Tu 1991)

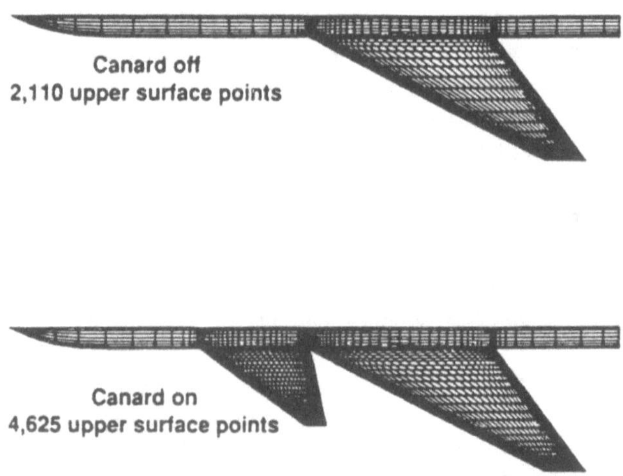

Fig. 9.37b. Surface grid for the wing-body configuration with and without canard (Tu 1991)

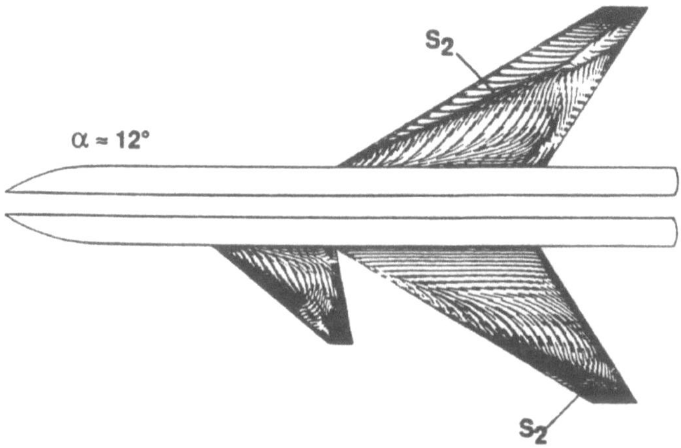

Fig. 9.38. Upper surface flow pattern for the wing-body configuration with and without canard for $M = 0.9$, Re=$1.52 \cdot 10^6$ and $\alpha = 12^o$ (Tu 1991)

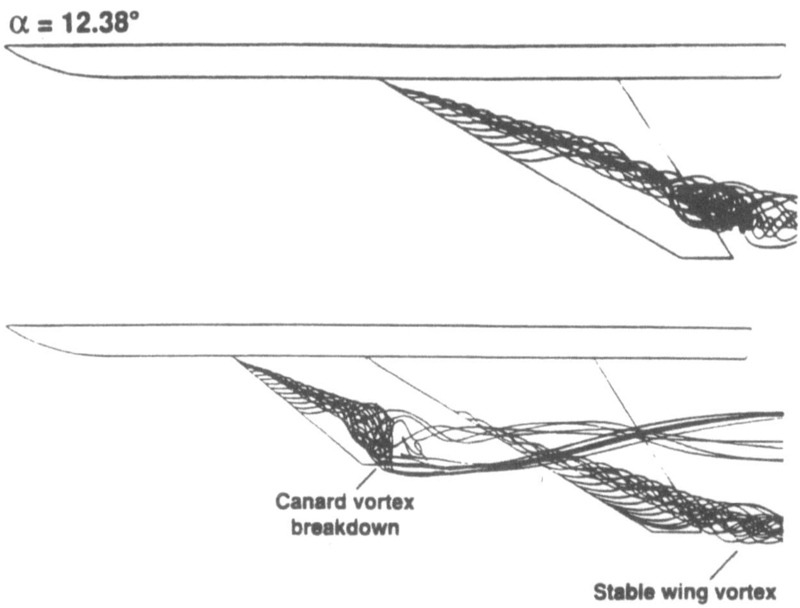

Fig. 9.39. Particle traces for the wing-body configuration with canard off and canard on at $M = 0.9$, Re=$1.52 \cdot 10^6$ and $\alpha = 12^o$ (Tu 1991)

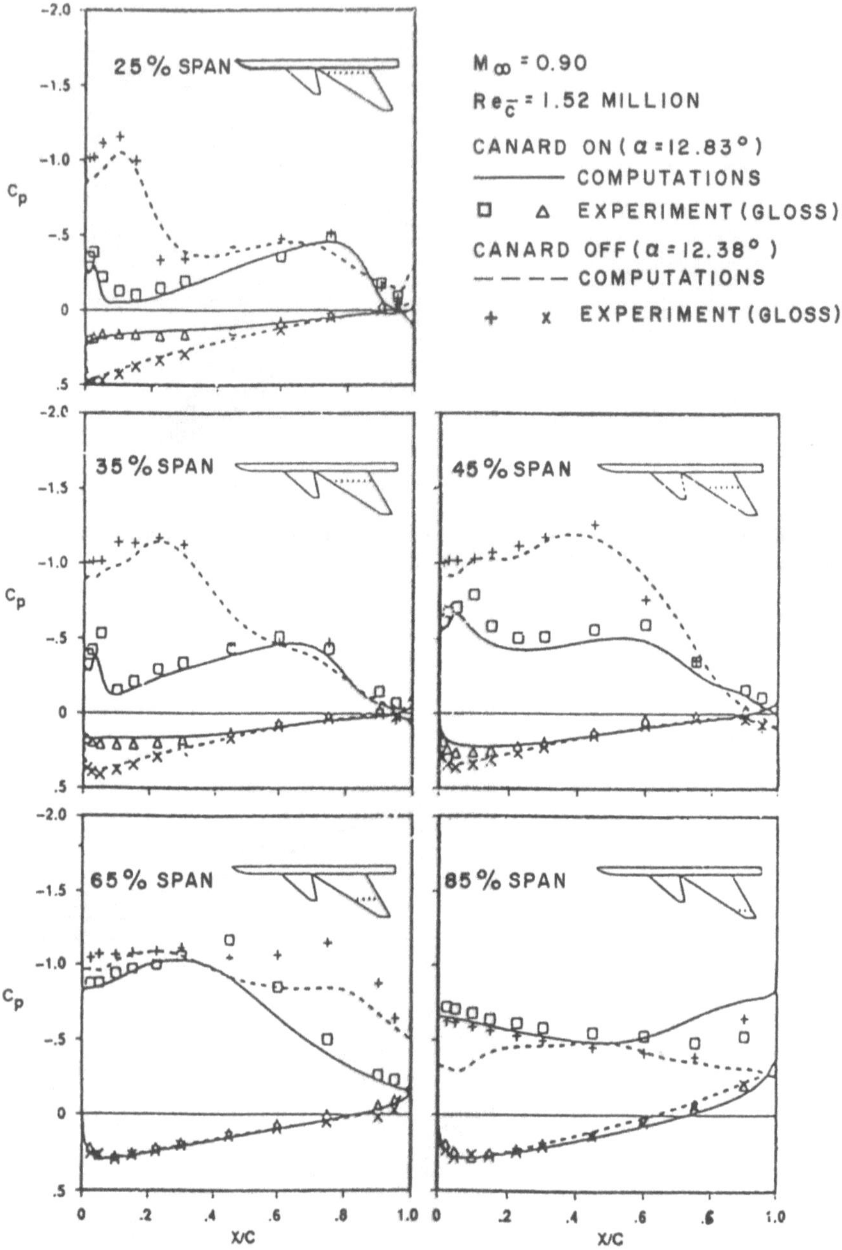

Fig. 9.40. Comparison of computed and experimental surface pressure coefficients for the refined grid with and without canard at $M = 0.9$, $\mathrm{Re}=1.52 \cdot 10^6$ and $\alpha \cong 12^\circ$ (Tu 1991)

the calculated and experimental results for the canard-on configuration for this range of angles of attack. It will be very interesting to examine the results of the calculations on the canard-wing-body configuration at higher angles of attack.

9.5.1.6 Calculations on Aircraft Configurations. It is the aspiration of the aerodynamic community to be able to use the computer as a numerical wind tunnel. It is hoped to make computations which will be able to simulate flight conditions and give results of comparable quality and accuracy to presently available wind tunnel data. The approximations needed to reduce the flow equations to solvable formulations, such as the various potential methods including the Euler solutions, cannot be expected to simulate all the physical phenomena obtained in the flow of real viscous fluids. The advancements in the last few years in the solutions of the Navier-Stokes equations with reasonably accurate boundary conditions on complete aircraft configurations have raised hopes that the new supercomputer laboratory may give results of quality comparable to those obtained in the wind tunnel laboratory. The results of the Navier-Stokes solutions on simple configurations, such as axisymmetric bodies and delta wings with and without strakes, are certainly close to those measured in the wind tunnel. We saw in the previous sections that the Navier-Stokes solutions of flow on these bodies are able to show effects of three-dimensional primary, secondary and higher order separations, free vortical flows and vortex breakdown. The calculated pressure distributions and aerodynamic coefficients on these simple geometric configurations are in reasonably close agreement with experimental data, so as to enable us to rely on the computed results with only a few verification tests. This step can represent a great savings in time and costs in aerodynamic design projects, making the Navier-Stokes computations faster and cheaper.

A major effort towards the development of a credible program for the solution of the Navier-Stokes equations on a fighter aircraft configuration is now underway as part of the NASA High Alpha Technology Program in the NASA laboratories. The program includes flight tests of the F-18 High Alpha Research Vehicle (HARV) combined with comparison with the results of the aircraft model testing in various wind tunnels, some of which are reported by Del Frate and Zuniga (1990). Very good agreement is found between the flight test measurements on the F-18 HARV aircraft, shown in Fig. 9.42, and the model testing in wind tunnels and visualization experiments in water tunnels. Therefore, the capability of the Navier-Stokes computations remains to be examined in order to simulate these experimental results at high angles of attack.

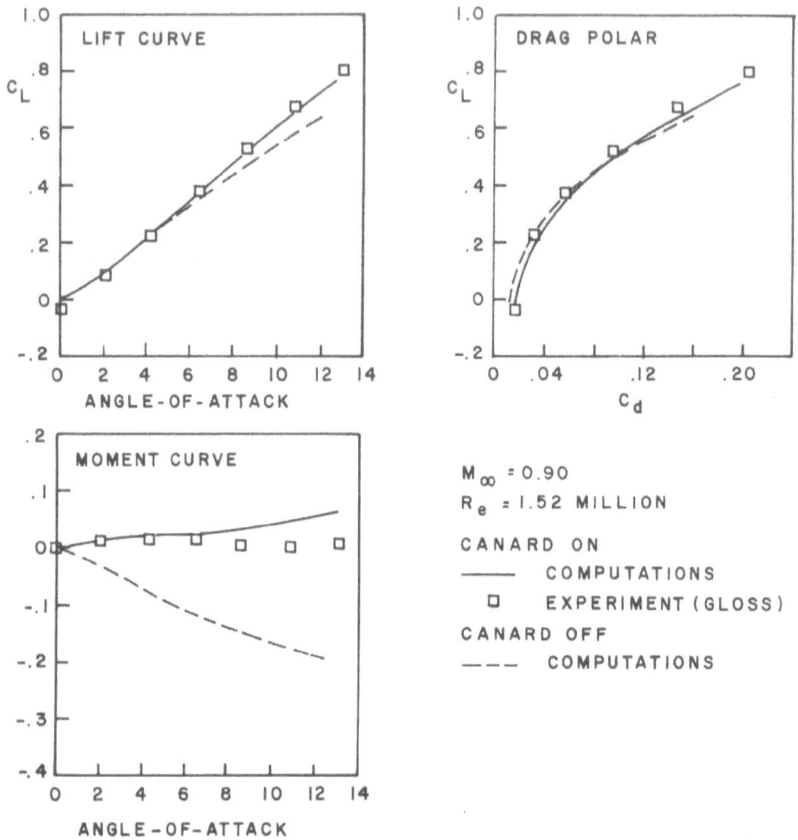

Fig. 9.41. Comparison of computed and experimental forces and moment coefficients on the wing-body configuration with and without canard at $M = 0.9$, Re=$1.52 \cdot 10^6$ (Tu 1991)

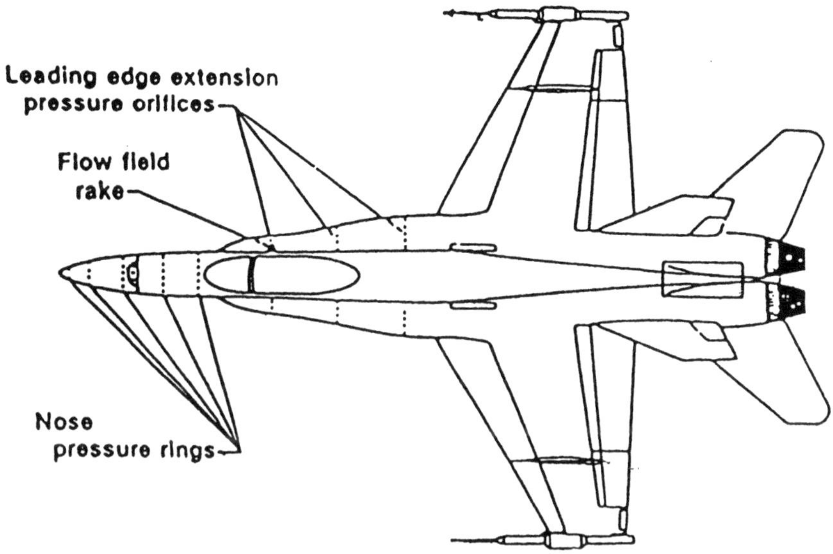

Fig. 9.42. Top view of the F-18 HARV aircraft instrumented for surface pressure and oil flow (Rizk, Schiff and Gee 1990)

It is by now well established that high angle of attack flows are characterized by regions of three-dimensional separated flows which leave the surfaces and roll up into concentrated vortices. Very high grid resolution is required to capture numerically the detailed structure of these three-dimensional separated flows close to the surfaces and in the free stream. The availability of the current generation of supercomputers, such as the Cray-2 and Cray-YMP, has permitted a significant increase in the size of the computational grids. Therefore, it is possible to compute high angle of attack flows over bodies and aircraft components with codes based on the Reynolds-averaged Navier-Stokes equations utilizing the large number of grid points which are required to resolve adequately the main features of the vortical flows at high angles of attack. In conjunction with the increased capability of supercomputers, more effective numerical methods are being developed. As part of the High Alpha Technology Project, a well directed effort is underway at NASA Ames and Langley Research Centers aimed at developing the numerical tools for the computations of the flow on the F-18 configuration. The numerical program being developed at NASA Ames is based on time-marching Navier-Stokes computations, utilizing the three-dimensional partially flux-split Navier-Stokes code, the F3D algorithm, as reported by Steger et al. (1986), and as previously described in Sec. 9.3. The Chimera overset grid approach, described in Sec. 9.4, is applied to fit the complex configurations such as the F-18 aircraft. The F-18 computation program carried out at NASA Langley uses an algorithm for the Navier-Stokes equations which incorporates an upwind-biased, flux-difference-splitting approach along with longitudinally-patched grids, as reported by Thomas and Newsome (1986).

Subsonic and transonic flows calculations. At first, calculations of the flow about the fuselage forebody and wing leading-edge extension (LEX) of the F-18 HARV aircraft at conditions matching those of the flight tests were performed using the F3D code, as reported by Schiff et al. (1989). In order to simulate well the LEX section, a double-block grid is used in preference to a single-block grid. The forward grid, with $36(axial) \times 51(circumferential) \times 50(radial)$ points, covers the forebody, while the rear grid, with $35 \times 83 \times 50$ points, starts at the origin of the LEX and extends downstream. The surface flow patterns were calculated for laminar and turbulent flows at $M = 0.2$, and an angle of attack $\alpha = 30^o$ and with Reynolds number based on the mean aerodynamic chord, $Re_c = 0.8 \cdot 10^6$ (for the laminar flow) and $Re_c = 11.52 \cdot 10^6$ (for the turbulent flow). The calculated laminar and turbulent flow patterns are shown in Figs. 9.43a and 9.43b, respectively.

In the laminar flow case, Fig. 9.43a, a primary separation line is seen extending from the nose tip to near the origin of the LEX with a secondary separation line further above on the forebody. On the rear part downstream of the LEX, the flow pattern is more complex. The flow separates circumferentially at the LEX leading-edge and forms the strong primary LEX vortices. The primary LEX vortex induces a flow spanwise outward on the leeside of the LEX, which separates at a secondary separation line extending the length of the LEX and rolls up into a secondary vortex. In addition, the primary LEX vortex induces a secondary crossflow separation on the side of the fuselage above the LEX. The flow on the fuselage below the LEX is also strongly affected. Primary and secondary separation lines are seen on the side of the fuselage under the LEX.

The calculations using the F3D code were also performed for the turbulent flow case with the Baldwin-Lomax eddy-viscosity model modified by Degani-Schiff for separation effects. The flow was assumed to be turbulent over the entire length of the forebody, and no transition model was used. The calculated flow pattern on the surface of the forebody with the LEX is presented in Fig. 9.43b. The distinct differences between the laminar and turbulent flows shown in the separation lines on the surface of the forebody, which are displaced towards the leeward side of the body in comparison with the corresponding lines in the laminar case. In addition, the secondary separation line on the leeward surface of the LEX is displaced outwards in comparison to that line in the laminar flow.

A parallel calculation of the flow over the forebody and LEX of the F-18 with comparison with the flight tests was performed at NASA Langley. The patched-grid algorithm for the F-18 configuration is described by Thomas et al. (1989), and some results of these calculations on the forebody LEX are presented by Ghaffari et al. (1989).

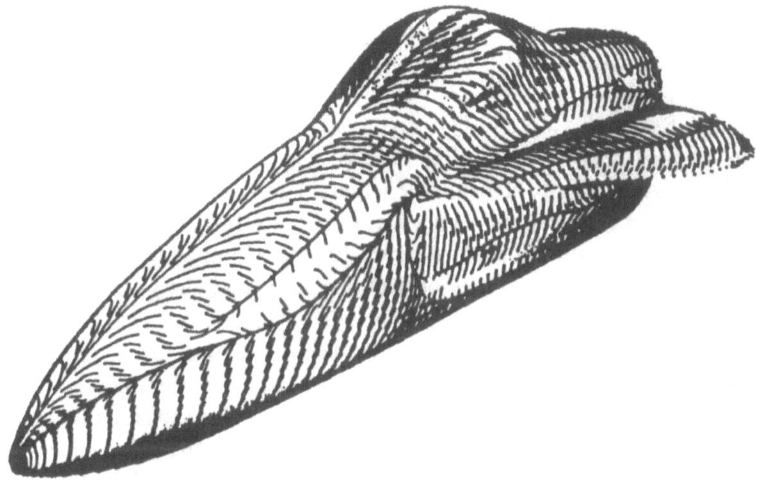

a. Laminar flow pattern, $M = 0.2$, $\mathrm{Re}_c = 0.8 \cdot 10^6$

b. Turbulent flow pattern, $M = 0.2$, $\mathrm{Re}_c = 11.52 \cdot 10^6$

Fig. 9.43. Calculated surface flow patterns on the forebody of the F-18 HARV aircraft at $\alpha = 30°$ (Schiff et al. 1989)

These surface flow patterns can be compared with both visualization experiments in wind tunnels and flight tests. The laminar calculation results have a good qualitative agreement with the experimental surface oil patterns obtained in the Basic Aerodynamic Research tunnel at the NASA Langley Research Center. The turbulent calculation results are compared with surface flow visualization photos taken on the F-18 HARV at the NASA Dryden Flight Research Facility. The flight tests results obtained under the same conditions as used in the computations were found to agree well with the calculated surface flow patterns. Comparison of the computed turbulent surface streamlines on the forebody and the apex of the LEX is shown in Fig. 9.44 and 9.45.

Further advances in the Navier-Stokes calculations of the flow field around the F-18 (HARV) aircraft were made by Cummings et al. (1990), by Rizk, Schiff and Gee (1990) and by Rizk and Gee (1991). The prediction of the flow field on the wing and fuselage (without the tail surfaces) of the F-18 (HARV) was presented by Cummings et al. (1990). Calculations of the flow over an improved configuration including the forebody, LEX, faired-over inlet, wing and deflected wing leading-edge flaps, was presented by Rizk, Schiff and Gee (1990). Computations of the steady and unsteady flows over the complete aircraft configuration, which is obtained by extending the previous configuration also to include both the horizontal and vertical tail surfaces, the leading- and trailing-edge flaps on the wing, the moveable control surfaces and the boundary layer divertor vent, was recently presented by Rizk and Gee (1991). The Reynolds-averaged Navier-Stokes equations are solved using the F3D code. The representation of the aircraft surface geometry required certain simplifications, but the essential features of the complex configuration were retained. The required multi-block grid is the Chimera overset grid scheme, allowing treatment of the closely coupled components of the configuration. A generalized overset/patch zonal grid interfacing method is added to the basic Chimera scheme to increase the range of application.

The complex geometry of the F-18 represents a challenge to grid generation. Some simplifications are made, such as the elimination of small perturbances (e.g. antenna fairings), as well as simulation of the inlet flow and the exhaust plumes. The simplified surface grid for the computational model of the F-18 is shown in Fig. 9.46. It consists of the forebody, LEX, faired-over inlet, wing, deflected leading-edge flaps, vertical and horizontal tails, and idealized boundary layer divertor vent.

The grid system for the F-18 consists of ten grids, shown in an overview in Fig. 9.47. The front fuselage grid is of a C-O type while the rear fuselage is of the H-O type. These relatively fine grids extend only two wing roots away from the body surface. A coarse outer grid of the C-O type is added, which extends to about seven chord lengths away from the body surface. Separate grids of the C-H type are used for the wing, the leading-edge flaps and the vertical and horizontal tails.

Computational results Flight test

Fig. 9.44. Comparison of computed turbulent surface streamlines with flight test – side view, at $M = 0.34$, Re=$13.5 \cdot 10^6$ and $\alpha = 19^o$ (Ghaffari et al. 1989)

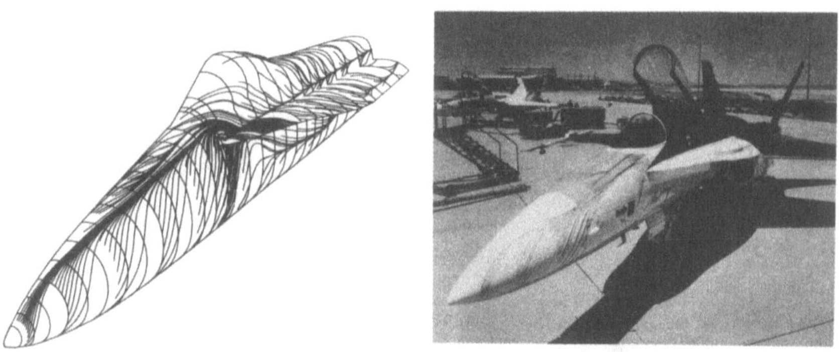

Computational results Flight test

Fig. 9.45. Comparison of computed turbulent surface streamlines with flight test at $M \cong 0.3$, Re=$10 \cdot 10^6$ and $\alpha = 30^o$ (Thomas et al. 1989)

Using the F3D code and assuming bilateral symmetry with ten grids consisting of 0.9 million points around half of the configuration allows for the computation of the aerodynamic characteristics of the F-18 aircraft. Conditions corresponding to the flight test data were selected for the turbulent flow calculations $M = 0.243$ and angle of attack of 30.3^o with Reynolds number of $11 \cdot 10^6$ (based on the mean aerodynamic chord of the wing). At this angle of attack, the inboard and outboard leading-edge flaps are deflected 33° nose down and the horizontal tail is deflected 7° nose down. The code was run on the Cray-YMP at NASA-Ames in thenontime-accurate mode for 1000 steps and in a time-accurate mode for an additional 4100

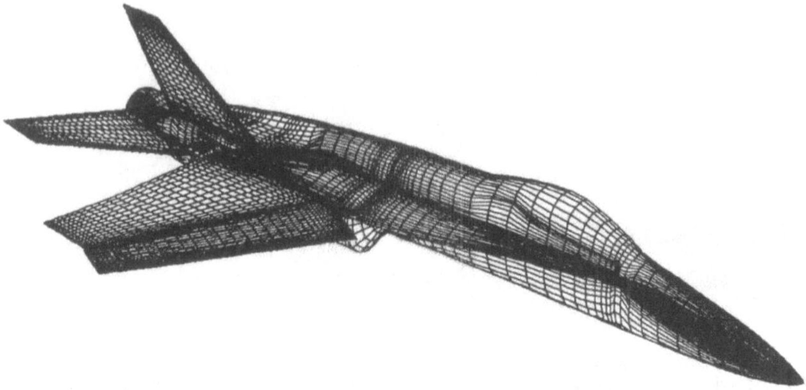

Fig. 9.46. Surface grid of the F-18 (Rizk and Gee 1991)

Fig. 9.47. Overall view of the grid system for the F-18 (Risk and Gee 1991)

time steps. Risk and Gee (1991) report that the solution required $8MW$ of core memory and took a total of about 55 hours of CPU time.

The computed pressure distributions at stations located on the forebody and LEX are compared to flight test data in Fig. 9.48. In general, there is a good qualitative and even a reasonable quantitative agreement between the computed and experimental results. Some of the discrepancies can be attributed to the lack of a transition zone in the present turbulent model (at stations 3 and 4) and the local effect of antenna (station 4). At station 8, there is a drop in suction predicted in the computation results, which is due to the LEX vortex breakdown at this station. Improved accuracy and better agreement with experimental data may probably be achieved by using a finer mesh in the computations.

The computed surface streamlines and the LEX vortex breakdown are presented in Figs. 9.49 and 9.50. These results are in good agreement with flight test visualization data.

The capability of this Navier-Stokes solution to predict the breakdown of the LEX vortex so well and to show the details of the three-dimensional separated flow over the various parts of the F-18 configuration so accurately is a very important breakthrough in the application of CFD methods in aerodynamic design technology.

In addition to the extensive F-18 calculations program, there are several calculations for other aircraft configurations. A computation of the Navier-Stokes equations of the flow at transonic speeds over a simplified F-16 aircraft configuration at low angles of attack is presented by Flores et al. (1987). Some calculations of steady and unsteady flows over a partial model of the F-15 aircraft are presented by Kwon and Sankar (1991). The calculated particle trace on the F-15 wing-body-inlet configuration is compared with experimental results in Fig. 9.51. Good qualitative agreement is once again found.

Supersonic and hypersonic flows calculations. The capabilities of the Navier-Stokes solutions to evaluate aerodynamic characteristics, including surface heat transfer coefficients are demonstrated in a paper by Narain, Muramoto and Lawrence (1991). The Navier-Stokes equations are solved using an adaptation of the three-dimensional code, UPS (Upwind Parabolized Solver), developed by Lawrence et al. (1987). The UPS code employs an upwind algorithm (based on the Chakravarthy and Szema (1985) modification of the Roe (1983) scheme) in both cross-flow directions for enhanced shock capturing characteristics. The UPS code has an implicit finite volume algorithm which is second-order accurate in the cross-flow direction and first-order accurate in the streamwise direction. The presence of a streamwise pressure gradient allows upstream propagation of the signals in the subsonic sublayer of the flow. The original UPS code has been modified in terms of step size, initiation of the marching process, redefinition of the outer boundary and asymmetric flow due to yaw, and for hypersonic applications, the addition of mass injection at the surface to simulate cooling

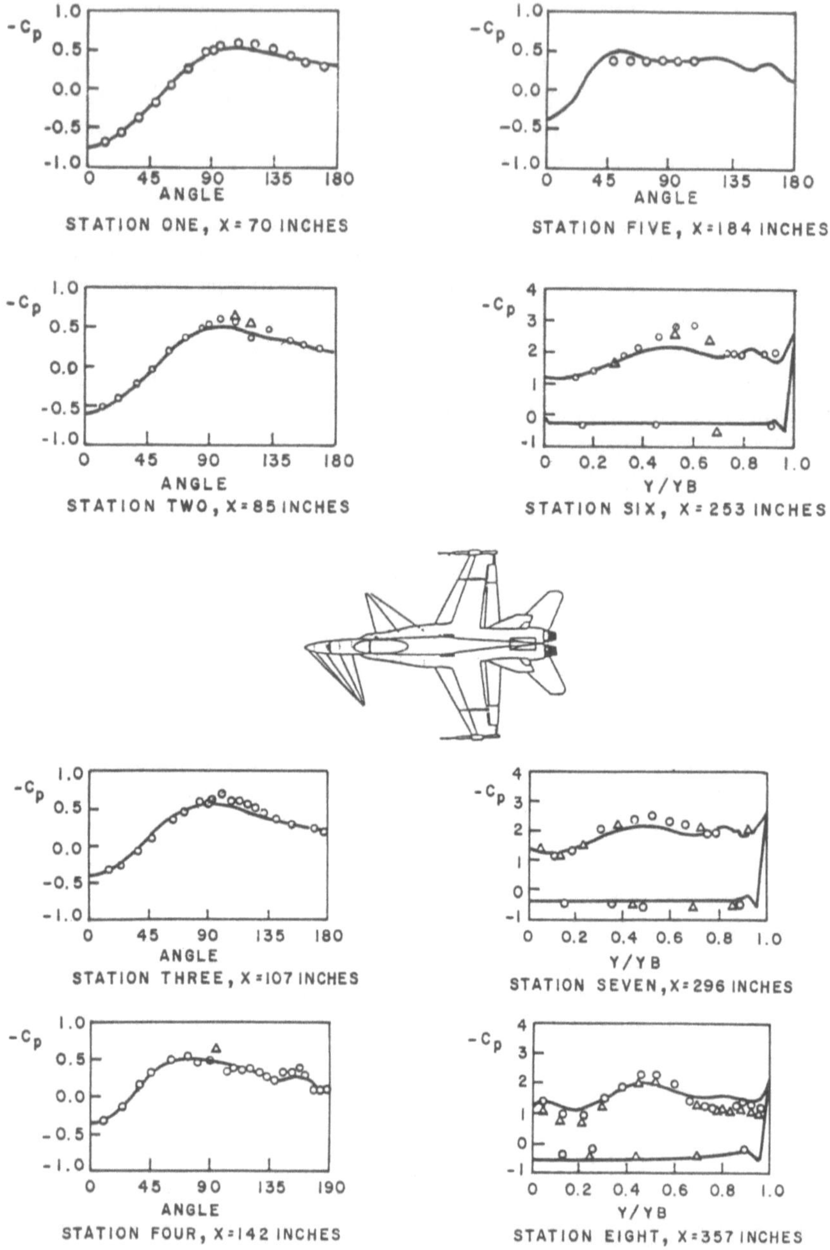

Fig. 9.48. Comparison of computations and flight test data of pressure distributions on F-18 aircraft at $M = 0.243$, Re=$11 \cdot 10^6$ and $\alpha = 30.3°$ (Rizk and Gee 1991)

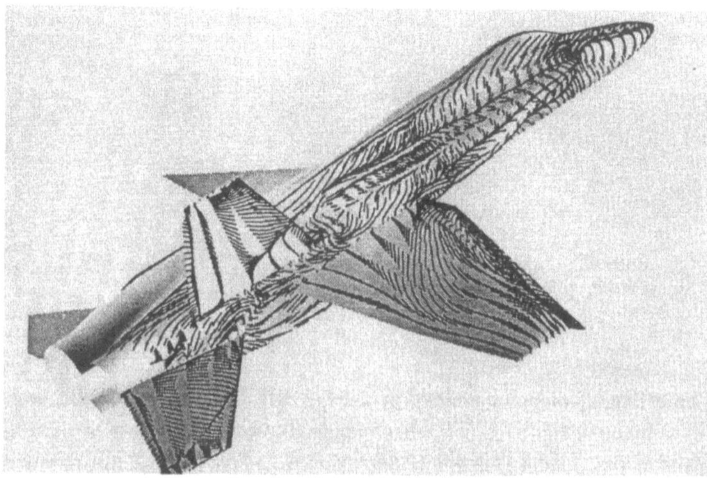

Fig. 9.49. Computed limiting surface streamlines on F-18 at $M = 0.243$, Re=$11 \cdot 10^6$ and $\alpha = 30.3^o$ (Rizk and Gee 1991)

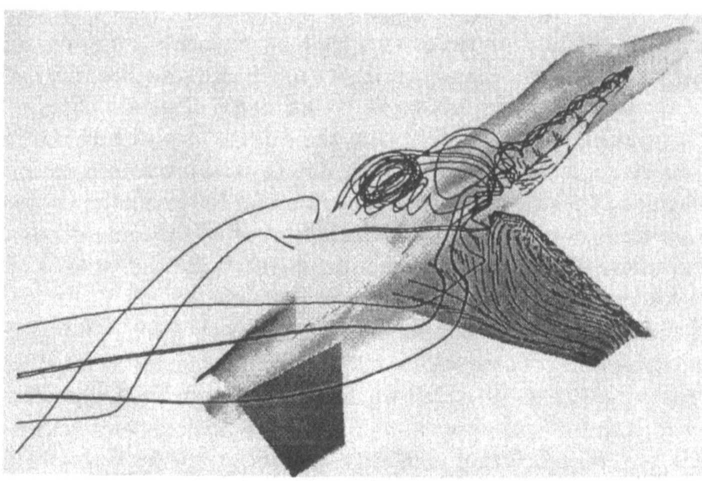

Fig. 9.50. Computed particle traces from the LEX of F-18 showing LEX vortex breakdown and surface flow pattern at $M = 0.243$, Re=$11 \cdot 10^6$ and $\alpha = 30.3^o$ (Rizk and Gee 1991)

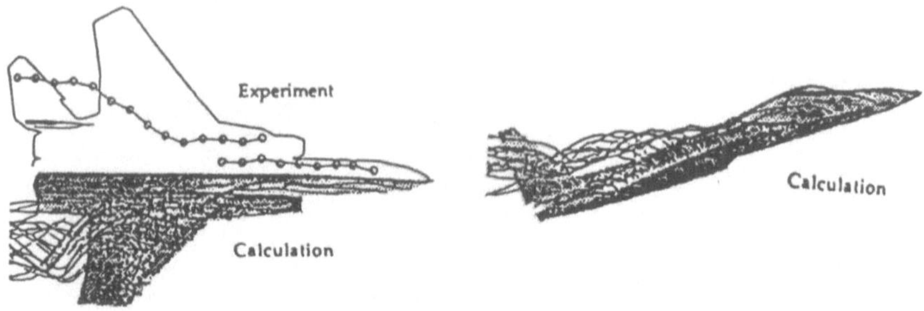

Fig. 9.51. Particle trajectories for flow over an F-15 wing-body-inlet configuration at $M = 0.15$, $\alpha = 20^o$ (Kwon and Sankar 1991)

due to ablation, film cooling etc.

The calculations are made on a high lift-to-drag ratio hypersonic vehicle known as FDL-5A, for which experimental results are available for code validation. The FDL-5A configuration, used in the computation, has a planform with a leading-edge sweep of 81°, as sketched in Fig. 9.52, (the vertical fin and a set of trailing elevons are omitted in the computations). The calculated pressure distributions on the centerline of the body with various rates of injection are presented in Fig. 9.52. The calculated results obtained with zero injection are in a good agreement with the experimental data at $M = 10.2$, Re=$2.1 \cdot 10^6$/ft and $\alpha = 5^o$. The comparison of the calculated lift-to-drag ratios and pitching moments variation with angle of attack at $M = 7.98$ and Re=$2.29 \cdot 10^6$/ft are presented in Fig. 9.53. Computational results are obtained for an angle of attack of up to 10°. Apparently, at higher angles of attack, the calculations with this UPS code do not converge, a fact which may be due to streamwise separation. It is then concluded that the UPS code is very useful for calculations of viscous hypersonic flows over vehicle configurations at small-up-to-moderate angles of attack, where flow separation is not experienced. Further studies of codes suitable for separated viscous hypersonic flows are needed.

We should expect new developments in the application of Navier-Stokes solutions to aircraft configurations, mainly involving improved accuracy and better computer economy, which would be a result of the high alpha technology project now underway at the various NASA centers. The present limitation on grid resolution manifests itself in errors in the calculation of pressure distributions, which may affect the accuracy of the evaluation of moment coefficients in the longitudinal and lateral directions. The present limitation on the grid resolution also requires imposing symmetry conditions on the flow. Therefore, the very important effects at high angles of at-

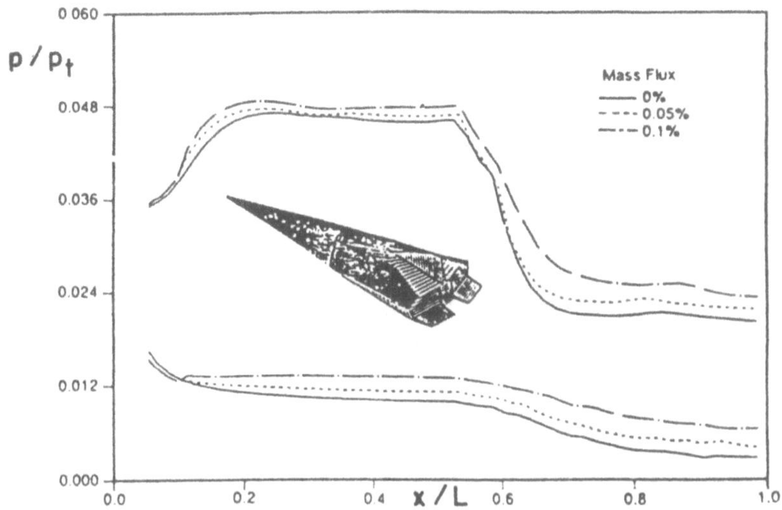

Fig. 9.52. Calculated axial pressure distribution for the FDL-5A configuration with various mass injection at $M = 10.2$, Re=$2.1 \cdot 10^6$/ft and $\alpha = 5°$ (Narain, Muramoto and Lawrence 1991)

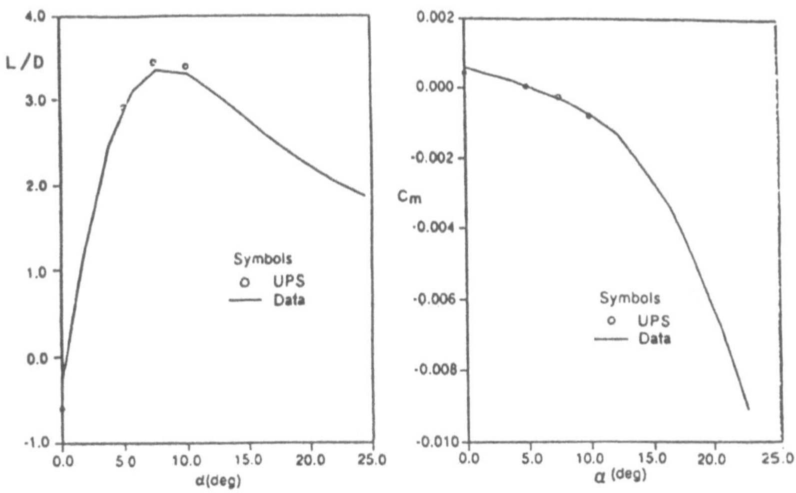

Fig. 9.53. Comparison of the calculated and measured lift-to-drag ratios and pitching moments on the FDL-5A configuration at $M = 7.98$, Re=$2.29 \cdot 10^6$/ft (Narain, Muramoto and Lawrence 1991)

tack of the appearance of asymmetric vortex structures with the associated lateral forces and moments are not calculated at present. It is clear that in order to transform the Navier-Stokes solvers into tools for aerodynamic designers we require a drastic reduction in supercomputer computation time by at least an order of magnitude lower than the tens of hours of the CRAY-YMP required for the present F-18 calculations. However, having set these high goals, we should admire the tremendous progress in the utilization of Navier-Stokes solutions that we have witnessed over the past few years and can expect that with the introduction of faster supercomputers and with parallel computing, it will be possible to make Navier-Stokes calculations in reasonable periods of time. We will then still be confronted with the selection of the appropriate turbulent model for the computation of that model, and we will need an even bigger and faster computing system.

References

1. Atkins, H.L. (1991), "A Multi-Block Multigrid Method for the Solution of the Euler and Navier-Stokes Equations for Three-Dimensional Flows", AIAA Paper 91-0101.

2. Baldwin, B.S. and Lomax, H. (1978), "Thin Layer Approximation and Algebraic Model for Separated Turbulent Flows", AIAA Paper 78-257.

3. Bannink, W.J. and Nebbeling, C. (1978), "Measurements of the Supersonic Flow Field Past a Slender Cone at High Angles of Attack", *High Angles of Attack Aerodynamics*, AGARD CP-247, Paper 22.

4. Barth, T.J. (1991), "Numerical Aspects of Computing Viscous High Reynolds Number Flows on Unstructured Meshes", AIAA Paper 91-0721.

5. Beam, R.M. and Warming, R.F. (1978), "An Implicit Factored Scheme for the Compressible Navier-Stokes Equations", AIAA Journal, Vol. 16, No. 4, pp. 393–402.

6. Benek, J.A., Buning, P.G. and Steger, J.L. (1985), " A 3-D Chimera Grid Embedding Technique", AIAA Paper 85-1523.

7. Benek, J.A., Donegan, T.L. and Suhs, N.E. (1987), "Extended Chimera Grid Scheme with Application to Viscous Flows", AIAA Paper 87-1126.

8. Bluford, G.S. (1979), "Numerical Solution of the Supersonic and Hypersonic Viscous Flow Around Thin Delta Wings", AIAA Journal, Vol. 17, No. 9, pp. 942–949.

9. Brenner, G., Da Costa, J.L., Niederdrenk, P. and Devezeaux, D. (1991), "Simulation of Viscous Hypersonic Flows Past Hyperboloid Configurations with Flare", AIAA Paper 91-0393.

10. Buelow, P. (1989), "Comparisons of TVD Schemes Applied to the Navier-Stokes Equations", AIAA Paper 89-0847.

11. Buelow, P.E., Tannehill, J.C., Ievalts, J.O. and Lawrence, S.L. (1990), "A Three-Dimensional Upwind Parabolized Navier-Stokes Code for Chemically Reacting Flows", AIAA Paper 90-0394.

12. Chaderjian, N.M. (1991), "Numerical Algorithm Comparison for the Accurate and Efficient Computation of High-Incident Vortical Flow", AIAA Paper 91-0175.

13. Chakravarthy, S.R. and Szema, K.Y. (1985), "An Euler Solver for Three-Dimensional Supersonic Flows with Subsonic Pockets", AIAA Paper 85-1703.

14. Chiang, S.T., Hoffmann, K.A. and Rutledge, W.H. (1991), "Comparison of Flux-Vector Splitting Schemes for Finite Difference and Finite Volume Techniques", AIAA Paper 91-0170.

15. Chieng, C.C. and Danberg, J.E. (1988), "Navier-Stokes Computations for a Transonic Projectile at Angle of Attack and Comparison with Experiment Data", AIAA Paper 88-2584.

16. Cummings, R.L., Risk, Y.M., Schiff, L.B. and Chaderjian, N.M. (1990), "Navier-Stokes Predictions of the Flowfield Around the F-18 (HARV) Wing and Fuselage at Large Incidence", AIAA Paper 90-0099.

17. Davis, R.L., Ni, R.H. and Bowley, W.W. (1984), "Prediction of Compressible Laminar Viscous Flows Using a Time-Marching Control Volume and Multiple-Grid Technique", AIAA Journal, Vol. 22, No. 11, pp. 1573–1581.

18. Deese, J.E. and Agarwal, R.K. (1988), "Navier-Stokes Calculations of Transonic Viscous Flow About Wing/Body Configurations", Journal of Aircraft, Vol. 25, No. 12, pp. 1106–1112.

19. Deese, J.E., Agarwal, R.K. and Johnson, J.G. (1991), "Calculation of Vortex Flowfields Around Forebodies and Delta Wings", AIAA Paper 91-0176.

20. Degani, D. and Steger, J.L. (1983), "Comparison Between Navier-Stokes and Thin-Layer Computations for Separated Supersonic Flow", AIAA Journal, Vol. 21, No. 11, pp. 1604–1606.

21. Degani, D. and Schiff, L.B. (1983), "Computation of Supersonic Viscous Flows Around Pointed Bodies at Large Incidence", AIAA Paper 83-0034.

22. Degani, D. and Schiff, L.B. (1986), "Computation of Turbulent Supersonic Flows Around Pointed Bodies Having Crossflow Separation", Journal Comp. Physics, Vol. 66, pp. 173–196.

23. Degani, D. and Schiff, L.B. (1989), "Numerical Simulation of the Effect of Spatial Disturbances on Vortex Asymmetry", AIAA Paper 89-0340.

24. Degani, D., Schiff, L.B. and Levy, Y. (1990), "Physical Considerations Governing Computation of Turbulent Flows over Bodies at Large Incidence", AIAA Paper 90-0096.

25. Degani, D. (1990), "Numerical Investigation of the Origin of Vortex Asymmetry", AIAA Paper 90-0593.

26. Del Frate, J.H. and Zuniga, F.A. (1990), "In-Flight Flow Field Analysis on the NASA F-18 High Alpha Research Vehicle with Comparisons to Ground Facility Data", AIAA Paper 90-0231.

27. Ekaterinaris, J.A. and Schiff, L.B. (1990a) , "Vortical Flows over Delta Wings and Numerical Prediction of Vortex Breakdown", AIAA Paper 90-0102.

28. Ekaterinaris, J.A. and Schiff, L.B. (1990b) , "Numerical Simulation of the Effects of Variation of Angle of Attack and Sweep Angle on Vortex Breakdown over Delta Wings", AIAA Paper 90-3000CP, pp. 59–67.

29. Ekaterinaris, J.A., Coutley, R.L., Schiff, L.B. and Platzer, M.F. (1991), "Numerical Investigation of the Flow over a Delta Wing at High Incidence", AIAA Paper 91-0753.

30. Flores, J. (1986), "Convergence Acceleration for a Three-Dimensional Euler Navier-Stokes Zonal Approach", AIAA Journal, Vol. 24, No. 9, pp. 1441–1442.

31. Flores, J., Resnick, S., Holst, T.L. and Gundy, K. (1987), "Transonic Navier-Stokes Solutions for Fighter-Like Configuration", AIAA Paper 87-0032, also Journal of Aircraft, Vol. 25, No. 10, pp. 875–881.

32. Flores, J. and Chaderjian, N.M. (1988), "The Numerical Simulation of Transonic Separated Flow About the Complete F-16A ", AIAA Paper 88-2506.

33. Flores, J. and Chaderjian, N.M. (1990), "Zonal Navier-Stokes Methodology for Flow Simulation About a Complete Aircraft", Journal of Aircraft, Vol. 27, No. 7, pp. 583–590.

34. Fujii, K. and Schiff, L.B. (1989), "Numerical Simulation of Vortical Flows over a Strake-Delta Wing", AIAA Journal, Vol. 27, No. 9, pp. 1153–1162.

35. Gee, K., Cummings, R. M. and Schiff, L. B. (1990), "The effect of Turbulence Models on the Numerical Prediction of the Flowfield about a Prolate Spheroid at High Angles of Attack", AIAA Paper 90-3106CP, pp. 109–123.

36. Ghaffari, F., Luckring, J.M., Thomas, J.L. and Bates, B.L. (1989), "Navier-Stokes Solutions About the F/A-18 Forebody-LEX Configuration", AIAA Paper 89-0338.

37. Govindan, T.R., Briley, W.R. and Chang, M.S. (1991), "Generalized Primary/Secondary Flow Analysis of Viscous Flow Around Bodies at Incidence", AIAA Paper 91-0186.

38. Hartwich, P.M., Hsu, C.H., Luckring, J.M. and Liu, C.H. (1988), "Numerical Study of the Vortex Burst Phenomenon for Delta Wings", AIAA Paper 88-0505.

39. Hartwich, P.M. and Hsu, C.M. (1988), "High Resolution Upwind schemes for the Three-Dimensional Incompressible Navier-Stokes Equations", AIAA Journal, Vol. 26, No. 11. pp. 1321–1328.

40. Hartwich, P.M., Hsu, C.H., Luckring, J.M. and Liu, C.H. (1988), "Aerodynamic Applications of an Efficient Incompressible Navier-Stokes Solver", ICAS Paper 88-5.9.1, pp. 1417–1427.

41. Hartwich, P.M. and Hall. R.M. (1990), "Navier-Stokes Solutions for Vortical Flows over a Tangent-Ogive Cylinder", AIAA Journal, Vol. 28, No. 7, pp. 1171–1179.

42. Harvey, A.D. III, Acharya, S. and Lawrence, S.L. (1991), "A Solution-Adaptive Grid Proceedure for the Three-Dimensional Parabolized Navier-Stokes Equations", AIAA Paper 91-0104.

43. Harvey, A.D. III, Acharya, S., Lawrence, S.L. and Cheung, S. (1990), "A Solution Adaptive Grid Procedure for an Upwind Parabolized Flow Solver", AIAA Paper 90-1567.

44. Holst, T.L., Kaynak, U., Gundy, K.L., Thomas, S.D. and Flores, J. (1987), "Numerical Solutions of Transonic Wing Flows Using an Euler/Navier-Stokes Zonal Approach", Journal of Aircraft, Vol. 24, No. 1, pp. 17–24.

45. Himansu, A. and Rubin, S.G. (1991), "Three-Dimensional RNS Computations with Multigrid Acceleration", AIAA Paper 91-0105.

46. Hsu, C.C., Shiau,N.H. and Reed, C.W. (1988), "Numerical Simulation of Transonic Turbulent Flow Past a Real Projectile", AIAA Paper 88-0218.

47. Jayaram, M. and Jameson, A. (1988), "Multigrid Solution of the Navier-Stokes Equations for Flow over Wings", AIAA Paper 88-0705.

48. Kaynak, U., Cantwell, B.J. and Holst, T.L. (1986), "Numerical Simulation of Transonic Separated Flows over Low-Aspect Ratio Wings", AIAA Paper 86-0508.

49. Kwon, O.J. and Sankar, L.N. (1991), "Viscous Flow Simulation of Fighter Aircraft", AIAA Paper 91-0278.

50. Lawrence, S.L., Tannehill, J.C. and Chaussee, D.S. (1986), "An Upwind Algorithm for the Parabolized Navier-Stokes Equations", AIAA Paper 86-1117.

51. Lawrence, S.L., Chaussee, D.S. and Tannehill, J.C. (1987), "Application of an Upwind Algorithm to the Three-Dimensional Parabolized Navier-Stokes Equations", AIAA Paper 87-1112, pp. 112–125.

52. Lawrence, S.L. (1989), "Calculation of Winged-Body Flow Field Using an Implicit Upwind Space-Marching Code", AIAA Paper 89-1826.

53. Levy, Y., Seginer, A. and Degani, D. (1988), "Graphical Representation of Three-Dimensional Vortical Flows by Means of Helicity Density and Normalized Helicity", AIAA Paper 88-2598.

54. Lin, H. and Chieng, C.C. (1991), "Comparisons of TVD Schemes for Turbulent Transonic Projectile Aerodynamics Computations", AIAA Paper 91-0171.

55. Lombard, C.K., Bardina, J., Venkatapathy, E. and Oliger, J. (1983), "Multi-Dimensional Formulation of CSCM – an Upwind Flux Difference Eigenvector Split Method for the Compressible Navier-Stokes Equations", AIAA Paper 83-1895.

56. MacCormack, R.W. (1969), "The Effect of Viscosity in Hypervelocity Impact Crating", AIAA Paper 69-354.

57. MacCormack, R.W. (1971), "Numerical Solution of the Interaction of a Shock Wave with a Laminar Boundary Layer", Proc. Second Intl. Conf. Num. Methods Fluid Dyn., *Lecture Notes in Physics,* Vol. 8, Springer-Verlag New-York, pp. 151–163.

58. MacCormack, R.W. and Paullay, A.J. (1972), "Computational Efficiency Achieved by Time Splitting of Finite Difference Operators", AIAA Paper 72-154.

59. MacCormack, R.W. and Baldwin, B.S. (1975), "A Numerical Method for Solving the Navier-Stokes Equations with Application to Shock-Boundary Layer Interactions", AIAA Paper 75-1.

60. MacCormack, R.W. (1976), "An Efficient Numerical Method for Solving the Time-Dependent Compressible Navier-Stokes Equations at High Reynolds Numbers", NASA TM X-73, and X-129.

61. MacCormack, R.W. (1981), "A Numerical Method for Solving the Equations of Compresssible Viscous Flow", AIAA Paper 81-0110.

62. MacCormack, R.W. (1985), "Current Status of Numerical Solutions of the Navier-Stokes Equations", AIAA Paper 85-0032.

63. MacCormack, R.W. and Candler, G.V. (1989), "The Solution of the Navier-Stokes Equations Using Gauss-Seidel Line Relaxation", Computers and Fluids, Vol. 17, No. 1, pp. 135–150.

64. Martinelli, L., Jameson, A. and Grasso, F. (1986), "A Multigrid Method for Navier-Stokes Equations", AIAA Paper 86-0208.

65. McDonald, H. and Briley, W.R. (1975), "Three-Dimensional Supersonic Flow of a Viscous or Inviscid Gas", Journal of Comp. Physics, Vol. 19, pp. 150–178.

66. McMillin, S.N., Thomas, J.L. and Murman. E.M. (1987), "Euler and Navier-Stokes Solutions for the Leeside Flow over Delta Wings at Supersonic Speeds", AIAA Paper 87-2270.

67. Meier, H.U., Kreplin, H.P. and Volimers, H. (1983), "Development of Boundary Layers and Separation Patterns on a Body of Revolution at Incidence", 2nd Symposium on Numerical and Physical Aspects of Aerodynamics Flows, Long Beach.

68. Narain, J.P., Muramoto, K.K. and Lawrence, S.L. (1991), "The Prediction of Viscous Hypersonic Flows About Complex Configurations Using an Upwind Parabolized Navier-Stokes Code", AIAA Paper 91-0394.

69. Pan, D. and Pulliam, T. H. (1986), "The Computation of Steady 3-D Separated Flows over Aerodynamic Bodies at Incidence and Yaw", AIAA Paper 86-0109.

70. Panaras, A.G. and Steger, J.L. (1988), "A Thin-Layer Solution of the Flow About a Prolate Spheroid", Z. Flugwiss, Weltraumforch., Vol. 12, pp. 173–180.

71. Radespiel, R. (1989), "A Cell-Vertex Multigrid Method for the Navier-Stokes Equations", NASA TM 101557.

72. Rainbird, J.R. (1968), "Turbulent Boundary-Layer Growth and Separation on a Yawed Cone", AIAA Journal, Vol. 6, No. 12, pp. 2410–2416.

73. Reklis, R.P. and Sturek, W.B. (1978), "Surface Pressure Measurements on Slender Bodies at Angles of Attack in Supersonic Flow", ARBRL -MR-02876, U.S. Army Ballistic Research Laboratory, ARRADCOM, Abardeen Proving Ground, Md.

74. Reznick, S. (1988), "Transonic Navier-Stokes Computations of Strake Generated Vortex Interactions for a Fighter-Like Configuration", NASA TM 100009.

75. Rizk, Y.M., Schiff, L.B. and Gee, K. (1990), "Numerical Simulation of the Viscous Flow Around a Simplified F/A-18 at High Angles of Attack", AIAA Paper 90-2999.

76. Rizk, Y.M. and Gee, K. (1991), "Numerical Prediction of the Unsteady Flowfield Around the F-18 Aircraft at Large Incidence", AIAA Paper 91-0020.

77. Roe, P.L. (1983), "Approximate Riemann Solvers, Parameter Vectors, and Difference Schemes", Journal of Computational Physics, Vol. 43, pp. 357–372.

78. Rosenfeld, M., Israeli, M. and Wolfstein, M. (1988), "Numerical Study of the Skin Friction on a Spheroid of Incidence", AIAA Journal, Vol. 26, No. 2, pp. 129–136.

79. Ruffin, S.M. and Murman, E.M. (1988), "Solutions for Hypersonic Viscous Flow over Delta Wings", AIAA Paper 88-0126.

80. Rumsey, C.L., Van Leer, B. and Roe, Ph. L. (1991), "A Grid-Independent Approximate Rieman Solver with Applications to the Euler and Navier-Stokes Equations", AIAA Paper 91-0239.

81. Schiff, L.B. and Steger, J.L. (1979), "Numerical Simulation of Steady Supersonic Viscous Flow", AIAA Paper 79-0130.

82. Schiff, L.B., Cummings, R.M., Sorenson, R.L. and Rizk, Y.M. (1989), "Numerical Simulation of High-Incidence Flow over the F-18 Fuselage Forebody", AIAA Paper 89-0339.

83. Shapiro, R.A. (1991), "Implementation of an Euler/Navier-Stokes Finite Element Algorithm on the Connection Machine", AIAA Paper 91-0438.

84. Shiau, N.H. and Hsu, C.C. (1988), "A Diagonalized TVD Scheme for Turbulent Transonic Projectile Aerodynamics Computation", AIAA Paper 88-0217.

85. Siclari, M.J. and Marconi, F. (1989), "The Computation of Navier-Stokes Solutions Exhibiting Asymmetric Vortices", AIAA Paper 89-1817.

86. Siclari, M.J. (1990), "Asymmetric Separated Flows at Supersonic Speeds", AIAA Paper 90-0595.

87. Stalnaker, J.F., Nicholson, L.A., Hanline, D.S. and McGraw, E.H. (1986), "Improvements to the AFWAL Parabolized Navier-Stokes Code Formulation", AFWAL-TR-86-3076.

88. Swanson, R.C. and Turkel, E. (1985), "A Multistage Time-Stepping Scheme for the Navier-Stokes Equations", AIAA Paper 85-0035.

89. Tannehill, J.C., Venkatapathy, E. and Rakich, J.V. (1981), "Numerical Solution of Supersonic Viscous Flow over Blunt Delta Wings", AIAA Paper 81-0049.

90. Tannehill, J.C., Buelow, P.E., Ievalts, J.O.and Lawrence, S.L. (1989), "A Three-Dimensional Upwind Parabolized Navier-Stokes Code for Real Gas Flows", AIAA Paper 89-1651.

91. Thomas, J.L. and Walters, W.W. (1985), "Upwind Relaxation Algorithms for the Navier-Stokes Equations", AIAA Paper 85-1501.

92. Thomas, J.L. and Newsome, R.W. (1986), "Navier-Stokes Computations of Lee-Side Flows over Delta Wings", AIAA Paper 86-1049.

93. Thomas, J.L., Taylor, S.L. and Anderson, W.K. (1987), "Navier-Stokes Computations of Vortical Flows over Low Aspect Ratio Wings", AIAA Paper 87-0207.

94. Thomas, J.L., Walter, R.W., Reu, T., Ghaffari, F., Weston, R.P. and Luckring, J.M. (1989), "A Patched-Grid Algorithm for Complex Configurations Directed Towards the F-18 Aircraft", AIAA Paper 89-0121.

95. Thomas, J.L., Krist, S.T. and Anderson, W.K. (1990), "Navier-Stokes Computations of Vortical Flows over Low Aspect Ratio Wings", AIAA Journal, Vol. 28, No. 2, pp. 205–212.

96. Thomas, P.D. and Neier, K.L. (1989), "Navier-Stokes Simulation of 3D Hypersonic Equilibrium Flows with Ablation", AIAA Paper 89-1650.

97. Tu, E.L. (1991), "Navier-Stokes Simulation of a Close-Coupled Canard-Wing-Body Configuration", AIAA Paper 91-0070.

98. Tysinger, T.L. and Caughey, D.A. (1991), "Implicit Multigrid Algorithm for the Navier-Stokes Equations", AIAA Paper 91-0242.

99. Vadysk, J. and Schuster, D.M. (1989), "Navier-Stokes Simulation of Burst Vortex Flowfields for Fighter Aircraft at High Incidence", AIAA Paper 89-2190CP, pp. 227–237.

100. Vasta, V.N., Thomas, J.L. and Wedan, B.W. (1987), "Navier-Stokes Computations of Prolate Spheroids at Angles of Attack", AIAA Paper 87-2627.

101. Vigneron, Y.C., Rakich, J.V. and Tannehill, J.C. (1978), "Calculation of Supersonic Viscous Flow over Delta Wings with Sharp Subsonic Leading-Edges", AIAA Paper 78-1137.

102. Ying, S.X., Steger, J.L., Schiff, L.B. and Baganoff, D. (1986), "Numerical Simulation of Unsteady, Viscous, High Angle of Attack Flows Using a Partially Flux Split Algorithm", AIAA Paper 86-2179.

103. Ying, S.X., Schiff, L.B. and Steger, J.L. (1987), "A Numerical Study of Three-Dimensional Separated Flow Past a Hemisphere Cylinder", AIAA Paper 87-1207.

Credits

The following figures were reprinted with permission from AIAA: 2.22b, 2.26b, 2.37, 2.46, 2.48, 2.49, 2.50, 2.51, 2.52, 2.53, 2.55, 2.57, 4.24, 4.25, 4.26, 4.27, 4.28, 4.29 – 4.36, 5.2, 5.5, 6.4, 6.5, 7.13, 7.16, 7.17, 7.38, 7.42, 7.43, 7.56, 8.5 – 8.7, 8.9, 8.12 – 8.23, 8.32 – 8.39, 9.2b, 9.4 – 9.21, 9.24, 9.26 – 9.32, 9.34 – 9.45, 9.47 – 9.53.

The following figures were reprinted with permission from AGARD: 2.22a, 2.23, 2.24, 2.31, 2.32, 2.34, 2.35, 2.36, 2.39 – 2.43, 2.45, 2.54.

The following figures were reprinted with permission from ICAS: 2.10, 7.1, 7.19 – 7.21, 7.39, 8.4, 8.24 – 8.31, 9.22, 9.23, 9.25, 9.33.

The following figures were reprinted with permission from NASA: 2.2, 2.5, 2.19, 2.21, 2.26, 2.28, 2.56, 2.58, 4.3, 4.4, 4.10 – 4.15, 4.20 – 4.23, 6.8, 6.14, 7.45 – 7.49, 7.52, 7.54, 7.57.

The following figures were reprinted with permission from ONERA: 2.3, 2.6 – 2.9, 2.12, 2.13, 2.16, 2.17, 2.44, 7.12, 7.14.

The following figures were reprinted with permission from Lockheed Missle and Space Company: 2.25 and 2.27.

The following figures were reprinted with permission from Oxford University Press: 3.4 and 3.5.

The following figures were reprinted with permission from Computer Methods in Applied Mechanics and Engineering: 4.5 – 4.9.

Figure 2.1 was reprinted with permission from Hemisphere Publishing Company.

Figure 2.4 was reprinted with permission from National Research Council of Canada.

Figure 2.11 was reprinted with permission from the Aeronautical Journal.

Figure 2.14 was reprinted with permission from ARA.

Figure 2.15 was reprinted with permission from DGLR.

Figures 2.29 and 2.30 were reprinted with permission from Beeman.

Figure 2.38 was reprinted with permission from AIAA and Hummel and Oelker.

Figure 5.1 was reprinted with permission from British A.R.C.

Figures 5.3, 6.6, and 6.7 were reprinted with permission from Ing. Arch.

Figure 6.2 was reprinted with permission from Betz.

Figures 6.12 and 6.13 were reprinted with permission from the Royal Society of London.

Figures 7.22–7.28 were reprinted with permission from ICAS and AIAA.

Figure 8.10 was reprinted with permission from J.F.M.

Figures 9.2a and 9.3 were reprinted with permission from Journal Comp. Physics.

Author Index

Subject Index

MIX
Papier aus verantwortungsvollen Quellen
Paper from responsible sources
FSC® C105338

If you have any concerns about our products,
you can contact us on
ProductSafety@springernature.com

In case Publisher is established outside the EU,
the EU authorized representative is:
Springer Nature Customer Service Center GmbH
Europaplatz 3, 69115 Heidelberg, Germany

Printed by Libri Plureos GmbH
in Hamburg, Germany